Lecture Notes in Computer Science 14350

Founding Editors

Gerhard Goos
Juris Hartmanis

The series Lecture Notes in Computer Science (LNCS), including its subseries Lecture Notes in Artificial Intelligence (LNAI) and Lecture Notes in Bioinformatics (LNBI), has established itself as a medium for the publication of new developments in computer science and information technology research, teaching, and education.

LNCS enjoys close cooperation with the computer science R & D community, the series counts many renowned academics among its volume editors and paper authors, and collaborates with prestigious societies. Its mission is to serve this international community by providing an invaluable service, mainly focused on the publication of conference and workshop proceedings and postproceedings. LNCS commenced publication in 1973.

Christian Wachinger · Beatriz Paniagua ·
Shireen Elhabian · Jianning Li · Jan Egger
Editors

Shape in Medical Imaging

International Workshop, ShapeMI 2023
Held in Conjunction with MICCAI 2023
Vancouver, BC, Canada, October 8, 2023
Proceedings

Editors
Christian Wachinger ⓘD
Technical University of Munich
Munich, Germany

Beatriz Paniagua ⓘD
University of North Carolina at Chapel Hill
Carrboro, NC, USA

Shireen Elhabian ⓘD
University of Utah
Salt Lake City, UT, USA

Jianning Li
Essen University Hospital
Essen, Germany

Jan Egger ⓘD
Essen University Hospital
Essen, Germany

ISSN 0302-9743   ISSN 1611-3349 (electronic)
Lecture Notes in Computer Science
ISBN 978-3-031-46913-8   ISBN 978-3-031-46914-5 (eBook)
https://doi.org/10.1007/978-3-031-46914-5

This Springer imprint is published by the registered company Springer Nature Switzerland AG
The registered company address is: Gewerbestrasse 11, 6330 Cham, Switzerland

Paper in this product is recyclable.

Preface

This volume contains the proceedings of the **International Workshop on Shape in Medical Imaging (ShapeMI 2023)** held in conjunction with the 26th International Conference on Medical Image Computing and Computer-Assisted Intervention (MICCAI 2023) on October 8, 2023, in Vancouver, Canada. ShapeMI 2023 was a continuation of the previous MICCAI ShapeMI 2020, ShapeMI 2018, SeSAMI 2016, and SAMI 2015 Workshops, as well as the Shape Symposium 2015 and 2014.

ShapeMI 2023 received 27 submissions via the CMT system. All submissions underwent single-blind peer review by at least two experts in the field. Based on these reviews, the program committee chairs accepted 23 full papers.

Shape and geometry processing methods have received significant attention as they apply in various fields, from medical image computing to paleontology, anthropology, and beyond. While imaging is the primary mechanism to acquire visual information, the underlying structures are usually 3D geometric shapes, often representing continuous or time-varying phenomena. Thus, 3D shape models better describe anatomical structures than voxels in a regular grid and can have a higher sensitivity to local variations or early disease/drug effects relative to traditional image-based markers such as the volume of a structure. Therefore, shape and spectral analysis, geometric learning and modeling algorithms, and application-driven research are the focus of the ShapeMI workshop.

In ShapeMI we strive to collect and present original methods and applications related to shape analysis and processing in medical imaging. The workshop provides a venue for researchers working in shape modeling, analysis, statistics, classification, geometric learning, and their medical applications to present recent research results, to foster interaction, and to exchange ideas. As a single-track workshop, ShapeMI also features excellent keynote speakers, technical paper presentations, and state-of-the-art software methods for shape processing.

We thank all the contributors for making this workshop such a huge success. We thank all authors who shared their latest findings and the Program Committee members who contributed quality reviews in a very short time. We especially thank our keynote speakers, who kindly accepted our invitation and enriched the workshop with their excellent presentations: Miaomiao Zhang (University of Virginia, USA) and Guido Gerig (New York University, USA).

October 2023

Christian Wachinger
Beatriz Paniagua
Shireen Elhabian
Jianning Li
Jan Egger

Organization

Program Committee

Constantin Seibold	Karlsruhe Institute of Technology, Germany
Ellen Gasparovic	Union College, USA
Fabian Bongratz	Technical University of Munich, Germany
Guido Gerig	New York University, USA
Ilkay Oksuz	Istanbul Technical University, Turkey
James Fishbaugh	Kitware, USA
Jens Kleesiek	UK Essen, Germany
Kathryn Leonard	Occidental College, USA
Miaomiao Zhang	University of Virginia, USA
Moritz Rempe	UK Essen, Germany
Stefan Sommer	University of Copenhagen, Denmark
Steve Pizer	University of North Carolina at Chapel Hill, USA
Tim Cootes	Manchester University, UK
Umberto Castellani	University of Verona, Italy
Veronika Zimmer	Technical University of Munich, Germany
Ilwoo Lyu	Ulsan National Institute of Science and Technology, South Korea
Sungmin Hong	Amazon Web Services, USA
Suyash Awate	Indian Institute of Technology, Bombay, India
Ye Han	Kitware, USA
Yitong Li	Technical University of Munich, Germany
Yonggang Shi	University of Southern California, USA

Contents

Anatomy Completor: A Multi-class Completion Framework for 3D Anatomy Reconstruction

Jianning Li[1]([⊠]), Antonio Pepe[2], Gijs Luijten[1,2], Christina Schwarz-Gsaxner[2], Jens Kleesiek[1], and Jan Egger[1,2]

[1] Institute for AI in Medicine (IKIM), University Hospital Essen, Essen, Germany
jianning.li@uk-essen.de
[2] Institute of Computer Graphics and Vision (ICG), Graz University of Technology, Graz, Austria

Abstract. In this paper, we introduce a completion framework to reconstruct the geometric shapes of various anatomies, including organs, vessels and muscles. Our work targets a scenario where one or multiple anatomies are missing in the imaging data due to surgical, pathological or traumatic factors, or simply because these anatomies are not covered by image acquisition. Automatic reconstruction of the missing anatomies benefits many applications, such as organ 3D bio-printing, whole-body segmentation, animation realism, paleoradiology and forensic imaging. We propose two paradigms based on a 3D denoising autoencoder (DAE) to solve the anatomy reconstruction problem: (i) the DAE learns a *many-to-one* mapping between incomplete and complete instances; (ii) the DAE learns directly a *one-to-one* residual mapping between the incomplete instances and the target anatomies. We apply a loss aggregation scheme that enables the DAE to learn the *many-to-one* mapping more effectively and further enhances the learning of the residual mapping. On top of this, we extend the DAE to a multi-class completor by assigning a unique label to each anatomy involved. We evaluate our method using a CT dataset with whole-body segmentations. Results show that our method produces reasonable anatomy reconstructions given instances with different levels of incompleteness (i.e., one or multiple random anatomies are missing). Codes and pre-trained models are publicly available at https://github.com/Jianningli/medshapenet-feedback/tree/main/anatomy-completor.

Keywords: Anatomical Shape Completion · Shape Reconstruction · Shape Inpainting · Whole-body Segmentation · Residual Learning · MedShapeNet · Diminished Reality

1 Introduction

3D anatomy reconstructions play important roles in medical applications and beyond, such as (1) 3D bio-printing and organ transplantation, where damaged/diseased organs from traumatic injuries or pathologies are replaced by

C. Wachinger et al. (Eds.): ShapeMI 2023, LNCS 14350, pp. 1–14, 2023.
https://doi.org/10.1007/978-3-031-46914-5_1

3D bio-printed artificial organs [23]; (2) paleoradiology and forensic imaging, in which the full anatomical structures are re-established based on the skeleton remains [13, 21, 31]; (3) whole-body segmentation, where pseudo labels of whole-body anatomies are generated given only sparse manual annotations [8, 26, 30]; (4) animation realism [2]; and (5) diminished reality, where the 3D view of an anatomy blocked by medical instruments is reconstructed. Such an anatomy reconstruction task is well aligned with the shape completion problem in computer vision, which is commonly solved based on the symmetry of geometric shapes [28] or using learning-based approaches, where auto-encoder and generative adversarial networks (GANs) [3, 25, 32, 33] are popular choices. Recent years have witnessed a growing interest in medical shape completion, with the rapid development of medical deep learning [4]. Nevertheless, existing works in this direction are mostly focused on reconstructing a pre-defined and geometrically simple bone structure, such as the cranium [10, 11, 14, 16–18, 22, 32], maxilla [34], spine [19] and teeth [29], which restricts their scope of application to implant and prosthetic design. Existing methods for medical shape completion are commonly based on variants of auto-encoder and U-Net [10] and statistical shape models (SSMs) [5, 24]. Reconstructing random anatomies with varied geometric complexity is significantly harder than when the reconstruction target is pre-defined as in prior works. To realize the former, a network learns not only to identify the targets (i.e., what are missing in the input) but to reconstruct them, a process analogous to object instance segmentation [6], where a network first identifies all objects in an image and then segments them. However, random anatomy reconstruction has not been covered by existing research, which only completes one fixed anatomy with missing part(s), and remains to be an open problem. The goal of this work is to extend medical shape completion to the whole body, covering the majority of anatomy classes, and to realize random anatomy reconstruction in a single shape completion framework. To achieve this goal, we derived a 3D anatomical shape dataset from a fully-segmented CT dataset and trained a 3D convolutional denoising auto-encoder on the dataset to learn a mapping relationship between the incomplete instances and the corresponding targets, i.e., the full segmentations or the missing anatomies. Both quantitative and qualitative evaluations have demonstrated the effectiveness of our proposed method towards solving the anatomical shape reconstruction problem.

2   Methods

2.1   Problem Formulation

Reconstructing random missing anatomies is formulated as a shape completion problem, where the goal is to learn a mapping $\mathcal{F}$ between the incomplete instances from N subjects $\mathcal{X} = \left\{x_n^m\right\}_{n=1,...,N}^{m=1,...,M}$ and the corresponding complete ground truth $\mathcal{Y} = \left\{y_n\right\}_{n=1}^{N}$ derived from whole-body anatomy segmentations. For subject x_n, there exist M instances i.e., $x_n^1, x_n^2, ..., x_n^m, ..., x_n^M$ with different

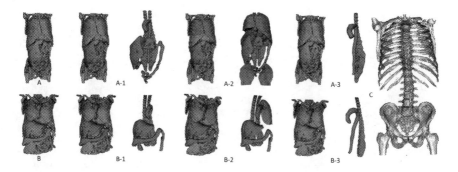

Fig. 1. Illustration of the pre-processed dataset. (A, B): the full anatomy segmentations from two subjects. (A-1, A-2, A-3) and (B-1, B-2, B-3): three incomplete instances with random missing anatomies (shown in red). (C): the skeleton in a CT scan. (Color figure online)

degrees of incompleteness, where one or multiple random anatomies are missing. Therefore, $\mathcal{F}$ is supposed to be a *many-to-one* mapping, i.e.,

$$\mathcal{F} : \left\{ x_n^m \right\}_{m=1}^M \to y_n, \, n = 1, 2, ..., N \qquad (1)$$

We use binary voxel grids to represent 3D anatomies, such that $x_n^m, y_n \in R^{L \times W \times H}$. The value of a voxel in x_n^m, y_n is '1' if the voxel belongs to an anatomy and '0' otherwise. Such a formulation extends existing medical shape completion methods that target only a single, pre-defined anatomy to random anatomies.

2.2   Denoising Auto-Encoder with Residual Connections

Given the notations in Sect. 2.1, the missing anatomies for subject x_n can be conveniently expressed in a residual form: $\left\{ y_n - x_n^m \right\}_{m=1}^M$. Therefore, apart from learning the full mapping $\mathcal{F}$, we can instead learn a residual mapping $\mathcal{F}_{res}$:

$$\mathcal{F}_{res} : \left\{ x_n^m \right\}_{m=1}^M \to \left\{ y_n - x_n^m \right\}_{m=1}^M, \, n = 1, 2, ..., N \qquad (2)$$

Unlike $\mathcal{F}$, the residual mapping $\mathcal{F}_{res}$ is obviously *one-to-one*, which can be straightforwardly realized based on deep residual learning [7]. Motivated by this observation, we propose to solve the shape completion problem using a 3D denoising auto-encoder (DAE) with a residual connection between the input and the output. The input x_n^m is treated as a corrupted version of y_n with random noise. The DAE denoises the input by restoring the anatomies missing in x_n^m. The DAE is trained in a supervised fashion, with the input being $\mathcal{X}$ and the ground truth being $\mathcal{Y}$. Even though both mappings are learnable by the DAE, we presume that a *one-to-one* mapping relationship is easier to learn than a *many-to-one* mapping, so that the DAE can reach a superior reconstructive performance by learning $\mathcal{F}_{res}$.

2.3   Loss Aggregation for Random Anatomy Completion

To learn the *many-to-one* mapping $\mathcal{F}$, we train the DAE by optimizing a Dice loss function $\mathcal{L}_{dice}$ aggregated over M versions of incomplete instances with random missing anatomies:

$$\mathcal{L}_{\mathcal{F}} = \sum_{m=1}^{M} \sum_{n=1}^{N} \mathcal{L}_{dice}(y_n, \tilde{y}_n^m) \tag{3}$$

where $\mathcal{L}_{dice} = \frac{2 \sum (y_n \odot \hat{y}_n^m)}{\sum (y_n \odot y_n) + \sum (\hat{y}_n^m \odot \hat{y}_n^m)}$ is the standard Dice loss [20]. $\hat{y}_n^m$ denotes the prediction for x_n^m given the mapping $\mathcal{F}$, and $\odot$ denotes the Hadamard product (i.e., element-wise multiplication between two matrices). $\sum$ denotes the summation of all the elements of a matrix. Optimizing such an aggregated loss function $\mathcal{L}_{\mathcal{F}}$ ensures that the DAE learns to reconstruct a complete set of anatomies regardless of the class and/or number of anatomies that are absent in the input. Similarly, to learn the *one-to-one* residual mapping $\mathcal{F}_{res}$, the following loss function is optimized:

$$\mathcal{L}_{\mathcal{F}_{res}} = \sum_{m=1}^{M} \sum_{n=1}^{N} \mathcal{L}_{dice}(y_n, \tilde{x}_n^m + x_n^m) \tag{4}$$

where $\tilde{x}_n^m$ denotes the reconstructed missing anatomies for x_n. Depending on the mapping to be learned, the respective loss function ($\mathcal{L}_{\mathcal{F}}$ or $\mathcal{L}_{\mathcal{F}_{res}}$) is used.

2.4   Multi-class Anatomy Completion

For the multi-anatomy completion task, compared to representing x_n^m and y^n as binary voxel grids in which different anatomies are not differentiated (Sect. 2.1), it is more desirable to assign a unique label to each anatomy in x_n^m and y_n. This extension can be easily achieved by setting the number of output channels of the penultimate layer of the DAE network to the number of anatomy classes. Each channel predicts the probability of occupancy of the voxel grids for an anatomy. The same Dice loss $\mathcal{L}_{dice}$ can be calculated between the output and the ground truth in one-hot encoding.

3   Experiments and Results

3.1   Dataset and Pre-processing

We validate our method using a public CT dataset with whole-body anatomy segmentations, which is publicly available at https://zenodo.org/record/6802614#.Y_YMwXbMIQ8. The dataset comprises 1024 CT images, each accompanied by a set of segmentation masks of 104 anatomies (organs, bones, muscles, vessels) [30]. After screening (discarding images with corrupted segmentations), 737 sets of segmentations are included in this work, which are further randomly split into a training (451) and test set (286). For each set of segmentations, we randomly remove anatomies accounting for at least 10%, 20% and 40% of the entire segmentation's volume to create the incomplete instances $\mathcal{X}$. The original segmentations serve as the ground truth $\mathcal{Y}$. Considering that anatomy ratios are subject-specific, different type and/or number of anatomies could have been removed for different subjects given the same threshold, as can be seen from Fig. 1. Thus, anatomy removal is analogous to inserting random noise to $\mathcal{Y}$. In general, using a 10% threshold (Fig. 1, A-2, B-2) removes more anatomies than using higher thresholds (20% and 40%), and using a threshold of 40% removes only large anatomies, such as the aorta and the autochthonous back muscles (Fig. 1, A-3, B-3). The small bones such as the individual ribs and vertebrae that form the skeleton (Fig. 1, C) enclosing the internal anatomies are generally not removed, providing a natural constraint for anatomy reconstruction. We use the ratio-based method to remove anatomies, so that each full segmentation yields three instances with random incompleteness in the training and test set. We denote the three test sets as D_{test1} (10%), D_{test2} (20%) and D_{test3} (40%). Besides random anatomy removal, we create another test set D_{test4} by removing only one specific anatomy from the full segmentations randomly selected from the test set. All the images are re-scaled to a uniform size of 128^3 ($L, W, H = 128$). We made the anatomical shape dataset used in this study publicly available through *MedShapeNet* [15].

3.2 Implementation Details

The DAE is comprised of four two-strided 3D convolutional (conv3D) and transposed convolutional (t_conv3D) layers for downsampling and upsampling. To increase the learning capacity, we add a single-strided conv3D layer after each t_conv3D layer, and further append four single-strided conv3D layers at the end of the DAE. We use *ReLu* activations and a kernel size of three for all layers, amounting to around 22M trainable parameters. The residual connection is implemented as an addition between the input and the output of the penultimate layer. The DAE is implemented using TensorFlow [1] and trained on an NVIDIA RTX 3090 GPU using the ADAM optimizer [9]. The learning rate is set to 0.0001 and the exponential decay rate for the first moment estimates is set to 0.3 for the ADAM optimizer.

3.3 Experimental Setup

Since, to our knowledge, our paper is the first to investigate random anatomy reconstruction, we adhere to the following steps to validate our methods: (i) A baseline is established by training the DAE without residual connection using a conventional Dice loss from existing single anatomy completion studies [16,18]; (ii) On top of the baseline, we train the DAE using the aggregated Dice loss (Eq. 3); (iii) We train the DAE with residual connection (Eq. 2) using a conventional Dice loss; (iv) We train the DAE with residual connection using the aggregated Dice loss (Eq. 4). For all experiments, the DAE is trained for 100 epochs. The baseline experiment evaluates the feasibility of realizing random anatomy reconstruction using a single shape completion framework, and experiments (ii-iv) verify the effectiveness of each proposed components (i.e., residual connection, loss aggregation) for the anatomy reconstruction task. We denote the trained DAE models from experiment (i-iv) as DAE_b, DAE_{agg}, DAE_{res} and $DAE_{agg+res}$, respectively. Dice similarity coefficient (DSC) is used for quantitative evaluation of the results on test set D_{test1}, D_{test2}, and D_{test3}. The output of the DAE is interpolated to the original size to calculate the DSC against the ground truth. On D_{test4}, we perform an empirical evaluation of our method in reconstructing one specific anatomy.

3.4   Results

Table 1. Mean (Standard Deviation) of DSC on D_{test1}, D_{test2}, D_{test3}.

Methods	D_{test1}	D_{test2}	D_{test3}
DAE_b	0.783 (0.075)	0.778 (0.061)	0.757 (0.058)
DAE_{agg}	0.789 (0.073)	0.803 (0.059)	0.812 (0.053)
DAE_{res}	**0.865** (0.069)	0.885 (0.046)	0.887 (0.047)
$DAE_{agg+res}$	**0.865** (0.074)	**0.904** (0.039)	**0.931** (0.030)

Quantitative Evaluation and Statistical Comparison. Table 1 presents the quantitative results of the ablation experiments, where the mean and standard deviations (SD) of DSC on test set D_{test1}, D_{test2} and D_{test3} are reported. The quantitative comparisons show that both loss aggregation (DAE_{agg}) and residual connection (DAE_{res}) help improve the anatomy reconstruction performance compared to the baseline (DAE_b). Furthermore, the comparison between DAE_{agg} and DAE_{res} demonstrates that the DAE is significantly better at learning the residual (Eq. 2) than the full anatomy (Eq. 1). Combining both components ($DAE_{agg+res}$) further improves the reconstructive performance of the DAE compared to using each component individually. Furthermore, DAE_{agg}, DAE_{res} and $DAE_{agg+res}$ also perform more stably across test instances (smaller SD) than the baseline, on all three test sets. Compared with D_{test1} and D_{test2}, we notice an obvious drop of mean DSC on D_{test3} for the baseline model, suggesting that DAE_b tends to perform worse when the combined ratio of all missing anatomies becomes smaller. The combined ratio of all missing anatomies in D_{test3} is likely to be lower, since fewer anatomies can be removed due to the higher ratio threshold. Higher sensitivity is required to detect and reconstruct smaller anatomies. Applying loss aggregation (DAE_{agg}) enforces the *many-to-one* mapping and therefore mitigates the low sensitivity issue. The residual mapping (DAE_{res}) overcomes the low-sensitivity issue even without loss aggregation. A statistical comparison of the DSC between different models on the three test sets is also performed based on a t-test, and the p values are reported in Table 2. $p < 0.05$ indicates a statistically significant improvement. Based on Table 1 and the statistical comparisons of $DAE_{agg} \leftrightarrow DAE_b$ and $DAE_{agg+res} \leftrightarrow DAE_{res}$, we can also conclude that loss aggregation does not significantly improve the results on D_{test1}, which has a very high combined ratio of missing anatomies.

Qualitative Evaluation. Fig. 2 illustrates the reconstruction results in 2D coronal planes. Multiple test instances with different degrees of incompleteness are presented. As seen from the ground truth (Fig. 2, second column), an ideal reconstruction covers 100% of the input and does not extend beyond the region enclosed by the ribs (Fig. 2, first column). The qualitative comparison shows that

Table 2. Statistical Comparison of DSC on Test Set D_{test1}, D_{test2}, D_{test3} Between Different Methods. The Table Reports the p Values From a T-test.

Methods	D_{test1}	D_{test2}	D_{test3}
$DAE_{agg} \leftrightarrow DAE_b$	0.328	4.301e-07	8.176e-29
$DAE_{res} \leftrightarrow DAE_b$	2.839e-35	2.147e-86	7.427e-114
$DAE_{agg+res} \leftrightarrow DAE_b$	2.295e-33	1.437e-110	2.985e-163
$DAE_{res} \leftrightarrow DAE_{agg}$	9.931e-32	3.051e-59	5.644e-57
$DAE_{agg+res} \leftrightarrow DAE_{res}$	0.989	2.866e-07	1.797e-34

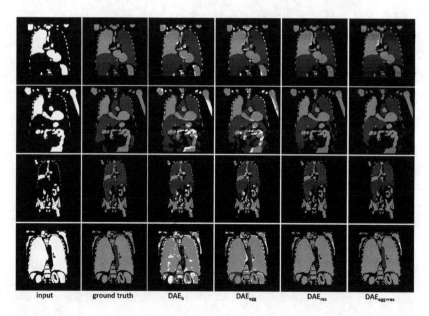

input   ground truth   DAE$_b$ DAE$_{agg}$ DAE$_{res}$ DAE$_{agg+res}$

Fig. 2. Qualitative comparison of anatomy reconstruction performance. ▨ indicates the overlap between the reconstruction and the input, and ▰ indicates the reconstructed missing anatomies. Small white blocks in the reconstructions indicate false negative predictions.

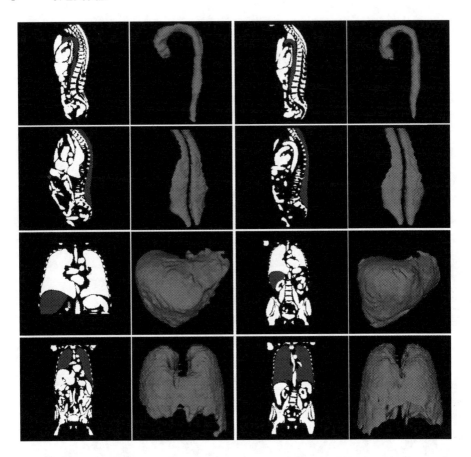

Fig. 3. The first to last row show the reconstructed aorta, autochthonous back muscles, liver and lung by $DAE_{agg+res}$. Two test instances are presented for each anatomy class.

the DAE models trained for full anatomy reconstruction (DAE_b and DAE_{agg}, Eq. 1) have a tendency to produce false negatives, i.e., they fail to fully reconstruct existing anatomies, as shown by the small white blocks in the third and fourth column of Fig. 2, as well as false positives, i.e., they generate a reconstruction beyond the missing anatomies. Resorting to residual learning (DAE_{res} and $DAE_{agg+res}$, Eq. 2) obviously mitigates the false prediction issue. Figure 3 shows the reconstruction results from the best performing model $DAE_{agg+res}$ for a single anatomy, specifically the aorta, the autochthonous back muscles, liver and lung. For single anatomy reconstruction, only one random anatomy is missing in the input (D_{test4}). For smaller anatomies like the kidney and spleen, these models are not sufficiently sensitive to detect their absence and produce a reasonable reconstruction. Only for relatively large anatomies, such as livers and lungs, single anatomy reconstruction is feasible (Fig. 3). Increasing the loss aggregation scope (i.e., the M in Eq. 3, 4) to explicitly cover the individual small anatomies during the training process is a promising solution to the low sensitivity problem. Appendix (A) provides preliminary results that support this observation regarding the reconstruction of small missing anatomies. In Appendix (B), we show that it is feasible to reconstruct the whole anatomies given only the skeleton (rib cage + spine). These findings are potentially useful for (semi-)supervised whole-body segmentations, in which a human annotator provides manual segmentations for only a few of the anatomies, while the anatomy completor generates the segmentation masks in 3D for the rest. Even though the quality of the generated segmentations might not be sufficient to serve as the ground truth, they could be used as the initial pseudo labels that can be iteratively refined [27]. Appendix (B) gives an extreme example where only the skeleton is given or annotated. It should be noted that the current results for such examples are not optimal, and serve only as a proof of concept.

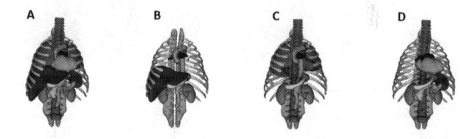

Fig. 4. Dataset for the multi-class anatomy completor. (A) the 12 anatomy segmentations. (B-D) three incomplete instances where some of the 12 anatomies are missing.

Multi-class Anatomy Completion. For the multi-class experiment, we choose 12 anatomies, including the lung, heart, spleen, stomach, pancreas, spine, rib cage, liver, kidney, aorta, a pair of autochthon muscles, and the pulmonary artery (Fig. 4 (A)). We extract the 12 above-mentioned anatomy segmentations from 18 whole-body segmentations randomly chosen from the training set. We create 10 incomplete instances for each case by randomly removing some of the 12 anatomies (e.g., Fig. 4 (B-D)), resulting in $18 \times 10 = 180$ training samples. Images are resized to $256 \times 256 \times 128$ ($L, W = 256, H = 128$) and the DAE_{agg} method is used for the experiment (i.e., to learn a *many-to-one mapping*). Figure 5 presents the multi-class anatomy completion results on test samples that are not involved during training. It is noticeable from the reconstructions that the long thin structures i.e., the ribs, are not well reconstructed (e.g., the last row of Fig. 5). Terracing artifacts are also obvious on the reconstructed anatomical shapes compared to the ground truth, which can be partly attributed to downsampling.

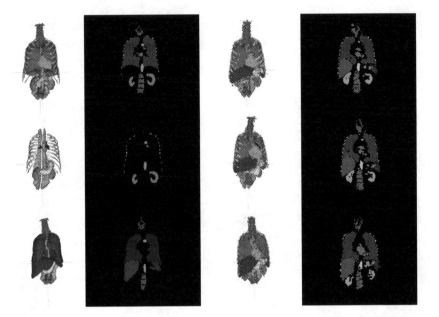

Fig. 5. Qualitative results of multi-class anatomy completion. The first and second column show three incomplete instances from the same subject in 3D and coronal views. The last two columns show the corresponding reconstruction results.

4   Discussion and Conclusion

In this paper, we demonstrated that multi-class anatomy reconstruction can be realized in a single shape completion framework. Given an incomplete instance with random missing anatomies, a DAE network reconstructs the missing anatomies specific to the instance, so that the new reconstructions geometrically align with existing anatomies. We further verified that residual learning and loss aggregation can significantly boost the performance of the DAE for the reconstructive task, and mitigate the low sensitivity and false prediction issues. Besides the baseline DAE, residual connection and loss aggregation can be easily implemented on top of more complicated network architectures. The models can not only reconstruct multiple missing anatomies simultaneously (Fig. 2) but also a specific anatomy, despite their sizes (Fig. 3 and Appendix A). There are several known limitations remaining to be addressed in future work: (i) Not all anatomy classes are covered by the segmentations of the CT dataset, such as the skull, full limb, brain, skins and soft tissues (e.g., facial soft tissues and most of the muscles); (ii) A quantitative evaluation for each specific anatomy is lacking (only qualitative results are provided in Fig. 3 and Appendix A); (iii) The reconstructions from the multi-class anatomy completer suffer from terracing artifacts and discontinuous ribs. A super-resolution procedure can be applied to refine the initial reconstructions using sparse convolutional neural networks [12]. An interesting direction for future work is to use the multi-class anatomy completor in whole-body segmentation, where it can be used to generate the initial pseudo labels of the organs given only skeletal annotations (e.g., the rib case and spine. See Appendix B).

Acknowledgement. The work is supported by the Plattform für KI-Translation Essen (KITE) from the REACT-EU initiative (EFRE-0801977, https://kite.ikim.nrw/) and "NUM 2.0" (FKZ: 01KX2121) FWF enFaced 2.0 (KLI 1044). The anatomical shape dataset used in this paper can be accessed through *MedShapeNet* at https://medshapenet.ikim.nrw/.

Appendix A. Reconstructing Small Anatomies

(see Fig. 6)

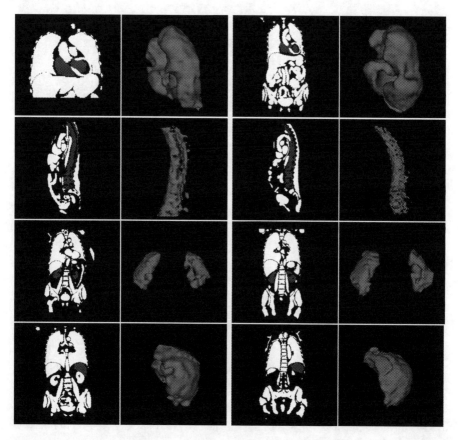

Fig. 6. Reconstruction results of individual, small anatomies by $DAE_{agg+res}$ trained with an increased loss aggregation scope (M). From the top: heart (2.4%), spine (4.3%), kidney (1.7%) and spleen (1.2%). The percentages in the brackets are the approximate volume ratio of the anatomy to the corresponding whole-body segmentations. The preliminary results demonstrate that increasing M (in Eq. 3, 4 in the main manuscript) increases also the sensitivity of the reconstructive model, which helps the model identify and reconstruct very small anatomies. Two test instances are presented for each anatomy class.

Appendix B. Anatomy Completion from Skeletons (rib cage + spine)

(see Fig. 7)

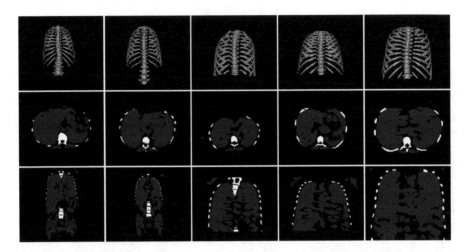

Fig. 7. The first row shows the input skeleton (ribs and spine), and the second to third row show the reconstruction results in axial and coronal views, respectively. The results are obtained by training DAE_{res} on 40 such 'skeleton-full' pairs for 200 epochs. The preliminary results demonstrate the feasibility of reconstructing the full anatomy based only on the skeleton.

References

1. Abadi, M., et al.: Tensorflow: large-scale machine learning on heterogeneous systems. https://www.tensorflow.org (2015)
2. Ali-Hamadi, D., et al.: Anatomy transfer. ACM Trans. Graph. (TOG) **32**(6), 1–8 (2013)
3. Chibane, J., Alldieck, T., Pons-Moll, G.: Implicit functions in feature space for 3D shape reconstruction and completion. In: Proceedings of the IEEE/CVF Conference on Computer Vision and Pattern Recognition, pp. 6970–6981 (2020)
4. Egger, J., et al.: Medical deep learning-a systematic meta-review. Comput. Methods Programs Biomed. **221**, 106874 (2022)
5. Goparaju, A., et al.: Benchmarking off-the-shelf statistical shape modeling tools in clinical applications. Med. Image Anal. **76**, 102271 (2022)
6. He, K., Gkioxari, G., Dollar, P., Girshick, R.: Mask R-CNN. In: Proceedings of the IEEE International Conference on Computer Vision (ICCV) (2017)
7. He, K., Zhang, X., Ren, S., Sun, J.: Deep residual learning for image recognition. In: Proceedings of the IEEE Conference on Computer Vision and Pattern Recognition, pp. 770–778 (2016)

8. Jaus, A., et al.: Towards unifying anatomy segmentation: automated generation of a full-body CT dataset via knowledge aggregation and anatomical guidelines. arXiv preprint arXiv:2307.13375 (2023)

9. Kingma, D.P., Ba, J.: Adam: a method for stochastic optimization. arXiv preprint arXiv:1412.6980 (2014)

10. Kodym, O., Španěl, M., Herout, A.: Skull shape reconstruction using cascaded convolutional networks. Comput. Biol. Med. **123**, 103886 (2020)

11. Kodym, O., Španěl, M., Herout, A.: Deep learning for cranioplasty in clinical practice: going from synthetic to real patient data. Comput. Biol. Med. **137**, 104766 (2021)

12. Kroviakov, A., Li, J., Egger, J.: Sparse convolutional neural network for skull reconstruction. In: Li, J., Egger, J. (eds.) AutoImplant 2021. LNCS, vol. 13123, pp. 80–94. Springer, Cham (2021). https://doi.org/10.1007/978-3-030-92652-6_7

13. La Cava, S.M., Orrù, G., Goldmann, T., Drahansky, M., Marcialis, G.L.: 3D face reconstruction for forensic recognition-a survey. In: 2022 26th International Conference on Pattern Recognition (ICPR), pp. 930–937. IEEE (2022)

14. Li, J., et al.: Towards clinical applicability and computational efficiency in automatic cranial implant design: an overview of the autoimplant 2021 cranial implant design challenge. Med. Image Anal. 102865 (2023)

15. Li, J., Pepe, A., Gsaxner, C., et al.: Medshapenet - a large-scale dataset of 3D medical shapes for computer vision. arXiv preprint arXiv:2308.16139 (2023)

16. Li, J., et al.: Autoimplant 2020-first MICCAI challenge on automatic cranial implant design. IEEE Trans. Med. Imaging **40**(9), 2329–2342 (2021)

17. Li, J., et al.: Automatic skull defect restoration and cranial implant generation for cranioplasty. Med. Image Anal. **73**, 102171 (2021)

18. Matzkin, F., et al.: Self-supervised skull reconstruction in brain CT images with decompressive craniectomy. In: Martel, A.L., et al. (eds.) MICCAI 2020. LNCS, vol. 12262, pp. 390–399. Springer, Cham (2020). https://doi.org/10.1007/978-3-030-59713-9_38

19. Meng, D., Keller, M., Boyer, E., Black, M., Pujades, S.: Learning a statistical full spine model from partial observations. In: Reuter, M., Wachinger, C., Lombaert, H., Paniagua, B., Goksel, O., Rekik, I. (eds.) ShapeMI 2020. LNCS, vol. 12474, pp. 122–133. Springer, Cham (2020). https://doi.org/10.1007/978-3-030-61056-2_10

20. Milletari, F., Navab, N., Ahmadi, S.A.: V-Net: fully convolutional neural networks for volumetric medical image segmentation. In: 2016 fourth international conference on 3D vision (3DV), pp. 565–571. IEEE (2016)

21. Missal, S.: Forensic facial reconstruction of skeletonized and highly decomposed human remains. In: Forensic Genetic Approaches for Identification of Human Skeletal Remains, pp. 549–569. Elsevier (2023)

22. Morais, A., Egger, J., Alves, V.: Automated computer-aided design of cranial implants using a deep volumetric convolutional denoising autoencoder. In: Rocha, Á., Adeli, H., Reis, L.P., Costanzo, S. (eds.) WorldCIST'19 2019. AISC, vol. 932, pp. 151–160. Springer, Cham (2019). https://doi.org/10.1007/978-3-030-16187-3_15

23. Parihar, A., Pandita, V., Kumar, A., Parihar, D.S., Puranik, N., Bajpai, T., Khan, R.: 3D printing: advancement in biogenerative engineering to combat shortage of organs and bioapplicable materials. Regenerative Engineering and Translational Medicine, pp. 1–27 (2021)

24. Pimentel, P., et al.: Automated virtual reconstruction of large skull defects using statistical shape models and generative adversarial networks. In: Li, J., Egger, J.

(eds.) AutoImplant 2020. LNCS, vol. 12439, pp. 16–27. Springer, Cham (2020). https://doi.org/10.1007/978-3-030-64327-0_3

25. Sarmad, M., Lee, H.J., Kim, Y.M.: RL-GAN-Net: a reinforcement learning agent controlled GAN network for real-time point cloud shape completion. In: Proceedings of the IEEE/CVF Conference on Computer Vision and Pattern Recognition (CVPR) (2019)

26. Seibold, C., et al.: Accurate fine-grained segmentation of human anatomy in radiographs via volumetric pseudo-labeling. arXiv preprint arXiv:2306.03934 (2023)

27. Seibold, C.M., Reiß, S., Kleesiek, J., Stiefelhagen, R.: Reference-guided pseudo-label generation for medical semantic segmentation. In: Proceedings of the AAAI Conference on Artificial Intelligence, vol. 36, pp. 2171–2179 (2022)

28. Shi, Y., Xu, X., Xi, J., Hu, X., Hu, D., Xu, K.: Learning to detect 3D symmetry from single-view RGB-D images with weak supervision. IEEE Trans. Pattern Analy. Mach. Intell. (2022)

29. Toscano, J.D., Zuniga-Navarrete, C., Siu, W.D.J., Segura, L.J., Sun, H.: Teeth mold point cloud completion via data augmentation and hybrid RL-GAN. J. Comput. Inf. Sci. Eng. **23**(4), 041008 (2023)

30. Wasserthal, J., Meyer, M., Breit, H.C., Cyriac, J., Yang, S., Segeroth, M.: Totalsegmentator: robust segmentation of 104 anatomical structures in CT images. arXiv preprint arXiv:2208.05868 (2022)

31. Wilkinson, C.: Facial reconstruction-anatomical art or artistic anatomy? J. Anat. **216**(2), 235–250 (2010)

32. Wodzinski, M., Daniol, M., Socha, M., Hemmerling, D., Stanuch, M., Skalski, A.: Deep learning-based framework for automatic cranial defect reconstruction and implant modeling. Comput. Methods Programs Biomed. **226**, 107173 (2022)

33. Yan, X., Lin, L., Mitra, N.J., Lischinski, D., Cohen-Or, D., Huang, H.: Shapeformer: transformer-based shape completion via sparse representation. In: Proceedings of the IEEE/CVF Conference on Computer Vision and Pattern Recognition, pp. 6239–6249 (2022)

34. Zhang, Y., Pei, Y., Guo, Y., Chen, S., Xu, T., Zha, H.: Cleft volume estimation and maxilla completion using cascaded deep neural networks. In: Liu, M., Yan, P., Lian, C., Cao, X. (eds.) MLMI 2020. LNCS, vol. 12436, pp. 332–341. Springer, Cham (2020). https://doi.org/10.1007/978-3-030-59861-7_34

C³Fusion: Consistent Contrastive Colon Fusion, Towards Deep SLAM in Colonoscopy

Erez Posner[✉], Adi Zholkover, Netanel Frank, and Moshe Bouhnik

Intuitive Surgical, Sunnyvale, CA 94086, USA
erez.posner@intusurg.com

Abstract. 3D colon reconstruction from Optical Colonoscopy (OC) to detect non-examined surfaces remains an unsolved problem. The challenges arise from the nature of optical colonoscopy data, characterized by highly reflective low-texture surfaces, drastic illumination changes and frequent tracking loss. Recent methods demonstrate compelling results, but suffer from: (1) frangible frame-to-frame (or frame-to-model) pose estimation resulting in many tracking failures; or (2) rely on point-based representations at the cost of scan quality. In this paper, we propose a novel reconstruction framework that addresses these issues end to end, which result in both quantitatively and qualitatively accurate and robust 3D colon reconstruction. Our SLAM approach, which employs correspondences based on contrastive deep features, and deep consistent depth maps, estimates globally optimized poses, is able to recover from frequent tracking failures, and estimates a global consistent 3D model; all within a single framework. We perform an extensive experimental evaluation on multiple synthetic and real colonoscopy videos, showing high-quality results and comparisons against relevant baselines.

Keywords: Colonoscopy · Coverage · 3D-Reconstruction

1   Introduction

The third most commonly diagnosed cancer worldwide is colorectal cancer (CRC) with over than 1.9 million incident cases in 2020 [3]. CRC is also among the most preventable cancers [28] and can be prevented from progressing if detected in it's primary stages by conducting screening and early detection measures [27,29]. Consequently, global incidence rates have been decreasing in the screening-eligible age group (50–75) due to the adoption of CRC screening [1]. The most common screening procedure is optical colonoscopy (OC) [20], which visually inspects the mucosal surface for abnormalities in the colon such as colorectal lesions. Nevertheless, performing a thorough endoscopic colon investigation solely from OC is very difficult. In practice, this means that not all regions of the colon will be covered and fully examined; consequently, tainting the polyp

C. Wachinger et al. (Eds.): ShapeMI 2023, LNCS 14350, pp. 15–34, 2023.
https://doi.org/10.1007/978-3-031-46914-5_2

detection rate. Lately, we are seeing a bloom in deep learning-based methods adapted to predict depth-maps from OC [6,33,35,42], aimed at providing a complete 3D geometric information of the colon including polyps. Thus, indicating the un-inspected surfaces during OC; as a result, increasing the polyp detection rate.

Despite the profusion of reconstruction solutions, a holistic solution for the problem of 3D colon reconstruction at scale that addresses real life issues during OC has yet to be seen. This is due to the numerous requirements that such system would have to support:

Accurate depth prediction - producing high-quality geometry-consistent depth estimation from a monocular video is imperative as well as challenging. *Scalability* - chosen representation should support extended scale environments while preserving global structure, and high local accuracy. *Global consistency* - the method should be robust to pose drifts and estimation error in order to enable the re-examination of previously scanned areas or loop closure. *Robust camera tracking* - tracking failure is extremely frequent in OC. Occlusions, fast motions, featureless regions [26] and deficient frames are a fraction of the reasons that contribute to loss of track. When these occur, the system should have the ability to re-localize the camera position.

There have been studies addressing specific parts of these problems [26,32, 35,38,40,43]. Direct SLAM systems optimize a photometric error which is susceptible to drastic illumination changes in OC imagery. Ma et al. [26] reconstructed fragments of the colon using Direct Sparse Odometry (DSO) [12] and a Recurrent Neural Network (RNN) for depth estimation. Zhang et al. [43] predicted gamma correction value to alleviate sudden illumination changes and [42] improved the depth estimation network using geometry-consistency losses. Indirect SLAM methods, like [9,30], usually utilize keypoints matching based on handcrafted descriptors. This kind of descriptors e.g., SIFT [24], are based on local gradients and hence not well suited to often texture-less and shadow prone OC imagery. Modern deep network based descriptors adopt CNN to predict both keypoint and descriptors for local feature matching. DeTone et al. [11], predicts keypoints and descriptors directly from a pre-trained DNN. However, it's localization accuracy is hampered due to the low dimensionality output. Moreover, as it's training is based on corner detection keypoints it is not optimized to OC cases, characterized with numerous occlusions. Although these aforementioned studies show promising results, there hasn't been a single solution to tackle all of these requirements up to date.

Our goal in this paper is to rigorously cope with *all* of these requirements in a single, end-to-end 3D reconstruction pipeline. At the core of our method is a robust positioning estimation scheme that utilizes contrastive deep-feature based correspondences. The proposed method globally optimizes the camera pose per-frame, taking into consideration all previously captured frames in an effective *local-to-global* hierarchical optimization framework.

In summary, the main contributions of our work are as follows: (1) A novel, deep-learning-driven global pose alignment SLAM system for OC which incorporates the complete sequence of input frames and removes the fuzzy nature of temporal tracking accuracy issues; (2) Large-scale colon 3D reconstruction, demonstrating model refinement in revisited areas, recovery from tracking failures, and robustness to drift and continuous loop closures; and (3) a novel method for local feature matching in low-texture areas, implicit loop closures in highly indistinguishable environments and highly-accurate fine-scale pose alignment.

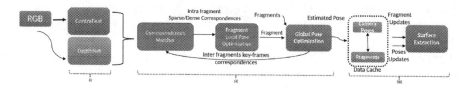

Fig. 1. Our novel, deep-learning-driven global pose alignment framework for colonoscopy SLAM system.

2   Method Overview

The main system pipeline (shown in Fig. 1) consists of three major parts: (i) depth estimation and deep feature extraction, (ii) hierarchical pose optimisation, and (iii) surface fusion. For each new frame, part (i) outputs a depth-map and keypoints with their deep descriptors, by inferencing DepthNet (Sect. 3.1) and ContraFeat (Sect. 3.2) respectively. Part (ii) starts with matching the new keypoints against previous frames and filtering mismatches (Sect. 4.1) to improve alignment and avoid false loop closures. To manage large scaled sequences comprised out of massive amount of frames and to make pose alignment fast, we carry out a hierarchical local-to-global pose optimization. This achieves robustness to frequent loss of tracking as we do not solely depend on temporal consistency. Thereby, enabling swift re-localization and allowing multiple visits of the same regions within the scene.

On the first (local) hierarchy level, $fragments$ are composed of sets of successive frames sharing similar spatial coverage. Each frame's pose is optimized by taking all of the $fragment's$ frames into account. On the second (global) hierarchy level, all $fragments'$ pose are optimized with respect to each other. In part (iii) the global 3D scene representation is acquired by fusing all fragments (Sect. 5) into a non-parametric surface represented implicitly by a scalable truncated signed distance function (TSDF) [8,45] following with marching cubes [22] applied to this volume to extract the final mesh.

3   Deep-Depth and Deep-Descriptors

3.1   Deep-Depth Self-supervised Training

Given as input an RGB image I_t, DepthNet predicts a depth map $\tilde{D}_t$. During training, for every frame I_t in a sequence of three sequential frames I_{t-1}, I_t, I_{t+1},

DepthNet predicts their corresponding depth maps $\tilde{D}_{t-1}, \tilde{D}_t, \tilde{D}_{t+1}$. PoseNet predicts relative camera poses between each image pair: $T_{ij} \forall (i,j) \in F_i = \{\forall i | j = i \pm 1\}$.

The network architecture is similar to the one used in Monodepth2 [13]. To reduce the impact of strong visual distortions (e.g. lens distortion) that characterized OC videos, we adopt deformable convolution [10] for the depth encoder, similar to [4,19]. We train in the same manner as [13], using auto-masking (μ) and per-pixel minimum photometric loss (L_{ph}). Since the photometric loss is not sufficiently informative for low-texture regions, common in OC imagery, and to enforce structural coherence, we apply extra regularization in the form of depth consistency loss [2], and extra spatio-temporal consistency losses. Considering that the additional regularization implicitly imposes depth smoothness, we discard the smoothness term (L_{ds}) used in [13].

To deal with specular reflections and occlusions by haustral folds we: (1) mask and in-paint specular reflections as in [32] and (2) remove outlier pixels having a loss greater than the 80-th percentile for the photometric and depth consistency errors. We compute the additional spatio-temporal consistency and depth consistency losses i.e., $L_{ph-extra}$ and L_{dc} respectively, between all image pairs in S.

$$L_{ph-extra}^{(i,j)} = \frac{1}{|V_\mu|} \sum_{p \in V_\mu} pe(I_i, I_{j \rightarrow i}), \tag{1}$$

$$L_{dc}^{(i,j)}(\tilde{D}_{j \rightarrow i}, \hat{D}_i) = \frac{\left| \tilde{D}_{j \rightarrow i} - \hat{D}_i \right|}{\tilde{D}_{j \rightarrow i} + \hat{D}_i}, \tag{2}$$

$$\forall (i,j) \in S = \{i = \{t-1, t+1\}, j = \{t-1, t, t+1\}, i \neq j\}$$

where pe is the photometric error from [13] containing L1 and SSIM losses, V_μ is a set comprises of all valid pixels in mask μ and the specular reflection mask based on image intensity threshold. $I_{j \rightarrow i}$ is a warped view of I_j to I_i pose, using the predicted depth $\tilde{D}_j$ and T_{ij}. $\tilde{D}_{j \rightarrow i}$ is the predicted depth map for image I_j in the coordinate system of image I_i. $\hat{D}_i$ is the interpolated depth map from the estimated depth map $\tilde{D}_i$. The final loss for the network is given in Eq. 3, where $\lambda_{ph-extra}, \lambda_{dc}$ are weights for the different loss components.

$$L = L_{ph} + \lambda_{ph-extra} \frac{1}{|S|} \sum_{(i,j) \in S} L_{ph-extra}^{(i,j)} + \lambda_{dc} \frac{1}{|S|} \sum_{(i,j) \in S} L_{dc}^{(i,j)} \tag{3}$$

3.2   Deep-Descriptors

Our deep feature descriptor block, $ContraFeat$, employs the detected keypoints from [24] in each frame, and extracts their deep feature representations $z = \phi(kp)$, where kp is a SIFT keypoint in 2D pixel coordinates. For the feature map ϕ, we use FPN [21] architecture in the bottom-up stream. The final feature map has the same spatial resolution as the original image and thus, retaining

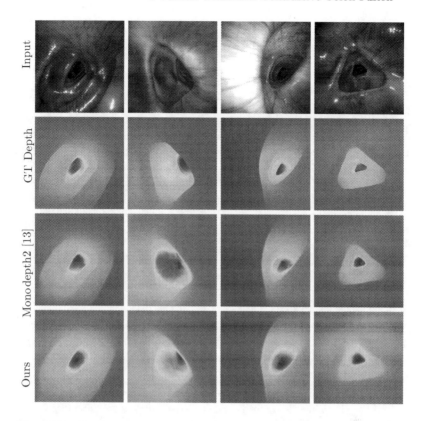

Fig. 2. Qualitative comparison of predicted depth-maps on synthetic data.

keypoint's pixel level accuracy. Each pixel is represented by a descriptor vector of length $c = 128$. Finally, to find correspondence sets between frames, descriptors are matched using cosine similarity. To train *ContraFeat* network, we use synthetic data [41](see Sect. 6). Accordingly, keypoints are extracted using [24]. Then, using known depth, pose and camera intrinsics, we collect ground-truth matches from corresponding 3D points and remove occluded points by filtering-out distant matches.

Contrastive Loss. Inspired by recent self-supervised learning methods based on contrastive losses [37,39], we use a loss similar to the InfoNCE loss [31] to train ContraFeat to learn discriminative representations of keypoints. The contrastive loss for an image pair (i, j) and a correct match k out of $M_{i,j}$ possible matches is given by

$$l_c^{i,j,k} = -\log \frac{\exp\left((z_i^k)^T \cdot z_j^k / \tau\right)}{\sum_{m=0}^{M_{i,j}} \exp\left((z_i^m)^T \cdot z_j^m / \tau\right)} \tag{4}$$

where z_i^m and z_j^m are descriptor vectors sampled at pixel coordinates kp_i^m and kp_j^m in the feature map of images i and j, respectively. τ represents a temperature parameter. We enforce $\|z_i^m\|_2 = 1$ via a L2-normalization layer. This loss is then averaged over all ground-truth matches.

4   Pose Alignment

Our system takes an RGB-D stream consisting of pairs $(I_{RGB}^i, I_{D_{pred}}^i)$, where i is the frame index, I_{RGB} is the 3-channel color image and $I_{D_{pred}}$ is the predicted depth map by the DepthNet network. Intrinsic and distortion parameters are assumed to be known. The goal of this stage is to estimate the ideal set of rigid camera poses $T_i = \{(R_i, t_i) | R_i \in SO(3), t_i \in \mathbb{R}^3\}_{i=1}^N$ in which all N frames align as best as possible, based on extracted 3D correspondences between all overlapping frames. The estimated transformations $T_i(p) = R_i p + t_i$ localizes all frames in the global coordinates system defined relative to the first frame and $p \in \mathbb{R}^3$.

4.1   Feature Matching

In order to find correspondence sets that result with a coherent and stable rigid transform between pair of frames f_i, f_j, we set to minimize outliers. To this end, we utilize the key point correspondence filter and the Surface Area Filter as in [9] to filter the sets of frame-pairwise matches based on geometric and feature-representation constancy. The transformation $T_{ij} \in SE(3)$ is constructed between f_i, f_j if minimal 10 matches are found with a re-projection error under 0.02cm and if it is valid. i.e. condition analysis results with a condition number less than ϵ_{cn}.

4.2   Hierarchical Pose Optimization

A colonoscopy procedure typically takes 30–60 minutes at 30 FPS. To be able to process such massive amount of frames in reasonable time we follow [9] and split the input sequence into fragments of consecutive frames that share similar coverage and apply two stage hierarchical optimization strategy. On the lower hierarchy level, we perform pose-graph optimization [15] to register all frames within a fragment. On the higher hierarchy level, we register all fragments with respect to each fragment's keyframe.

Fragment Construction Conditions. We keep track of one active fragment at all times. A new frame will either be appended to the active fragment, or will trigger the creation of a new fragment as it keyframe. There are two conditions which determine whether a new fragment should be constructed. (1) Structural affinity between the last two consecutive frames (i.e., minimal number of correspondences found is less than 100) (2) The new frame and the active fragment keyframe view frustums overlap is less than 85%.

Inter vs. Intra Fragment Registration. The two hierarchies fragment registration processes are similar. In Intra-fragment (local) registration the pose-graph optimization (see Sect. 4.3) is applied for all fragment's inhabited frames to align as best as possible with respect to the fragment's keyframe. Whereas in Inter-fragment (global) registration, we estimate the best global registration for all fragments using solely theirs keyframes.

Note that for the inter-fragment registration, we do not discard keyframes that have no correspondences with past keyframes. Instead, we keep them as a candidates, as they could share correspondences with future fragments. This enables incorporating the lone fragment later on in the sequence.

4.3   Registration as Pose-Graph Optimization

The goal of the pose-graph optimization is to estimate the ideal set of rigid transforms $\mathbb{T} = \{T_i\}$ such that all set of input frames F (which depends on the hierarchy level) align as best as possible. The process uses [14, 18] to estimate the relative rigid transforms T_{ij} $\forall (i, j) \in F$ based on the matched features and their predicted depth value. Given $\{T_{ij}\}$, we construct a pose-graph [15] with vertices $\{f_i\}$ and edges T_{ij}. As in [7], we set to minimize the inconsistency measure g between poses T_i, T_j and the relative pose T_{ij}, defined as the sum of squared distances between corresponding points in $T_i P_i$ and $T_j P_j$:

$$g(T_i, T_j, T_{ij}) = \sum_{(i,j)} \|T_j^{-1} T_i p_i - T_{ij} p_i\|^2 \tag{5}$$

Additional outlier removal filter in the form of edges pruning is applied to further improve the algorithm's robustness against false correspondences.

5   Scene Reconstruction

The colons 3D model is reconstructed by carefully fusing all RGB images, their predicted depth maps and the optimized global poses into an implicit scalable TSDF representation. The TSDF's unique features enable us to alleviate any further inconsistencies in successive depth maps predictions. The fusing scheme is based on the premise that the endoscope is slowly being withdrawn during the procedure; consequently, inspected regions won't be visited again. We fuse fragments when enough time has passed since last inspected ($> \epsilon_{f_{na}}$), and when the current camera position is far enough ($> \epsilon_{cf_d}$). This approach scales well to non-fixed scenes common in colonoscopy sequences, as demonstrated in Sect. 6.2.

6   Results

In this section, we analyze the results of our framework on 3 different data-sets: A colon simulator data-set, a CT colon rigid print, and real optical colonoscopy videos.

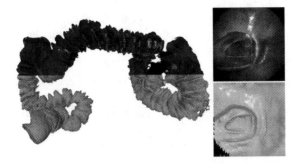

Fig. 3. Left: Full endoscopic colon reconstruction result on the synthetic data-set. Top right: the captured RGB images, Bottom right: The re-rendered reconstructed model.

We start by evaluating our results using a realistic synthetic colonoscopy simulator [41]. Like [42], we used the simulator to create 8 sequences of endoscopic procedures. For every frame, the ground-truth depth and pose are known. Each sequence contains on average 2000 frames with a trajectory length of 125cm, at a resolution of 512×512 with a field of view (FOV) of $125°C$. Different 'Material', 'Light and 'Wetness' properties were set to best resemble real imagery (See appendix A.1 for details) . We split the data-set sequences, similarly to [42], into training/validation/testing sets, containing 9.5 k/0.5 k/2.5 k frames. The metrics below are reported on the test set.

We also captured an endoscopic sequence of a rigid 3D-print CT based colon, using a calibrated Olympus CF-H185L/I colonoscope, along with the ground-truth camera trajectory that was captured by an EM tracker. The colon 3D model was fabricated as follows: a CT colon scan from [36] was segmented in order to extract the 3D surface of the colon, following with post processing re-modeling operations (cleaning, re-topology, skinning etc.) and texturing. The extracted mesh was then printed using a 3D printer. We intend to elaborate on this data-set creation in a future paper.

Lastly, to qualitatively test our framework on real optical colonoscopy sequences, we use Colon10K data-set [25]. Colon10K data-set contains 20 sequences cropped from full colonoscopies, each contains on average 500 frames. We split them to training/testing sets at a ratio of 80/20% such that the entire test sequences are not seen during training. For training and implementation details, please see Appendix A.2

6.1 Quantitative Results

In this section we start by evaluating our framework's components performance. First, our monocular depth estimation is compared to relevant baselines, showing it's strengths in predicting consistent depth-maps. Following with an assessment and comparison of our ContraFeat deep-descriptor in terms of recall/precision. Concluding with dense tracking evaluation, in which we analyze our full pipeline reconstruction trajectory accuracy.

Fig. 4. Left: Full endoscopic colon reconstruction result on 3D colon print. Right: zoomed in segments with visible holes (regions that were not covered during the scan).

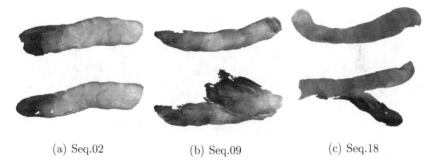

(a) Seq.02   (b) Seq.09   (c) Seq.18

Fig. 5. 3D reconstruction results on Colon10K data-set. Our proposed framework (top) outperforms the mesh reconstructed from depth and pose predictions by Godard et al. [13] (bottom).

Monocular Depth Estimation. We train our network and [13] on the synthetic data-set with the same data split paradigm as described above and compare the results. To accommodate different scales of depth-maps, we use per-image median ground-truth scaling as introduced by [46]. As shown in Table 1, our monocular method outperforms existing state-of-the-art self-supervised approaches. Specifically, our Sq Rel error has improved substantially (15.5% relatively) over [13] which corresponds to improved depth consistency.

Keypoints Correspondence Evaluation. We estimate the keypoint matching performance as precision (i.e., the percentage of **correct** correspondence from all correspondences found) and recall (i.e., the percentage of valid correspondences from all ground-truth correspondences) on the synthetic data-set with ground truth depth and pose. To determine if a keypoint pair correspondence is correct, we test if it lies in the ground truth set. We evaluate our contrastive deep-feature (*CDF*) performance by comparing it to SIFT-based and SuperPoint [11], with and without the addition of the key points filter (*kpf*) described in Sect.4.1. Table 2 shows that SIFT matches suffers from a low recall and extremely low precision which can impact the overall reconstruction. The

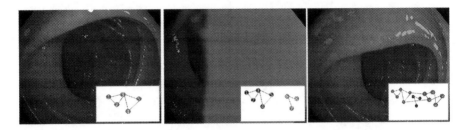

Fig. 6. Tracking failure recovery on real data: keyframes and their connectivity graph. (middle) recovery is lost (indicated as an additional disjoint sub-graph), (right) recovery is enabled as a connection is formed between frames 5 <-> 9. Active keyframes are highlighted (in blue). (Color figure online)

Fig. 7. Comparison of our frame-to-model approach with (left), and without (right) loop closure. The point-clouds are color coded based on their timestamp.

additional *kpf* increases the precision drastically to around 99.5% on average, at the cost of lowering the recall as can be expected. Although SuperPoint matches start with higher precision/recall than SIFT, after applying the key points filter, the precision greatly improves at the cost of drastically decreasing the recall. This could indicate that the localization error of the matched keypoints is high and therefore get filtered out. The *CDF* matches starts with much higher recall & precision, combined with *kpf* it achieves outstanding results of 74.9% recall and 100% precision on average. The results show that our hybrid approach i.e., using SIFT keypoint with contrastive deep-features, is well suited to OC domain. See Appendix A.3 for examples of predicted correspondences between frames.

Table 1. Quantitative results on the synthetic data.

Method	Abs Rel	Sq Rel	RMSE	RMSE log	$\delta < 1.25$	$\delta < 1.25^2$	$\delta < 1.25^3$
MonoDepth2 [13]	0.089	0.058	0.540	0.134	0.931	**0.982**	0.992
ColDE [42] †	0.077	0.079	0.701	0.134	0.935	0.975	0.989
Ours	**0.075**	**0.049**	**0.521**	**0.126**	**0.94**	0.980	**0.992**

† Results taken from paper since training data-set and code are not publicly available

Table 2. Keypoints recall/precision on the synthetic data-set: SIFT, SuperPoint (SP) and our ContraFeat deep feature (CDF) with/without our key points filter (kpf).

Seq	SIFT[%]	SIFT+kpf[%]	SP[%]	SP+kpf[%]	CDF[%]	CDF+kpf[%]
Seq1	31.4/12.1	24.2/100	62.4/78.9	22.4/100	74.2/86.1	**70.2/100**
Seq2	44.4/32.1	29.9/98.4	59.2/68.2	17.3/100	82.3/89.2	**79.2/100**
Seq3	37.8/21.3	26.8/100	55.4/64.8	16.2/100	78.1/83.5	**75.2/100**

Full Endoscopic Reconstruction. We evaluate the trajectory accuracy over the synthetic data-set and the 3D colon print sequence, and compare it to Direct Sparse Odometry (DSO) [12]. We also provide results for the SIFT-based feature descriptors instead of our suggested deep feature descriptor. Table 3 summarizes the ATE [44] results. As can be seen from the rmse and std values, our method surpasses DSO even using SIFT-based feature descriptors. When incorporating the deep features, our final approach is 46% better than DSO on the synthetic datast and 79% on the 3D colon print. This shows the major contribution of our framework to the reconstruction. It is important to note that, although ContraFeat was only trained on synthetic data, the ATE has improved by 10% relatively over SIFT on the 3D print model. Thus, demonstrating ContraFeat's robustness. See Appendix A.4 for visual trajectories comparison.

Table 3. Avg. ATE statistics over synthetic and 3D colon print data-sets (avg. trajectory of 125cm).

Method	synthetic data-set		3D colon print data-set	
	RMSE[cm]	std[cm]	RMSE[cm]	std[cm]
DSO	53.4	12.3	37.1	13.9
Ours(SIFT)	36.1	3.3	8.72	**3.53**
Ours	**28.9**	**2.7**	**7.88**	3.75

6.2 Qualitative Results

Monocular Depth Estimation. In Figure 2 we qualitatively compare our predicted depth maps with monodepth2 [13]. Note that our method produces high quality depth-maps characterized with consistent depth around the colon's surface while maintaining sharp boundaries around haustral folds. Furthermore, our method is more robust to specular reflection artifacts. Extra depth-maps can be seen in Appendix A.5.

Full Endoscopic Reconstruction Figure 3 and Fig. 4 depict the qualitative results of our method; showing the reconstructions of a fully endoscopic colon investigation on the synthetic data-set and the 3D colon print respectively. Figure 3 demonstrates that our approach produces high quality scans with negligible camera drift and high local quality of the reconstructions in both geometry and texture. Note in Fig. 3, the clear resemblance between the re-rendered mesh of the reconstructed model (in gray) and the captured RGB images. We are able to successfully capture the geometric curvatures of the colon while keeping the missing regions visible. In addition, we show in Fig. 5 a qualitative comparison of the reconstructed surfaces based on real optical colonoscopy videos from Colon10K. The results from MonoDepth2 [13] were produced using their pre-trained network on ImageNet, which was fine-tuned with semi-supervision over the data-set. To be comparable, we additionally fused [13] outputs as described in Sec.5 in order to generate a mesh. Extra images from different point of views are shown in Appendix A.6.

Our novel hierarchical global pose optimization framework implicitly handles loop closure, recovers from tracking failures, and reduces geometric drift. Our method is able to support multiple loop closure as it does not rely on explicit loop closure detection; thus, scales better.

Tracking Failure Recovery. When a new keyframe cannot be aligned with any of the previous keyframes, tracking is assumed to be lost. Sensibly, this means that the keyframe won't have any edges connecting it to any previous keyframes in the pose-graph optimization. Thus, an additional lone fragment will be created in the fragment's connectivity graph and the predicted surface won't be included in the output reconstruction. Based on our approach, recovery is enabled to any previously scanned areas as we don't require temporal nor spatial coherence. A common tracking loss sequence is shown in Fig. 6 in which the camera is occluded by a Haustral fold (colon wall protrusions). As our method globally matches new keyframes against all existing keyframes, tracking can be lost and recovered at a completely different place.

Loop Closure Operation. Our global pose optimization continuously operates in the background; detects and handle loop closures seamlessly thus mitigating camera pose errors and evidently preventing geometric drifts over time. See Fig. 7 where the synthetic sequence is played forward and backward in order to create a loop. Notice how the forward and backwards trajectories align in our method versus the vanilla frame-to-model approach.

7   Conclusion

We have presented a novel deep learning 3D reconstruction approach that provides a robust tracking with negligible geometric-drift and implicitly solves the tracking loss problem frequent during OC. The proposed approach was evaluated on multiple data-sets, showing outstanding reconstruction quality and completeness compared to previously suggested methods. Additional experiments were

conducted to illustrate the proficiency of the suggested method in several difficult cases common in colonoscopic sequences not supported by previous methods. The reconstruction can be used to indicate un-inspected surfaces that could
contain colorectal lesions and decrease the miss rates of polyps. Although our
framework deals with real life issues common in OC, some still remains unat-
tended: non-rigidity of the colon can be suppressed within the TSDF only up to
some extent and dynamic objects like stool and degenerate frames needs to be
discarded. We leave these further explorations to future work.

A   Appendix

A.1   Synthetic Data Generation

For the purpose of reproducibility, we state the parameters that were used to
build each synthetic sequence using the synthetic colonoscopy simulator [41].
The parameters are summarised in Table. 4, where RP stands for *Random Path*.
It is worth mentioning that the user does not have the ability to set the seed of
the random number generator for the random path chosen.

A.2   Depth Training and Implementation Details

We use AdamW optimizer [23], with $\beta_1 = 0.9$, $\beta_2 = 0.999$. We train the synthetic
and the Colon10k models for 40 epochs, with a batch size of 16 on a 24GB
Nvidia 3090 RTX. The initial learning rate is 10^{-4}; we reduce it by half on
each of the 16th, 24th and 32nd epochs. As for the 3D colon print model, we
train for 200 epochs; reduce the learning rate by half on each of the 80th, 120th
and 170th epochs. We center-crop the synthetic images to 400×400 to remove
vignetting effects. The Colon10k images are provided in an un-distorted and
center-cropped version of 270×216 pixels. Finally, the cropped image is scaled
to 224×224 before feeding to the network. For the 3D colon print, we employ
test time training due to the scarcity of the data and the fact that the training
process is completely self-supervised.

To generate the specular reflection mask for each frame, we convert the input
frames to YUV color-space and apply a threshold of 90% on the Y channel and
dilate the resulting binary mask with a kernel of 13 pixels.

We use MMLab's [5] implementation of ResNet [16], deformable convolutions
and FPN. All ResNet encoders and the FPN were pre-trained on ImageNet [34].
We use ResNet50 for the depth encoder. For the pose encoder and FPN, we use
ResNet18. Deformable convolution layers are applied in the depth encoder stages
of conv3, conv4 and conv5. We set $\lambda_{ph-extra} = 0.1$ and $\lambda_{dc} = 0.1$, $\tau = 0.01$.

A.3   Correspondence Matching Qualitative Results

Matching examples of ContraFeat, SIFT [24] and SuperPoint [11] are shown
in Fig. 8. ContraFeat incline to produce more correct matches and spread out
evenly throughout the image, and is less susceptible against drastic illumination
changes.

SIFT [24]   SuperPoint [11]   ContraFeat

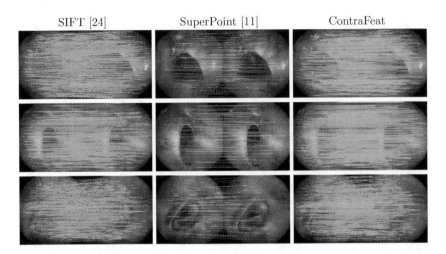

Fig. 8. Matching qualitative comparison on the synthetic data. Correct matches are green lines and mismatches are red lines. Mismatches defined when correspondence re-projection error is greater than 1% of colons diameter. (Color figure online)

A.4   Comparison of the Estimated Trajectories and Ground Truth Trajectories

Fig 9 compares the estimated trajectory and ground truth trajectory on the 3D colon print between DSO [12], our framework using SuperPoint [11] and our proposed method. The pose estimation from the network is of arbitrary scale. Therefore, we first align the two trajectories using similarity transform [18] following with first-frame alignment for better visualization and comparison. Note that the estimated trajectories by our framework is more accurate with loops of similar shape as compared to the ground truth trajectory.

DSO [12]   Ours + SuperPoint [11]   Ours

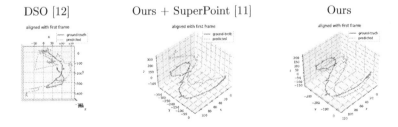

Fig. 9. Comparison of the estimated trajectories and ground truth trajectories on the 3D colon print sequence.

A.5   Extra Qualitative Depth-Map Predictions Results

In Fig. 10 and Fig. 11 we show extra depth-map predictions of the 3D colon print and Colon10K [25].

A.6   Extra Qualitative 3D Reconstruction Results on Colon10K

In Fig. 12 and Fig. 13 we show extra points-of-view of the 3D reconstructions by our proposed framework on Colon10K [25] and 3D colon print data.

Table 4. Synthetic data creation parameters.

Parameters	Sequence							
	Seq. 1	Seq. 2	Seq. 3	Seq. 4	Seq. 5	Seq. 6	Seq. 7	Seq. 8
Path	A	RP	B	RP	RP	RP	RP	RP
Trip Duration	150	150	150	150	150	150	150	150
Shots per sec	30	30	30	30	30	30	30	30
Shots Resolution	512×512	512×512	512×512	512×512	512×512	512×512	512×512	512×512
Hue	0	0	3	3	100	100	100	100
Saturation	72	72	90	90	78	78	78	78
Value	100	100	100	100	52	52	52	52
Wetness	88	88	56	56	40	40	40	40
Vessel Size	60	60	60	60	60	60	60	60
Vessel Opacity	30	30	30	30	30	30	30	30
Angle	150	150	150	150	150	150	150	150
Intensity	74	74	59	59	62	62	62	62
Distance	0	0	0	0	0	0	0	0
Shadow	On	On	On	On	On	On	On	On
Dynamic	Off	Off	Off	Off	Off	Off	Off	Off
Field of View	110	110	110	110	110	110	110	110
Vignette	On	On	On	On	On	On	On	On
Bloom	On	On	On	On	On	On	On	On
Grain	0	0	1	1	3	3	3	3

A.7   SuperPoint Training

SuperPoint [11] was trained using [17] Pytorch implementation with their suggested improvements that enable end to end training using a softargmax at the detector head and a sparse descriptor loss that allows an efficient training. Photo-metric augmentations were adapted to the colon data-set by lowering the contrast, blur and noise levels to values that enabled the extraction of features even from deeper shadowed areas of the colon. the network was trained for about 100 epochs, with a batch size of 10 and learning rate of 0.0003. The best checkpoint was chosen based on validation set precision and recall.

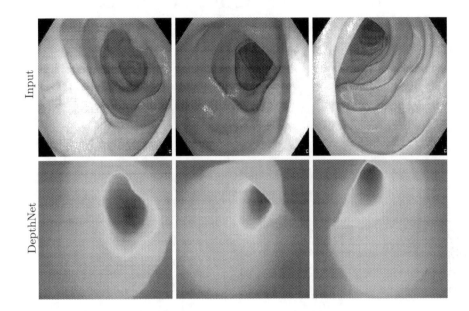

Fig. 10. Extra depth map prediction results on 3D colon print.

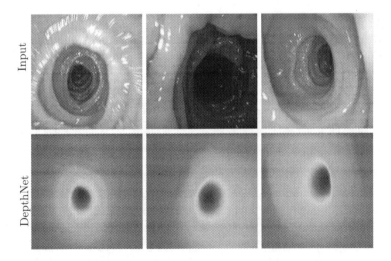

Fig. 11. Extra depth map prediction results on Colon10K [25]. Left image exhibits a highly specular area with strong motion blur. The middle image exhibits strong illumination differences. The right image exhibits low texture images. In all three examples, our depth network produces detailed and artifact-free depth maps.

Colon10K [25]   3D colon print

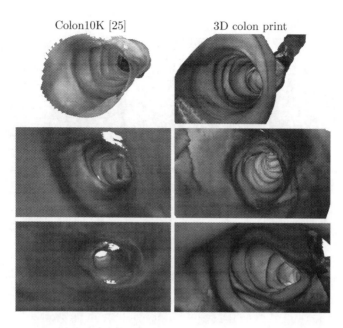

Fig. 12. Extra reconstruction qualitative results on Colon10K [25] and 3D colon print.

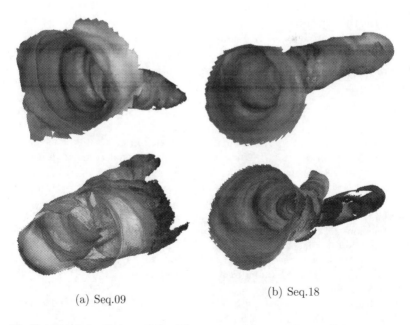

(a) Seq.09   (b) Seq.18

Fig. 13. Extra points-of-view of the 3D reconstruction results on Colon10K data-set. proposed framework (top), mesh reconstructed from depth and pose predictions by Godard et al. [13] (bottom).

RGB Image   Re-rendered reconstructed
                             model (with texture)

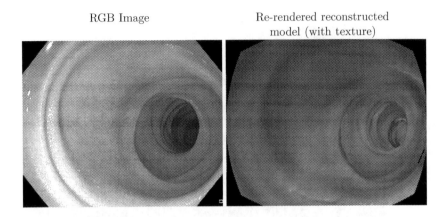

Fig. 14. Captured video and re-rendered reconstruction model similarity.

A.8 Supplementary Video Results

In the supplementary video, labeled as *rgb_tex_geo.mp4*, we show the fully endoscopic investigation of the 3D colon print while comparing the resemblance between the reconstructed model and the captured RGB images. This is accomplished by re-rendering the reconstructed model using the camera intrinsics, camera predicted pose and framework's output mesh. In the video *rgb_tex_geo.mp4* we visualise the captured video (Left) next to the re-rendered reconstructed model with texture (right). An example can be seen in Fig. 14. An additional camera fly-through video is available, labeled as *fly_through.mkv*, showing the final reconstruction of the 3D colon print.

References

1. Alyabsi, M., Algarni, M., Alshammari, K.: Trends in colorectal cancer incidence rates in Saudi Arabia (2001–2016) using Saudi national registry: Early- versus late-onset disease. Front. Oncol. **11**, 3392 (2021)
2. Bian, J., et al.: Unsupervised scale-consistent depth and ego-motion learning from monocular video. In: NeurIPS (2019)
3. International Agency for Research on Cancer: Globocan 2020: Cancer fact sheets-colorectal cancer". https://gco.iarc.fr/today/data/factsheets/cancers/10_8_9-Colorectum-fact-sheet.pdf
4. Xiang, C., H., Li, K., Fu, Z., Liu, M., Chen, Z., Guo, Y.: Distortion-aware monocular depth estimation for omnidirectional images. IEEE Signal Process. Lett. **28**, 334–338 (2021)
5. Chen, K., et al.: MMDetection: open mmlab detection toolbox and benchmark. arXiv preprint arXiv:1906.07155 (2019)
6. Chen, R.J., Bobrow, T.L., Athey, T., Mahmood, F., Durr, N.J.: Slam endoscopy enhanced by adversarial depth prediction. In: KDD Workshop on Applied Data Science for Healthcare 2019 (2019)

7. Choi, S., Zhou, Q.Y., Koltun, V.: Robust reconstruction of indoor scenes. In: IEEE Conference on Computer Vision and Pattern Recognition (CVPR) (2015)
8. Curless, B., Levoy, M.: A volumetric method for building complex models from range images. Proceedings of the 23rd Annual Conference on Computer Graphics and Interactive Techniques (1996)
9. Dai, A., Nießner, M., Zollhöfer, M., Izadi, S., Theobalt, C.: Bundlefusion: real-time globally consistent 3d reconstruction using on-the-fly surface re-integration. CoRR (2016)
10. Dai, J., et al.: Deformable convolutional networks. In: 2017 IEEE International Conference on Computer Vision (ICCV), pp. 764–773 (2017)
11. DeTone, D., Malisiewicz, T., Rabinovich, A.: Superpoint: self-supervised interest point detection and description. CoRR (2017). http://arxiv.org/abs/1712.07629
12. Engel, J., Koltun, V., Cremers, D.: Direct sparse odometry. IEEE Trans. Pattern Anal. Mach. Intell. (2018)
13. Godard, C., Mac Aodha, O., Firman, M., Brostow, G.J.: Digging into self-supervised monocular depth prediction (October 2019)
14. Gower, J.: Generalized procrustes analysis. Psychometrika **40**(1), 33–51 (1975)
15. Grisetti, G., Kümmerle, R., Stachniss, C., Burgard, W.: A tutorial on graph-based slam. IEEE Intell. Transp. Syst. Mag. **2**(4), 31–43 (2010)
16. He, K., Zhang, X., Ren, S., Sun, J.: Deep residual learning for image recognition. In: Proceedings of the IEEE Conference on Computer Vision and Pattern Recognition (CVPR) (June 2016)
17. Jau, Y.Y., Zhu, R., Su, H., Chandraker, M.: Deep keypoint-based camera pose estimation with geometric constraints. In: 2020 IEEE/RSJ International Conference on Intelligent Robots and Systems (IROS), pp. 4950–4957 (2020). https://doi.org/10.1109/IROS45743.2020.9341229
18. Kabsch, W.: A solution for the best rotation to relate two sets of vectors. Acta Crystallogr. A **32**(5), 922–923 (1976)
19. Kumar, V.R., et al.: Fisheyedistancenet: self-supervised scale-aware distance estimation using monocular fisheye camera for autonomous driving. 2020 IEEE International Conference on Robotics and Automation (ICRA), pp. 574–581 (2020)
20. Liang, Z., Richards, R.: Virtual colonoscopy vs optical colonoscopy. Expert Opinion Med. Diagn. 4(2), 159–169 (2010), 20473367[pmid]
21. Lin, T.Y., Dollár, P., Girshick, R.B., He, K., Hariharan, B., Belongie, S.J.: Feature pyramid networks for object detection. In: 2017 IEEE Conference on Computer Vision and Pattern Recognition (CVPR), pp. 936–944 (2017)
22. Lorensen, W.E., Cline, H.E.: Marching cubes: a high resolution 3d surface construction algorithm. SIGGRAPH Comput. Graph. **21**(4), 163–169 (1987)
23. Loshchilov, I., Hutter, F.: Decoupled weight decay regularization. In: ICLR (2019)
24. Lowe, D.G.: Object recognition from local scale-invariant features. In: Proceedings of the seventh IEEE International Conference on Computer Vision, vol. 2, pp. 1150–1157. IEEE (1999)
25. Ma, R., et al.: Colon10k: a benchmark for place recognition in colonoscopy. In: 2021 IEEE 18th International Symposium on Biomedical Imaging (ISBI), pp. 1279–1283 (2021). https://doi.org/10.1109/ISBI48211.2021.9433780
26. Ma, R., et al.: Rnnslam: reconstructing the 3d colon to visualize missing regions during a colonoscopy. Med. Image Anal. **72**, 102100 (2021)
27. Mirzaei, H., Panahi, M., Etemad, K., GHanbari-Motlagh, A., Holakouie-Naini, K.A.: Evaluation of pilot colorectal cancer screening programs in iran. Iranian J. Epidem. **12**(3) (2016)

28. Mohaghegh, P., Ahmadi, F., Shiravandi, M., Nazari, J.: Participation rate, risk factors, and incidence of colorectal cancer in the screening program among the population covered by the health centers in arak, iran. Inter. J. Cancer Manag. **14**(7), e113278 (2021)
29. Moshfeghi, K., Mohammadbeigi, A., Hamedi-Sanani, D., Bahrami, M.: Evaluation the role of nutritional and individual factors in colorectal cancer. Zahedan J. Res. Med. Sci. **13**(4), e93934 (2011)
30. Mur-Artal, R., Montiel, J.M.M., Tardós, J.D.: ORB-SLAM: a versatile and accurate monocular SLAM system. CoRR (2015). http://arxiv.org/abs/1502.00956
31. van den Oord, A., Li, Y., Vinyals, O.: Representation learning with contrastive predictive coding. ArXiv (2018)
32. Ozyoruk, K.B., et al.: Endoslam dataset and an unsupervised monocular visual odometry and depth estimation approach for endoscopic videos. Med. Image Anal. **71**, 102058 (2021)
33. Rau, A., et al.: Implicit domain adaptation with conditional generative adversarial networks for depth prediction in endoscopy. Inter. J. Comput. Assisted Radiol. Surgery**4** (2019)
34. Russakovsky, O., et al.: Imagenet large scale visual recognition challenge. CoRR (2014). http://arxiv.org/abs/1409.0575
35. Shao, S., et al.: Self-supervised monocular depth and ego-motion estimation in endoscopy: Appearance flow to the rescue. Med. Image Anal., 102338 (2021)
36. Smith, K., et al.: Data from ct colonography. the cancer imaging archive (2015). https://doi.org/10.7937/K9/TCIA.2015.NWTESAY1
37. Sohn, K.: Improved deep metric learning with multi-class n-pair loss objective. In: Proceedings of the 30th International Conference on Neural Information Processing Systems, NIPS 2016, pp. 1857–1865. Curran Associates Inc., Red Hook, NY, USA (2016)
38. Widya, A.R., Monno, Y., Okutomi, M., Suzuki, S., Gotoda, T., Miki, K.: Learning-based depth and pose estimation for monocular endoscope with loss generalization. CoRR abs/ arXiv: 2107.13263 (2021)
39. Wu, Z., Xiong, Y., Yu, S.X., Lin, D.: Unsupervised feature learning via non-parametric instance-level discrimination. ArXiv (2018)
40. Yao, H., Stidham, R.W., Gao, Z., Gryak, J., Najarian, K.: Motion-based camera localization system in colonoscopy videos. Med. Image Anal. **73**, 102180 (2021)
41. Zhang, S., Zhao, L., Huang, S., Ye, M., Hao, Q.: A template-based 3d reconstruction of colon structures and textures from stereo colonoscopic images. IEEE Trans. Med. Robotics Bionics **3**(1), 85–95 (2021)
42. Zhang, Y., et ak.: Colde: a depth estimation framework for colonoscopy reconstruction (2021)
43. Zhang, Y., Wang, S., Ma, R., McGill, S.K., Rosenman, J.G., Pizer, S.M.: Lighting enhancement aids reconstruction of colonoscopic surfaces (2021)
44. Zhang, Z., Scaramuzza, D.: A tutorial on quantitative trajectory evaluation for visual(-inertial) odometry. In: 2018 IEEE/RSJ International Conference on Intelligent Robots and Systems (IROS).,pp. 7244–7251 (2018)
45. Zhou, Q.Y., Koltun, V.: Dense scene reconstruction with points of interest. ACM Trans. Graph. 32 (2013)
46. Zhou, T., Brown, M., Snavely, N., Lowe, D.G.: Unsupervised learning of depth and ego-motion from video. CoRR (2017). http://arxiv.org/abs/1704.07813

Anatomy-Aware Masking for Inpainting in Medical Imaging

Yousef Yeganeh[1,2]([✉]), Azade Farshad[1,2], and Nassir Navab[1,3]

[1] Technical University of Munich, Munich, Germany
[2] Munich Center for Machine Learning, Munich, Germany
y.yeganeh@tum.de
[3] Johns Hopkins University, Baltimore, USA

Abstract. Inpainting has recently been employed as a successful deep-learning technique for unsupervised model discovery in medical image analysis by taking advantage of the strong priors learned by models to reconstruct the structure and texture of missing parts in images. Even though the learned features depend on the masks as well as the images, the masks used for inpainting are typically random and independent of the dataset, due to the unpredictability of the content of images, i.e., different objects and shapes can appear in different locations in images. However, this is rarely the case for medical imaging data since they are obtained from similar anatomies. Still, random square masks are the most popular technique for inpainting in medical imaging. In this work, we propose a pipeline to generate, position and sample the masks to efficiently learn the shape and structures of the anatomy and generate a myriad of diverse anatomy-aware masks, aiding the model in learning the statistical shape prior to the topology of the organs of interest. We demonstrate the impact of our approach compared to other masking mechanisms in the reconstruction of anatomy. We compare the effectiveness of our proposed masking approach over square-shaped masks, which are traditionally used in medical imaging, and irregular shape masks, which are used in SOTA inpainting literature. Project Page: https://anatomyaware.github.io/.

Keywords: Inpainting · Mask Generation · Superpixel · Self-supervised

1 Introduction

Recent advances in deep learning and particularly generative models have made it possible to take advantage of these methods for image generation, completion, or manipulation tasks. A commonly used image manipulation method is image inpainting [4,5,10,11], which reconstructs partially masked images. The main application of inpainting is to replace corrupted or unwanted objects in an image

---

Y. Yeganeh and A. Farshad—Equal Contribution.

using normal distribution priors [7]. Early medical image inpainting methods relied on classical techniques such as Mumford-Shah function [14,29] or manual pixel modification [28] to remove erroneous parts. However, recently, most of the works incorporate deep learning techniques [34]. Some works utilize inpainting to remove anomalies to improve model robustness in other tasks [16,27]. Inpainting has also shown to be effective in removing brain lesions and improving brain atrophy detection performance [16]. Armanious et al. proposed two networks for inpainting - one for MR images [3] and another with a more general direction for arbitrary regions [2]. To preserve the structure and edges in the image, some works enforce edge continuity during inpainting [14,35,37]. For example, [37] uses the edge and structure information to address distortion in medical images, including COVID CT, Abdominal CT, and Abdominal MR reconstruction. Multi-scale masks for medical images are used to handle homogeneous areas and structure details [15]. Bukas et al. [6] use inpainting to reconstruct and straighten broken vertebrae. In 3D brain MR scans, the inpainting of sparse 2D scans can be utilized to decrease the run time [20]. Kim et al. [22] generate tumors in the healthy brain using inpainting to visualize the tumor progression.

Masking Approaches. Liu et al. [26] introduced the irregular masks dataset consisting of binary images extracted from video clips, as a substitute for rectangular masks. Elharrouss et al. [8] found that models can effectively reconstruct holes up to 20 percent of the original image but struggle with larger holes. To address this, Rojas introduced mask shapes as an additional input to improve model performance [31].

Superpixels for Masking. Superpixels have never been utilized for training, but Isogawa et al. [18] used superpixels to shape the masks during testing, leveraging their shape as prior information for reconstruction. Similarly, Li et al. [25] utilized superpixels to generate holes in the image during inference. They then identified significant differences between the reconstructed and original images as anomalies.

Mask Placement. Previous works [8,18,24,33,35] have shown that the size and placement of masks can affect the performance of inpainting models. However, finding the optimal mask for each image is challenging due to the high dimensionality of images. In computer vision, irregular-shaped masks are commonly used, while previous work on inpainting with medical imaging relies on random square-shaped masks. We argue that irregular-shaped masks are suitable for non-medical images, but for medical images with predictable shapes and textures, anatomy-based masks are more appropriate to capture the structure of the target organ or anatomical region. We investigate the effect of the inpainting mask's position and shape on image reconstruction performance. We propose a novel anatomy-aware mask generation and positioning method that conforms to the image's structures and shapes. Segmentation maps can provide strong

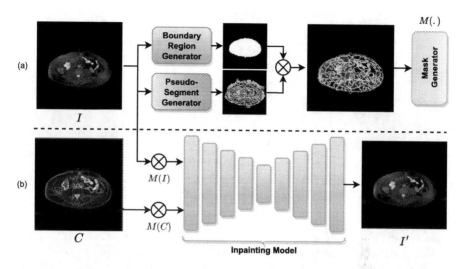

Fig. 1. Here, we show an overview of (a) our mask generation method, and (b) the inpainting framework. The organ bounded region and pseudo-segments are generated by morphological opening and the superpixel algorithm, respectively. Finally, our anatomy-aware masks are sampled using the mask generator and used for the inpainting model.

shape prior information for mask generation, but they are hard to obtain and require a lot of supervised training data for segmentation models [9,12,40,41]. To overcome this challenge, we use superpixel over-segmentation maps, which do not need any training or manual annotation. Several over-segmentation algorithms [1,19,39] have been proposed; however, in this work we use a fast and straightforward off-the-shelf method [13] used in medical imaging [30]. We extensively evaluate different masking strategies using image reconstruction metrics. To summarize our contributions:

- We propose a self-supervised anatomy-aware mask localization, generation, and sampling strategy for the more anatomy-friendly inpainting.
- To the best of our knowledge, this is the first work that uses shape priors to generate masks in medical imaging.
- We verify our hypothesis that the masks' location and shape are essential aspects of mask generation for image reconstruction in inpainting.
- Our proposed method outperforms existing mask shapes (square and irregular) and arbitrary sampling and positioning of the mask in inpainting.

2 Methodology

Our anatomy-aware mask generation strategy for inpainting is explained in this section. First, we define the inpainting framework and then present our proposed mask localization and mask shape generation techniques. An overview of our method is presented in Fig. 1 and Algorithm 1.

Algorithm 1. Mask Generation

1: $I \leftarrow$ Image
2: $M(.)$: Mask sampling function
3: $B(.)$: Boundary Region Generator
4: $G(.)$: Pseudo-Segment Generator
5: S: Set of s_i, configurations for $G(s_i, I)$
6: **for** s_i in S **do**
7:      $PS_i = G(s_i, I) * B(I)$
8:      $m_I \leftarrow (ps_{(i,j)}, \text{area}(ps_{(i,j)}), \text{center}(ps_{(i,j)})), \forall ps_j \in PS_i$
9:      $\omega_I \leftarrow \text{area}(ps_{(i)}) / \sum \text{area}(ps_{(i,j)}) , \forall ps_{(i,j)} \in PS_i$
10: **end for**
11: $M = \{m' | m' \sim P(m_I | \omega_I)\}$ ▷ where P is the probability density function

---

Definitions. Given an input image $I \in \mathbb{R}$, the goal of our inpainting network $f(.)$ is to reconstruct pixels of the image which are masked by $M(.)$ before feeding it to $f(.)$. In addition to reconstructing the masked image, the inpainting model also reconstructs the masked edges C'. Therefore, the output of our network, which is the reconstructed image I' becomes $f(M(I))$.

Pseudo-segment Generation. Inspired by [30], we generate pseudo-segments for our images using the Felzenszwalb superpixel over-segmentation algorithm [13], which maps image pixels into a graph defined by (V, E). Each edge $\epsilon_{i,j}$ connects neighboring pixels ν_i, ν_j and has a distance measure $\delta_{i,j} \in \Omega$ that defines the dissimilarity between ν_i and ν_j. Our approach is not bound to a specific algorithm, but we demonstrate the effectiveness of our approach through image reconstruction metrics. To generate the pseudo-segments for mask generation, dissimilarity distances $\delta_{i,j}$ are computed for each pixel's eight neighboring pixels. These distances are then merged using the dissimilarity measure Ω to create the segments. The size of the segments is determined by a threshold value τ based on the distance between their vertices.

Shape and Sampling. We created a variety of shapes and sizes of pseudo-segments using different settings $s_i \in S$ (as it can be seen in Fig. 2) for our pseudo-segment generator $G(s_i, I)$, and intentionally increased the frequency of smaller segments to enhance their importance in the model's objective function. We achieved this by creating a weight-biases matrix ω that paired each pseudo-segment with a value calculated by the inverse number of pixels, resulting in pseudo-segments PS_n for each image I_n used by the sampler to generate masks. Our approach prioritizes smaller segments that may hold important information, as traditional objective functions for image reconstruction can be less sensitive to them because the number of pixels are an impacting factor for them.

Masking Position. We place a set of sampled pseudo-segments in specific regions to generate our desired mask. The placement regions are identified based on the centers of the pseudo-segments. Three approaches are outlined for the placement of the sampled pseudo-segments: 1) random locations in the image, 2) random locations inside the bounded region of the organ (IBR), 3) exact location of the pseudo-segment (OPS):

- Inside Bounded Region (**IBR**): To obtain the bounded region of the organ, we apply morphological opening to the input image I_n. This provides us with a pseudo-segmentation mask that defines the whole region of the organ. The mask coordinates are then randomly sampled from the area inside the bounded region.
- On Pseudo-segments (**OPS**): We suggest that the location of masked regions in image inpainting plays a significant role in modeling the dataset. Therefore, in our approach we sample m random pseudo-segments from each image and use the center of sampled pseudo-segments as the positioning anchor in the mask layout.

Inpainting Framework: Our inpainting framework is based on CTSDG [17], which is an state-of-the-art architecture for inpainting. The reason behind this choice lies in CTSDG relying both on the texture and the structure for image reconstruction, and taking advantage of the edges in the image. With that we show that the proposed masking strategy is the main contributing factor in the improvement of the model performance, since the model already receives the extracted edges of the image as an extra input and has an additional independent objective function to ensure the continuity of the edges.

Objective Functions: Our model is trained with a combination of reconstruction loss, perceptual loss, adversarial loss, and style loss similar to [17]:

$$\mathcal{L} = \|I' - I\| + \sum \|\phi(I') - \phi(I)\| + \sum \|\psi(I') - \psi(I)\| + \mathcal{L}_{adv} + \mathcal{L}_{int} \quad (1)$$

where ϕ corresponds to the activation map of the image from a pre-trained VGG [32] network and ψ is the Gram matrix of the activation map ϕ to preserve the style. The adversarial and intermediate losses are defined as $\mathcal{L}_{adv}$ and $\mathcal{L}_{int}$ respectively:

$$\mathcal{L}_{adv} = \min_{f} \max_{D} \mathbb{E}[\log(D(I,C))] + \mathbb{E}[\log(1 - D(I',C'))] \quad (2)$$

$$\mathcal{L}_{int} = BCE(C,C'') + \|I - I''\| \quad (3)$$

where C'', I'' are intermediate reconstructions of the edge and image and $D(.)$ is the discriminator in our inpainting network.

3   Experiments

In this section, we present the experiments that validate our hypothesis. To evaluate our mask generation and positioning method, we train and test it on

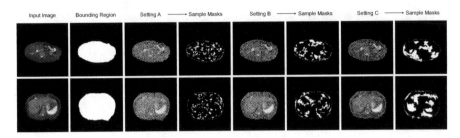

Fig. 2. Variety of generated anatomy-dependent masks by our proposed method.

the recently published CHAOS [21] (Combined (CT-MR) Healthy Abdominal Organ Segmentation) dataset. The CHAOS dataset includes healthy abdominal MR and CT scans from 80 patients. We train our model on T1-DUAL In-phase MR scans from 40 patients. The scans from 75% of the patients were randomly chosen for training the model and the scans from the remaining 25% patients for evaluation.

3.1 Experimental Setup

In all the experiments on CHAOS [21], we follow the same protocols and hyper-parameters as [17] for the inpainting framework unless explicitly specified. All our models were trained for $10K$ iterations, batch size of 6, and a learning rate of $2e-4$ for the inpainting network and 0.1 for the discriminator. All networks were trained with the Adam optimizer [23]. For the superpixel over-segmentation, we used the scikit-image [36] library with the scale factor of 2, Gaussian filtering with $\sigma = 0.5$, and a minimum size of 9 for each segment. To merge the pseudo-segments, we set the threshold value $\tau = 10$ and refine the pseudo-segments based on their mean color and their distance to their neighbors. In the experiments on CHAOS [21], we evaluate the model using three different masking techniques to show the model's robustness to masks of other shapes during testing. Since we did not observe any commonly used or publicly available benchmark for evaluating inpainting masks in prior works, we define our own evaluation protocol, including the ground truth segmentation maps, and square-shaped masks, which are the standard masks used in medical image inpainting.

Evaluation Masks. In the experiments on CHAOS [21], we used three evaluation masks: 1) segmentation shaped mask + OPS, which explicitly masks a segment that correlates with the shape of the organs, 2) multiple small squares, with similar total size to the area of the segments scattered in the image, and, 3) one large random sized and randomly located square that masks 20% to 60% of the image. Since progressive mask reconstruction becomes more challenging [8,26], we evaluate the methods on large square masks (over 20 percent of the image area) as well. In the inpainting literature, the capability of the model to reconstruct *larger masks* is a main challenge. This is due to the fact that in

Table 1. A comparison of various mask locations for training the model tested in 3 different evaluation settings. **IBR**: Inside Bounded Region, **OPS**: On Pseudo-Segments. The mask shape in these experiments is square.

Position	Evaluation Mask											
	Segments				Multiple Small Squares				Single Large Square			
	L1↓	PSNR↑	SSIM↑	LPIPS↓	L1↓	PSNR↑	SSIM↑	LPIPS↓	L1↓	PSNR↑	SSIM↑	LPIPS↓
Random	0.049	16.66	0.60	0.2885	0.052	16.78	0.63	0.32	0.56	**14.08**	0.64	0.29
IBR (Ours)	0.048	18.04	0.67	0.2686	0.05	18.13	0.67	**0.2586**	0.057	13.65	0.65	**0.2670**
OPS (Ours)	**0.047**	**18.07**	**0.68**	**0.2676**	**0.049**	**18.45**	**0.69**	0.2888	**0.055**	13.49	**0.67**	0.2685

the larger masks, the model has less neighboring information, that would make the task more challenging. On the other hand, *small square masks* have closer correlation with the tasks and the size of the organs.

Evaluation Metrics. We evaluate our method and the baselines on four different similarity metrics: 1) PSNR (Peak Signal-to-Noise Ratio), 2) SSIM [38] (Structural Similarity Index Measure), 3) LPIPS [42] (Learned Perceptual Image Patch Similarity), and 4) the L1 distance. All the metrics were calculated only on the region of interest (RoI) where the masks are applied.

3.2 Results

Our results are presented in three sections: Table 1 shows the effect of the masks locations on the performance of the model. In Table 2, we discuss the shapes of the masks, and in Table 3 we ablate the source and locations of the shape distributions and how our weight-biased sampling improves the results.

Finally, in Table 3 we show that the reconstruction performance of our proposed masking strategy on both IBR and OPS outperforms even similar settings. We also used masks that are sampled from another image to investigate

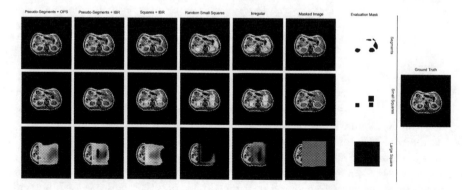

Fig. 3. Comparison of reconstructed masked image with different masking techniques.

the model performance's dependency on the domain of the mask shapes; however, the number of pseudo-segments are not the same, the masks from the other image are sampled based on their own weight-biases (ω).

Table 2. A comparison of various mask shapes used for training the model tested in 3 different evaluation settings. The positioning of the masks for all shapes is based on OPS. **PS**: Pseudo-Segments.

Shape	Evaluation Mask											
	Segments				Multiple Small Squares				Single Large Square			
	L1↓	PSNR↑	SSIM↑	LPIPS↓	L1↓	PSNR↑	SSIM↑	LPIPS↓	L1↓	PSNR↑	SSIM↑	LPIPS↓
Square	0.047	18.07	0.68	0.2676	0.049	18.45	0.69	0.2888	0.055	13.49	0.67	0.2685
Irregular	0.046	16.44	0.61	0.3078	0.047	16.63	0.64	0.3174	0.053	14.13	0.66	0.2781
PS (Ours)	0.045	19.38	0.77	0.2655	0.044	19.58	0.78	0.2630	0.052	14.4	0.74	0.2663

For the experiments in Table 1, we first evaluate different mask localization approaches using the square masks to demonstrate that regardless of the shape, appropriate mask positioning improves the reconstruction performance. It can be seen that bounding the masking region to the organ region brings the most significant improvement. Furthermore, if we enforce the mask centers to fall into the predicted pseudo-segments, the performance slightly improves. To ensure comparable results, we set the size of the squares the masks in all of the experiments to be the same

Table 3. IBR: Inside Bounded Region, **OPS**: On Pseudo-Segments. **Weighted**: weighted sampling.

Distribution	Weighted	Position	Evaluation Mask								
			Segments			Multiple Small Squares			Single Large Square		
			PSNR ↑	SSIM ↑	LPIPS ↓	PSNR ↑	SSIM ↑	LPIPS ↓	PSNR ↑	SSIM ↑	LPIPS ↓
Same Image	-	OPS	18.25	0.6456	0.29	18.5	0.6767	0.2955	14.02	0.6865	0.2862
Other Image	✓	OPS	18.31	0.6295	0.3009	18.48	0.6552	0.3054	14.11	0.6878	0.2871
Same Image	✓	IBR	19.16	0.6836	0.2544	19.18	0.7064	0.2599	13.42	0.6619	0.29665
Same Image	✓	OPS	19.38	0.77	0.2655	19.58	0.78	0.263	13.5	0.74	0.2853

In Table 2, we evaluate and compare our proposed masking strategy against the commonly used irregular [26] and square masks. As it can be seen, the model trained with pseudo-segment shaped masks (PS) achieves the best performance in most of the evaluation settings and metrics. Some qualitative results of reconstruction with different models and evaluated on various tasks is shown in Fig. 3.

We demonstrate the importance of the mask's shape and position in Table 3. Masks sampled from another image perform better than square masks, but masks sampled from the same image and positioned based on IBR perform even better. Moreover, masks that use OPS achieve the best performance.

4   Discussion and Conclusion

We proposed a novel approach for anatomy-aware mask generation that enforce geometric priors captured from the the images to generate masks for context encoders. Our mask generator and sampler considers shape, location and the size of anatomies, which proved to be more anatomy friendly than the commonly used square masks. Our experiments demonstrate that positioning the masks in the region of the organ or on pseudo-segments has a significant effect on the model's reconstruction performance. Therefore, as shown in Table 3, especially in smaller dataset, our method is preferred. Finally, we showed that using pseudo-segments from images other than the current image for mask generation has a higher performance than commonly used square shaped and irregular masks and therefore a dataset of our generated pseudo-segment masks can be considered a substitute to square and irregular masks for the medical image inpainting or pretext tasks.

References

1. Achanta, R., Shaji, A., Smith, K., Lucchi, A., Fua, P., Süsstrunk, S.: SLIC superpixels compared to state-of-the-art superpixel methods. IEEE Trans. Pattern Anal. Mach. Intell. **34**(11), 2274–2282 (2012). https://doi.org/10.1109/TPAMI.2012.120
2. Armanious, K., Kumar, V., Abdulatif, S., Hepp, T., Gatidis, S., Yang, B.: ipA-MedGAN: inpainting of arbitrary regions in medical imaging. In: 2020 IEEE International Conference on Image Processing (ICIP). IEEE (2020)
3. Armanious, K., Mecky, Y., Gatidis, S., Yang, B.: Adversarial inpainting of medical image modalities. In: ICASSP 2019–2019 IEEE International Conference on Acoustics, Speech and Signal Processing (ICASSP). IEEE (2019)
4. Astaraki, M., et al.: Autopaint: A self-inpainting method for unsupervised anomaly detection. arXiv preprint arXiv:2305.12358 (2023)
5. Bertalmio, M., Sapiro, G., Caselles, V., Ballester, C.: Image inpainting. In: Proceedings of the 27th annual conference on Computer graphics and interactive techniques (2000)
6. Bukas, C., et al.: Patient-specific virtual spine straightening and vertebra inpainting: an automatic framework for osteoplasty planning. In: de Bruijne, M., et al. (eds.) Medical Image Computing and Computer Assisted Intervention – MICCAI 2021: 24th International Conference, Strasbourg, France, September 27–October 1, 2021, Proceedings, Part IV, pp. 529–539. Springer, Cham (2021). https://doi.org/10.1007/978-3-030-87202-1_51
7. Dhamo, H., et al.: Semantic image manipulation using scene graphs. In: Proceedings of the IEEE/CVF Conference on Computer Vision and Pattern Recognition, pp. 5213–5222 (2020)
8. Elharrouss, O., Almaadeed, N., Al-Maadeed, S., Akbari, Y.: Image inpainting: a Review. Neural Process. Lett. **51**(2), 2007–2028 (2020). https://doi.org/10.1007/s11063-019-10163-0

9. Farshad, A., Makarevich, A., Belagiannis, V., Navab, N.: MetaMedSeg: volumetric meta-learning for few-shot organ segmentation. In: Kamnitsa, K., et al. (eds.) Domain Adaptation and Representation Transfer: 4th MICCAI Workshop, DART 2022, Held in Conjunction with MICCAI 2022, Singapore, September 22, 2022, Proceedings, pp. 45–55. Springer, Cham (2022). https://doi.org/10.1007/978-3-031-16852-9_5

10. Farshad, A., Yeganeh, Y., Chi, Y., Shen, C., Ommer, B., Navab, N.: SceneGenie: scene graph guided diffusion models for image synthesis. arXiv preprint arXiv:2304.14573 (2023)

11. Farshad, A., Yeganeh, Y., Dhamo, H., Tombari, F., Navab, N.: DisPositioNet: disentangled pose and identity in semantic image manipulation. In: 33rd British Machine Vision Conference 2022, BMVC 2022, London, UK, November 21–24, 2022. BMVA Press (2022)

12. Farshad, A., Yeganeh, Y., Gehlbach, P., Navab, N.: Y-Net: a Spatiospectral dual-encoder network for medical image segmentation. In: Wang, L., Dou, Q., Fletcher, P.T., Speidel, S., Li, S. (eds.) Medical Image Computing and Computer Assisted Intervention – MICCAI 2022: 25th International Conference, Singapore, September 18–22, 2022, Proceedings, Part II, pp. 582–592. Springer, Cham (2022). https://doi.org/10.1007/978-3-031-16434-7_56

13. Felzenszwalb, P.F., Huttenlocher, D.P.: Efficient graph-based image segmentation. Int. J. Comput. Vision **59**(2), 167–181 (2004). https://doi.org/10.1023/B:VISI.0000022288.19776.77

14. Feng, Z., Chi, S., Yin, J., Zhao, D., Liu, X.: A variational approach to medical image inpainting based on mumford-shah model. In: 2007 International Conference on Service Systems and Service Management (2007)

15. Gapon, N.V., Voronin, V.V., Sizyakin, R.A., Bakaev, D., Skorikova, A.: Medical image inpainting using multi-scale patches and neural networks concepts. IOP Confer. Ser.: Mater. Sci. Eng. **680**(1), 012040 (2019). https://doi.org/10.1088/1757-899X/680/1/012040

16. Guizard, N., Nakamura, K., Coupé, P., Fonov, V.S., Arnold, D.L., Collins, D.L.: Non-local means inpainting of MS lesions in longitudinal image processing. Front. Neurosci. **9** (2015). https://doi.org/10.3389/fnins.2015.00456

17. Guo, X., Yang, H., Huang, D.: Image inpainting via conditional texture and structure dual generation. In: ICCV (2021)

18. Isogawa, M., Mikami, D., Iwai, D., Kimata, H., Sato, K.: Mask optimization for image inpainting. IEEE Access **6**, 69728–69741 (2018). https://doi.org/10.1109/ACCESS.2018.2877401

19. Jampani, V., Sun, D., Liu, M.-Y., Yang, M.-H., Kautz, J.: Superpixel Sampling Networks. In: Ferrari, V., Hebert, M., Sminchisescu, C., Weiss, Y. (eds.) Computer Vision – ECCV 2018: 15th European Conference, Munich, Germany, September 8–14, 2018, Proceedings, Part VII, pp. 363–380. Springer, Cham (2018). https://doi.org/10.1007/978-3-030-01234-2_22

20. Kang, S.K., et al.: Deep learning-Based 3D inpainting of brain MR images. Sci. Rep. **11**(1), (2021). https://doi.org/10.1038/s41598-020-80930-w

21. Kavur, A.E., et al.: CHAOS Challenge - combined (CT-MR) healthy abdominal organ segmentation. Med. Image Anal. **69**, 101950 (2021). https://doi.org/10.1016/j.media.2020.101950

22. Kim, S., Kim, B., Park, H.W.: Synthesis of brain tumor multicontrast MR images for improved data augmentation. Med. Phys. **48**(5), 2185–2198 (2021). https://doi.org/10.1002/mp.14701

23. Kingma, D.P., Ba, J.: Adam: a method for stochastic optimization. arxiv.org:1412.6980 (2014)
24. Li, W., Lin, Z., Zhou, K., Qi, L., Wang, Y., Jia, J.: Mat: mask-aware transformer for large hole image inpainting. In: CVPR (2022)
25. Li, Z., et al.: Superpixel masking and inpainting for self-supervised anomaly detection. In: BMVC (2020)
26. Liu, G., Reda, F.A., Shih, K.J., Wang, T.-C., Tao, A., Catanzaro, B.: Image inpainting for irregular holes using partial convolutions. In: Ferrari, V., Hebert, M., Sminchisescu, C., Weiss, Y. (eds.) Computer Vision – ECCV 2018: 15th European Conference, Munich, Germany, September 8-14, 2018, Proceedings, Part XI, pp. 89–105. Springer, Cham (2018). https://doi.org/10.1007/978-3-030-01252-6_6
27. Manjón, J.V., et al.: Blind MRI brain lesion inpainting using deep learning. In: Burgos, N., Svoboda, D., Wolterink, J.M., Zhao, C. (eds.) Simulation and Synthesis in Medical Imaging: 5th International Workshop, SASHIMI 2020, Held in Conjunction with MICCAI 2020, Lima, Peru, October 4, 2020, Proceedings, pp. 41–49. Springer, Cham (2020). https://doi.org/10.1007/978-3-030-59520-3_5
28. Arnold, M., Ghosh, A., Ameling, S., Lacey, G.: Automatic segmentation and inpainting of specular highlights for endoscopic imaging. EURASIP J. Image Video Process. **2010**, 1–12 (2010). https://doi.org/10.1155/2010/814319
29. Mumford, D., Shah, J.: Optimal approximations by piecewise smooth functions and associated variational problems. Commun. Pure Appl. Math. **42**(5), 577–685 (1989). https://doi.org/10.1002/cpa.3160420503
30. Ouyang, C., Biffi, C., Chen, C., Kart, T., Qiu, H., Rueckert, D.: Self-supervision with Superpixels: training few-shot medical image segmentation without annotation. In: Vedaldi, A., Bischof, H., Brox, T., Frahm, J.-M. (eds.) Computer Vision – ECCV 2020: 16th European Conference, Glasgow, UK, August 23–28, 2020, Proceedings, Part XXIX, pp. 762–780. Springer, Cham (2020). https://doi.org/10.1007/978-3-030-58526-6_45
31. Rojas, D.J.B., Fernandes, B.J.T., Fernandes, S.M.M.: A review on image inpainting techniques and datasets. In: 2020 33rd SIBGRAPI Conference on Graphics, Patterns and Images (SIBGRAPI). IEEE (2020)
32. Simonyan, K., Zisserman, A.: Very deep convolutional networks for large-scale image recognition. arXiv:1409.1556 (2014)
33. Suvorov, R., et al.: Resolution-robust large mask inpainting with Fourier convolutions. In: WACV (2022)
34. Tran, M.T., Kim, S.H., Yang, H.J., Lee, G.S.: Deep Learning-Based Inpainting for Chest X-Ray Image. Association for Computing Machinery, New York, NY, USA (2020)
35. Tran, M.-T., Kim, S.-H., Yang, H.-J., Lee, G.-S.: Multi-task learning for medical image inpainting based on organ boundary awareness. Appl. Sci. **11**(9), 4247 (2021). https://doi.org/10.3390/app11094247
36. van der Walt, S., et al.: scikit-image: image processing in Python. PeerJ **2**, e453 (2014). https://doi.org/10.7717/peerj.453
37. Wang, Q., Chen, Y., Zhang, N., Gu, Y.: Medical image inpainting with edge and structure priors. Measurement **185**, 110027 (2021). https://doi.org/10.1016/j.measurement.2021.110027
38. Wang, Z., Bovik, A.C., Sheikh, H.R., Simoncelli, E.P.: Image quality assessment: from error visibility to structural similarity. IEEE Trans. Image Process. **13**(4), 600–612 (2004). https://doi.org/10.1109/TIP.2003.819861
39. Yang, F., Sun, Q., Jin, H., Zhou, Z.: Superpixel segmentation with fully convolutional networks. In: CVPR (2020)

40. Yeganeh, Y., et al.: Scope: structural continuity preservation for medical image segmentation. arXiv preprint arXiv:2304.14572 (2023)
41. Yeganeh, Y., Farshad, A., Weinberger, P., Ahmadi, S.A., Adeli, E., Navab, N.: Transformers pay attention to convolutions leveraging emerging properties of vits by dual attention-image network. In: Proceedings of the IEEE/CVF International Conference on Computer Vision. pp. 2304–2315 (2023)
42. Zhang, R., Isola, P., Efros, A.A., Shechtman, E., Wang, O.: The unreasonable effectiveness of deep features as a perceptual metric. In: CVPR (2018)

Particle-Based Shape Modeling for Arbitrary Regions-of-Interest

Hong Xu$^{(\boxtimes)}$, Alan Morris, and Shireen Y. Elhabian

Scientific Computing and Imaging Institute, School of Computing,
University of Utah, Salt Lake City, UT, USA
{hxu,amorris,shireen}@sci.utah.edu
http://www.sci.utah.edu

Abstract. Statistical Shape Modeling (SSM) is a quantitative method for analyzing morphological variations in anatomical structures. These analyses often necessitate building models on targeted anatomical regions of interest to focus on specific morphological features. We propose an extension to particle-based shape modeling (PSM), a widely used SSM framework, to allow shape modeling to arbitrary regions of interest. Existing methods to define regions of interest are computationally expensive and have topological limitations. To address these shortcomings, we use mesh fields to define free-form constraints, which allow for delimiting arbitrary regions of interest on shape surfaces. Furthermore, we add a quadratic penalty method to the model optimization to enable computationally efficient enforcement of any combination of cutting-plane and free-form constraints. We demonstrate the effectiveness of this method on a challenging synthetic dataset and two medical datasets.

1 Introduction

Statistical Shape Modeling (SSM) is a widespread method used to analyze shape variation across 3D anatomical samples within a population. These analyses are crucial in detecting common morphological pathologies and advancing the understanding of different diseases by studying the form-function relationships between anatomies [2,5–8,13,16–18,21]. While building SSMs, certain biomedical and clinical applications require a focus on specific anatomical regions of interest (ROIs) to tailor the analysis to precise morphological features (e.g. [1–4,11,14–16]). Such applications might require excluding certain surface aspects, modeling certain regions in isolation, or a mix of these. ROI definition without altering the input shape has been achieved using *constraints*, mathematical delimiters that restrict model construction to certain surface areas [11]. Our approach focuses on redesigning the constraint application method to improve its functionality, flexibility, and efficiency during SSM construction.

To construct such SSMs, two distinct families of shape representations can be used to allow for statistical analysis, *deformation fields* and *landmarks*. Whereas the former encodes *implicit* transformations between cohort samples and a predefined (or learned) atlas, the latter uses *explicit* landmark points spread on

C. Wachinger et al. (Eds.): ShapeMI 2023, LNCS 14350, pp. 47–54, 2023.
https://doi.org/10.1007/978-3-031-46914-5_4

shape surfaces that correspond across the population [19,20]. We focus on the latter approach given its extensive use due to its simplicity, computational efficiency, and interpretability for statistical analysis [19,22]. Although landmarks used to be manually placed on specific anatomical features of interest, the modern convention uses dense automatically-placed landmarks obtained through computational methods, such as minimum description length (MDL) [12], and particle-based shape modeling (PSM) [9,10]). We utilize PSM, an efficient and robust entropy-based optimization method that creates a system of dense landmarks or *particles*, which conform to all population shape surfaces while maintaining correspondence across them.

A previous attempt to constrain PSM particle distributions uses geometric primitives in the form of spheres or cutting planes to exclude regions [11]. This exclusion is achieved by projecting *virtual particles* onto these geometric primitives (represented as parametric constraints), relying on the entropy objective to repel landmark particles away from these areas. Such an approach has the advantage of not altering input surfaces, which can otherwise distort morphology or necessitate manual expert-driven reprocessing of data. However, it falters when arbitrary regions of interest cannot be expressed via geometric primitives, limiting the topologies to be modeled. It also exhibits poor scaling due to it requiring an additional set of projected virtual particles per constraint. Thereby, to address these shortcomings in the existing literature, we propose the use of the quadratic penalty method in the optimization to allow the simultaneous and scalable application of cutting-plane, spheres, other primitive constraints, as well as a proposed method of defining arbitrary surface constraints, or *free-form constraints* (FFCs). This method provides both flexibility in the definition of constraints to define ROIs and scalability with large-scale or heavily constrained populations without the need to reprocess data.

2   Method

The aforementioned automatic landmark placement methods take in a population of $I-$ shapes $\mathcal{S} = \{\mathbf{S}_i\}_{i=1}^{I}$ (binary segmentations, meshes, or n-dimensional contours), and obtain particles $\mathcal{P} = \{\mathbf{P}_i\}_{i=1}^{I}$ where the $i-$th shape point distribution model (PDM) is denoted by $J-$particles $\mathbf{P}_i = [\mathbf{p}_{i,1}, \mathbf{p}_{i,2}, \cdots, \mathbf{p}_{i,J}]$, where $\mathbf{p}_{i,j} \in \mathbb{R}^3$. Such particles are obtained by optimizing an objective $f(\mathcal{P})$, which give

$$f(\mathcal{P}) = H(\mathcal{P}) - \sum_{j=1}^{J} H(\mathbf{P}_i), \qquad (1)$$

where H is an estimation of the differential entropy. The particles enable quantifying subtle differences and computing shape statistics (e.g., by performing the principal component analysis (PCA) on corresponding particles) by providing a population-specific anatomical mapping across the given cohort.

We constrain each shape $\mathbf{S}_i$ by M_i−inequality constraints in the form $g_{i,m}(\mathbf{p}) \leq 0$, where $g_{i,m}(\mathbf{p})$ is a differentiable function. These parametric constraints can be in the form of cutting planes or spheres as showcased in [11] (by using the equations of planes or spheres), other parametric delimiters, or free-form constraints, which allow arbitrary surface region definition. These constraints limit the distribution of particles to regions that satisfy the inequality, a region more easily demarcated using parametric constraints in some anatomies, and/or free-form *surface-painting* in others.

In this section, we describe the use of a quadratic penalty method to allow efficient and simultaneous enforcement of an arbitrary number of parametric constraints, and the use of signed mesh vector fields to build free-form constraints that allow arbitrary surface region isolation. We will also showcase a friendly graphical interface to define these constraints.

2.1   Quadratic Penalty for Efficient Constrained PDM Construction

We define an extended objective function to express this constrained optimization problem in an unconstrained form. For each constraint function in the form $g_{i,m}(\mathbf{p}) \leq 0$, we add a quadratic penalty term $g_{i,m}^+(\mathbf{p}) = \max(0, g_{i,m}(\mathbf{p}))$ to the optimization objective, yielding

$$F(\mathcal{P}) = f(\mathcal{P}) + \sum_{i=1}^{I} \sum_{m=1}^{M_i} \sum_{j=1}^{J} g_{i,m}^+(\mathbf{p}_{i,j}). \tag{2}$$

We optimize this objective function using a Gauss-Seidel gradient descent scheme, with the second term preventing particles from violating constraints, hence restricting their movement exclusively to feasible regions. This method scales linearly with respect to the number of particles per shape, whereas the virtual particle model [11] scales quadratically.

2.2   Free-Form Constraints

We express free-form constraints in the same form $g_{i,m}(\mathbf{p}) \leq 0$ for each shape $\mathbf{S}_i$ by attributing a distance and gradient field onto each vertex of a mesh $\mathbf{M}_i$. Any feasible region on the surface of $\mathbf{M}_i$ can be delineated by a set of surface boundaries $\mathcal{B}_i = [\mathbf{B}_{i,1}, \mathbf{B}_{i,2}, \cdots, \mathbf{B}_{i,B}]$, which are represented as vertex loops on the mesh surface. A distance field query for a particle $\mathbf{p}$, denoted $\mathbf{M}_i^d(\mathbf{p})$, provides the signed geodesic distance to the closest constraint boundary $\mathbf{B}_{i,*}$ from the projection of $\mathbf{p}$ onto M_i, illustrated in Fig. 1 (b). Similarly, a gradient field query, denoted $\mathbf{M}_i^g(\mathbf{p})$, would provide the gradient direction, shown in Fig. 1 (c). Ultimately, the mesh $\mathbf{M}_i$ together with its fields $\mathbf{M}_i^d$ and $\mathbf{M}_i^g$, can approximate the distance and first-order gradients over near-surface points, effectively simulating a differentiable function $g_{i,m}(\mathbf{p})$. When integrated into the aforementioned penalty method 2.1, this approach can enforce arbitrary surface constraints.

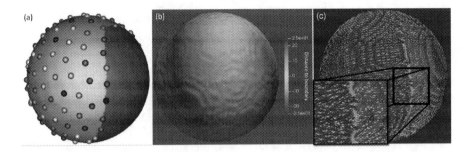

Fig. 1. (a) Constrained particle distribution on a sphere, where yellow illustrates the feasible region of the constrained area where particles are allowed to be distributed, the gray is the infeasible region where if particles were to be there, they would be violating the constraint. (b) Distance field $\mathbf{M}_i^d(\mathbf{p})$ of signed geodesic distances to the surface of every mesh vertex. (c) Gradient field $\mathbf{M}_i^g(\mathbf{p})$ on the mesh surface at every mesh vertex represented using white arrows and the blue surface as the feasible region. (Color figure online)

2.3 Graphical Interface Tool

We include a graphical interface tool that can define cutting planes and FFCs and can roughly propagate these to all shapes in the population. Cutting planes are defined by prompting 3 points that the user can pick that are on the shape surface, and can be copy-pasted into all other shapes. FFCs are defined using a "painting" tool that can define included and excluded areas with an adjustable brush size. This tool allows precise and arbitrarily customizable definition of constraints. An FFC on a single shape can be propagated to others using deformation parameters computed from image registration. This functionality is also included. All the graphical interface functionality is illustrated in Fig. 2.

3 Results

We demonstrate our results by integrating our method into an open-source implementation of the particle-based shape modeling (PSM) framework, Shape-Works [9], and produce SSMs from three datasets. The first is a synthetic dataset of ellipsoids that vary between values of 10, 20, 30, and 40, in each of their three major axes, totaling 64 ellipsoids. These ellipsoids are constrained by a free-form boundary that divides each ellipsoid into upper and lower halves by a full period of a sine wave projected onto the surface, providing a challenging but uniformly delimited population of shapes. Figure 3 shows a few examples and the modes of variation from the SSM. The constraints have the desired effect, and the modes of variation meet expectations as they mimic the variation in the three major axes.

The second is a dataset of 25 computerized tomography (CT) femurs, where the region of interest is the proximal femur sans the lesser trochanter (femoral head, neck, and greater trochanter). For each shape, we use a cutting plane

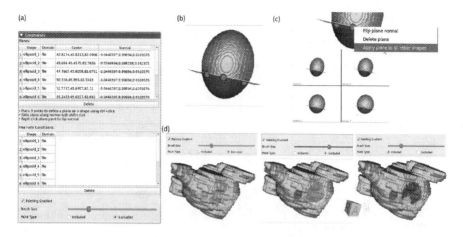

Fig. 2. (a) The constraint panel shows the constraints that have been defined and the tools to define the constraints. (b) Cutting-plane constraints are defined by ctrl-clicking 3 points on the shape surface. (c) Constraints can be flipped or applied to all other shapes via the right-click menu. (d) FFCs are defined with a painting tool with different brush sizes and options to customize included and excluded areas. We show how the painting of excluded areas of different sizes applied to a segmentation of a left atrium.

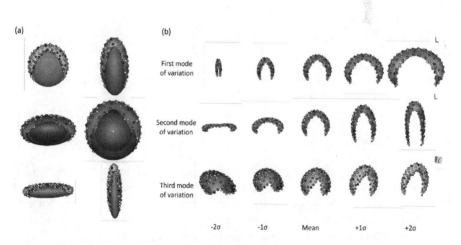

Fig. 3. (a) Sample ellipsoids from the dataset with feasible regions in yellow and restricted regions in grey. (b) The first three modes of variation in the dataset, which show the variation in corresponding major axes. (Color figure online)

constraint to exclude the shaft and a free-form constraint to exclude the lesser trochanter. Figure 4 illustrates a few examples and the first two modes of variation. We observe that a cutting plane allows a more straightforward exclusion of the shaft whilst the FFC precisely excludes the lesser trochanter. The constraints restrict the movement of particles to the feasible region as expected, and the modes of variation meet expectation as well.

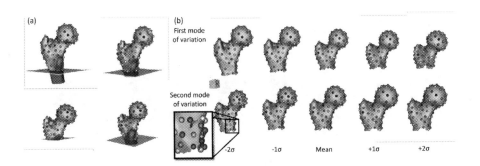

Fig. 4. (a) Example of defined constraints. The feasible region is shown in yellow and the constrained region in grey. (b) The first two modes of variation in the dataset. Notice that particles are excluded from the lesser trochanter. (Color figure online)

The third dataset comprises 21 segmentation of left atria models obtained from MRIs. The pulmonary veins represent the area of greatest variation both in anatomical structure (e.g. number of veins, common veins, etc.) as well as greatest variability in segmentation by expert observers (e.g. length into vein to segment). While the position of veins may be important from a shape modeling perspective, their exact shape is not particularly relevant to LA shape morphology. Thus, we paint a free-form constraint exclusion area around the veins. Figure 5 showcases some examples of the shape and the first three modes of variation. The models meet expectations.

4   Conclusion

We demonstrate a flexible and more scalable approach to define regions of interest in fully-groomed shapes for landmark-based statistical shape modeling by allowing arbitrary definition of surface constraints via FFCs and incorporating mixed constraint types into the optimization. This significantly improves the usability of PSM methods, obviating the need for reprocessing datasets. Future work includes the automatic propagation of constraints to the entire cohort given manual definitions on certain representative shapes.

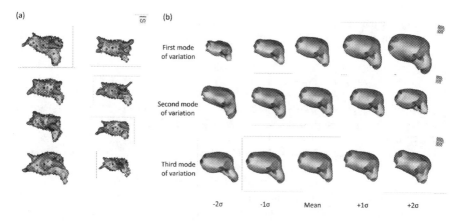

Fig. 5. (a) Example of defined constraints where the left atrium (in yellow) segmentations have the pulmonary veins excluded (in grey). (b) The first three modes of variation in the dataset. Notice how the pulmonary vein areas remain hollow. (Color figure online)

References

1. Atkins, P.R., et al.: Prediction of femoral head coverage from articulated statistical shape models of patients with developmental dysplasia of the hip. J. Orthop. Res. **40**(9), 2113–2126 (2022). https://doi.org/10.1002/jor.25227
2. Atkins, P.R., et al.: Quantitative comparison of cortical bone thickness using correspondence-based shape modeling in patients with cam femoroacetabular impingement. J. Orthop. Res. **35**(8), 1743–1753 (2017)
3. Atkins, P.R., et al.: Which two-dimensional radiographic measurements of cam femoroacetabular impingement best describe the three-dimensional shape of the proximal femur? Clin. Orthop. Relat. Res. **477**(1), 242 (2019)
4. Audenaert, E.A., Pattyn, C., Steenackers, G., De Roeck, J., Vandermeulen, D., Claes, P.: Statistical shape modeling of skeletal anatomy for sex discrimination: Their training size, sexual dimorphism, and asymmetry. Front. in Bioeng. Biotechnol. **7** (2019). DOI: https://doi.org/10.3389/fbioe.2019.00302,https://www.frontiersin.org/articles/10.3389/fbioe.2019.00302
5. Bhalodia, R., Dvoracek, L.A., Ayyash, A.M., Kavan, L., Whitaker, R., Goldstein, J.A.: Quantifying the severity of metopic craniosynostosis: a pilot study application of machine learning in craniofacial surgery. J. Craniofac. Surg. **31**(3), 697–701 (2020). https://doi.org/10.1097/SCS.0000000000006215
6. Bruse, J.L.: A statistical shape modelling framework to extract 3D shape biomarkers from medical imaging data: assessing arch morphology of repaired coarctation of the aorta. BMC Med. Imaging **16**, 1–19 (2016)
7. Carriere, N., et al.: Apathy in Parkinson's disease is associated with nucleus accumbens atrophy: a magnetic resonance imaging shape analysis. Mov. Disord. **29**(7), 897–903 (2014)
8. Cates, J., et al.: Computational shape models characterize shape change of the left atrium in atrial fibrillation. Clin. Med. Insights: Cardiol. **8s1**, CMC.S15710 (2014). https://doi.org/10.4137/CMC.S15710

9. Cates, J., Elhabian, S., Whitaker, R.: ShapeWorks. In: Statistical Shape and Deformation Analysis, pp. 257–298. Elsevier (2017). https://doi.org/10.1016/B978-0-12-810493-4.00012-2

10. Cates, J., Fletcher, P.T., Styner, M., Shenton, M., Whitaker, R.: Shape modeling and analysis with entropy-based particle systems. In: Karssemeijer, N., Lelieveldt, B. (eds.) Information Processing in Medical Imaging: 20th International Conference, IPMI 2007, Kerkrade, The Netherlands, July 2-6, 2007. Proceedings, pp. 333–345. Springer, Berlin, Heidelberg (2007). https://doi.org/10.1007/978-3-540-73273-0_28

11. Datar, M., Cates, J., Fletcher, P.T., Gouttard, S., Gerig, G., Whitaker, R.: Particle based shape regression of open surfaces with applications to developmental neuroimaging. In: Yang, G.-Z., Hawkes, D., Rueckert, D., Noble, A., Taylor, C. (eds.) Medical Image Computing and Computer-Assisted Intervention – MICCAI 2009, pp. 167–174. Springer, Berlin, Heidelberg (2009). https://doi.org/10.1007/978-3-642-04271-3_21

12. Davies, R.H., Twining, C.J., Cootes, T.F., Waterton, J.C., Taylor, C.J.: A minimum description length approach to statistical shape modeling. IEEE Trans. Med. Imaging $21(5)$, 525–537 (2002)

13. Harris, M.D., Datar, M., Whitaker, R.T., Jurrus, E.R., Peters, C.L., Anderson, A.E.: Statistical shape modeling of cam femoroacetabular impingement. J. Orthop. Res. $31(10)$, 1620–1626 (2013). https://doi.org/10.1002/jor.22389

14. Jacxsens, M., et al.: Thinking outside the glenohumeral box: Hierarchical shape variation of the periarticular anatomy of the scapula using statistical shape modeling. J. Orthop. Res. $38(10)$, 2272–2279 (2020). https://doi.org/10.1002/jor.24589

15. Jacxsens, M., Elhabian, S.Y., Brady, S.E., Chalmers, P.N., Tashjian, R.Z., Henninger, H.B.: Coracoacromial morphology: a contributor to recurrent traumatic anterior glenohumeral instability? J. Shoulder Elbow Surg. $28(7)$, 1316–1325 (2019)

16. Lenz, A.L.: Statistical shape modeling of the talocrural joint using a hybrid multi-articulation joint approach. Sci. Rep. $11(1)$,(2021). https://doi.org/10.1038/s41598-021-86567-7

17. Merle, C., et al.: How many different types of femora are there in primary hip osteoarthritis? an active shape modeling study. J. Orthop. Res. $32(3)$, 413–422 (2014)

18. Merle, C., et al.: High variability of acetabular offset in primary hip osteoarthritis influences acetabular reaming-a computed tomography-based anatomic study. J. Arthroplasty $34(8)$, 1808–1814 (2019)

19. Sarkalkan, N., Weinans, H., Zadpoor, A.A.: Statistical shape and appearance models of bones. Bone 60, 129–140 (2014)

20. Thompson, D.W., et al.: On growth and form. On growth and form. (1942)

21. van Buuren, M., et al.: Statistical shape modeling of the hip and the association with hip osteoarthritis: a systematic review. Osteoarthritis and Cartilage $29(5)$, 607–618 (2021). https://doi.org/10.1016/j.joca.2020.12.003,https://www.sciencedirect.com/science/article/pii/S106345842031219X

22. Zachow, S.: Computational planning in facial surgery. Facial Plast. Surg. $31(05)$, 446–462 (2015)

Optimal Coronary Artery Segmentation Based on Transfer Learning and UNet Architecture

Belén Serrano-Antón[1,2,3](✉) ⓘ, Alberto Otero-Cacho[1,2,3],
Diego López-Otero[4,5], Brais Díaz-Fernández[4,5], María Bastos-Fernández[4,5],
Gemma Massonis[1], Santiago Pendón[1], Vicente Pérez-Muñuzuri[3,6],
José Ramón González-Juanatey[4,5,7], and Alberto P. Muñuzuri[2,3]

[1] FlowReserve Labs S.L., 15782 Santiago de Compostela, Spain
[2] Galician Center for Mathematical Research and Technology (CITMAga), 15782 Santiago de Compostela, Spain
belenserrano.anton@usc.es
[3] Group of Nonlinear Physics. University of Santiago de Compostela, 15782 Santiago de Compostela, Spain
[4] Cardiology and Intensive Cardiac Care Department, University Hospital of Santiago de Compostela, 15706 Santiago de Compostela, Spain
[5] Centro de Investigación Biomédica en Red de Enfermedades Cardiovasculares (CIBERCV), 28029 Madrid, Spain
[6] Institute CRETUS, Group of Nonlinear Physics, University of Santiago de Compostela, 15705 Santiago de Compostela, Spain
[7] Instituto de Investigación Sanitaria de Santiago de Compostela (IDIS), 15706 Santiago de Compostela, Spain

Abstract. Recent results demonstrated that the use of AI to perform complicated segmentation of medical images becomes very useful when the coronary arteries are considered. Nevertheless, the different segments of the coronary arteries (distal, middle and proximal) exhibit singularities, mostly linked to section changes and image visibility, that point in the direction to consider each in a singular way. In the present contribution we thoroughly analyse the quality of the segmentation obtained using different neural networks, based on the UNet architecture, applied to the three segments of the coronary arteries.

We observe that for proximal segments any of the AI considered provides acceptable segmentations while for distal segments the 3D UNet is not able to recognise the coronary structures. In addition, in the distal region there is a noticeable improvement in the 2D UNet without pre-training compared to the 2D networks with pre-training.

Supported by the Spanish Ministerio de Economía y Competitividad and European Regional Development Fund under contract RTI2018-097063-B-I00 AEI/FEDER, UE, by Xunta de Galicia under Research Grant No. 2021-PG036 and by the Spanish Ministerio de Ciencia e Innovación MCIN/AEI/10.13039/501100011033 and European Union NextGenerationEU/PRTR. Research grant No: DIN2020-011068. All these programs are co-funded by FEDER (UE).

C. Wachinger et al. (Eds.): ShapeMI 2023, LNCS 14350, pp. 55–64, 2023.
https://doi.org/10.1007/978-3-031-46914-5_5

Keywords: Coronary · Artery · Segmentaion · CT · Neural Network

1   Introduction

The use of artificial intelligence as a diagnostic aid in clinical practice is becoming increasingly widespread. Despite its limitations in terms of computational resources or explainability, this tool is of great help in reducing costs, time and, above all, risk for patients [7].

In particular, invasive procedures such as catheterisation can be avoided by CT angiography (CCTA). Based on the CT images, the coronary arteries can be reconstructed. These geometries are then used to calculate, by means of computational fluid simulation techniques (CFD), parameters of clinical interest, such as FFR [4,5,9]. Extracting the geometries from the images manually is a tedious, time-consuming and error-prone task. Therefore, some studies have developed artificial intelligence techniques capable of accurately extracting these geometries [1,2,6,8].

In most of the medical studies on coronary arteries, an additional degree of differentiation is considered depending on the location of the artery considered. Three segments are typically analyzed: proximal (close to the aorta), middle and distal (corresponding with the artery segments farther from the aorta). It is known that each segment has differences basically linked to the artery diameter and visibility due to contrast loss [3,10], thus, it is likely that different AI might be more suitable for each region.

In this study we will focus on evaluating the performance of different UNet networks developed in [8] for the different segments of the coronary arteries. Specifically, we will analyse 2D networks with and without pre-training (with ImageNet data) and a 3D network. Our case study is based on the analysis in the three coronary regions: proximal, middle and distal.

2   Methods

2.1   Dataset

The paper on which this contribution is based, [8], uses a dataset of up to 88 patients, which is divided into training, validation and test sets. Both the manual segmentations (ground truth) and the network predictions considered in this study are from the test set, which consists of 10 patients. In the following they will be named from T001 to T010.

In [8], training is done with different numbers of patients and coronary structures (aorta and coronary arteries or coronary arteries only). For this study, we will just consider network weights of the training with the maximum number of patients (N=65) and coronary arteries only, since we want to evaluate the best performance of the networks only in this region. The influence of the number of patients used for training is demonstrated in [8]. When data is scarce, performance is worse, especially in networks without pre-training. Note that the availability of segmented coronary arteries is always a problem so it is important to develop techniques that can provide accurate answers without the need of extensive training data set.

In addition, the prediction segments of the neural networks have been cleaned of all false positives located in a separate connected component of the coronary arteries. That is, only those pixels that are in contact (or their neighbors are in contact) with a true positive have been selected. This allows us to evaluate the results only in the region of interest.

2.2 Network Implementation

The neural networks of the study are described in [8]. We have 4 UNet with different backbones. Specifically, 2D pre-trained UNet, 2D UNet and a Effcient-Net and a 3D UNet without pre-training. The UNet architecture is divided into two parts: an encoder, which extracts features from the image by increasing the number of channels and reducing its size, and a decoder, which performs the opposite process and generates a prediction of the same size as the original image. The networks using transfer learning (pre-trained) keep the pre-trained weights in the encoder part and only train the decoder part with the CTCA images.

2.3 Separation of the Coronary Tree into Three Regions: Proximal, Middle and Distal

In order to evaluate the performance of the different neural networks, the coronary tree is divided into three regions: proximal, middle and distal.

The separation into three regions is done using 3D Slicer. First, the ground truth (GT) mask segment for each patient is obtained. Then a curve is manually drawn along the main vessels (usually the right coronary artery, the left coronary artery and the circumflex artery) of the manual segmentation of each patient, from the ostium to the last bifurcation in the most distal part (see Fig. 1). The proximal part corresponds to 30% of the length, measured from the ostium, the distal part corresponds to 30% of the length from the distal end of the curve, and finally the middle part corresponds to the remaining 40% between the proximal and distal parts. We also distinguish between right coronary tree (RCT) and left coronary tree (LCT).

2.4 Evaluation Metrics

The parameters used to assess network performance are well known in the field of segmentation. They include true positives (TP), false positives (FP), true negatives (TN) and false negatives (FN). Following with the above parameters, we define:

- Sensitivity or recall: $TP/(TP+FN)$. Value in the interval $[0,1]$. Measures the number of true positives out of the total number of positives of the manual segmentation (GT). The closer to 1, the more vessel has been detected.

Proximal   Middle   Distal

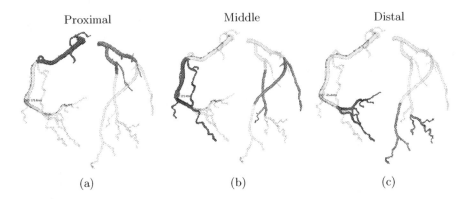

(a)   (b)   (c)

Fig. 1. Hand-made segmentation of the test patient T002. In red the part corresponding to the right coronary tree (RCT). In orange the part corresponding to the left coronary tree (LCT). Blue shows a curve along the RCT with its associated length in mm. (a) Proximal region. (b) Middle region. (c) Distal region. (Color figure online)

- F_1 score or dice similarity coefficient (DSC): $TP/(TP + 0.5 * (FP + FN))$. It is the harmonic mean of precision and sensitivity. It can also be expressed as $2*(Y \cap \widehat{Y})/(|Y| + |\widehat{Y}|)$. Where Y and $\widehat{Y}$ represent ground truth and prediction, respectively, and $Y, \widehat{Y} \in \{0, 1\}$. The closer to 1, the better is the result of our prediction.
- False negative rate (FNR) or miss rate: $FN/(FN + TP)$. Value in the interval $[0, 1]$. Measures the number of false negatives out of the total number of positives of the manual segmentation (GT). The closer to 1, he more vessel that has not been detected.
- Critical success index (CSI): $TP/(TP + FN + FP)$. Similar to F_1 score but does not give as much weight to true positives, thus it is more influenced by prediction errors.

3   Results

Figure 2 shows the segmentations of each region, proximal, middle and distal, obtained with each of the neural networks, for the patient T002. A first inspection of the pictures reveals that proximal segments are well-reproduced independently of the network used. On the contrary, there is a large diversity of results when distal segments are considered. In order to quantify this, for each of these regions, the DSC, FNR, sensitivity and CSI parameters are analysed.

Parameter values for the proximal segment are shown in Fig. 3. The DSC (Fig. 3.(a)) shows values above 0.8 for all 2D networks. However, the 3D network has values below 0.8. This fact is more accentuated in the left branch. This is easily explained with the help of the FNR (Fig. 3.(b)), since a reduction in vessel detection is observed (false negatives increase, true positives decrease (sensitivity, Fig. 3.(c))).

Proximal   Middle   Distal

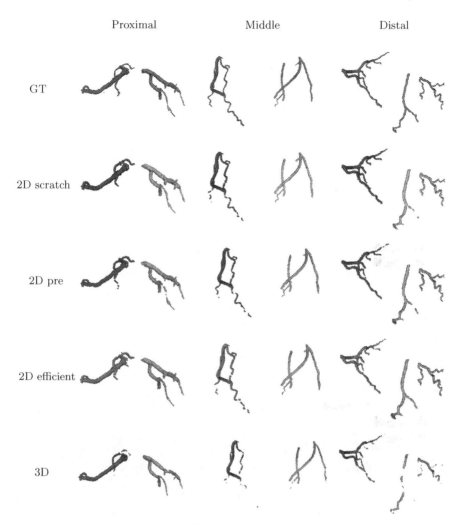

Fig. 2. Segmentations of test patient T002. In columns the proximal, middle and distal regions. In rows the manual segmentation (*GT*), the 2D UNet without pre-training (2D scratch), the 2D UNet pre-trained, the 2D efficient net pre-trained (2D efficient) and the 3D UNet without pre-training (3D).

For the middle region (Fig. 4) we observe the same pattern as in the proximal region. However, worse results are shown for the left coronary tree. This may be due to the fact that, in this region, the left coronary artery has more branches and complexity than the right coronary artery. The network that is least affected by this fact is the 2D efficient net, since the DSC and CSI values (Fig. 4.(a) and Fig. 4.(d)) are practically identical for both branches.

The fact that the 3D UNet achieves similar DSC and CSI values (Fig. 4.(a) and Fig. 4.(d)) to the 2D nets, despite detecting less vessel quantity (Fig. 4.(b)

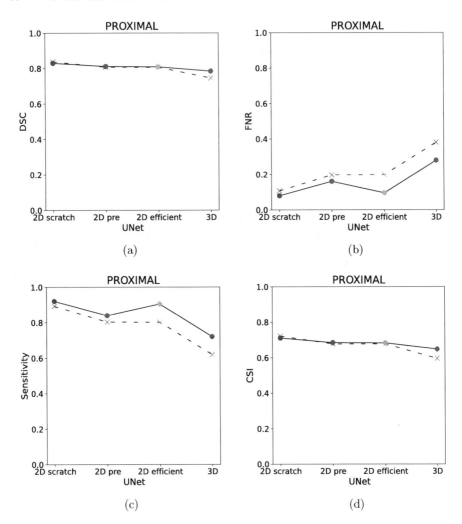

Fig. 3. Parameter values in proximal region in y-axis, network type in x-axis. (a) Dice Similarity coefficient (DSC) parameter. (b) False Negative Rate (FNR) parameter. (c) Sensitivity parameter. (d) Critical Success Index (CSI) parameter. Dot represents right coronary tree (RCT). Cross represents left coronary tree (LCT).

and Fig. 4.(c)) could indicate a significant reduction in the number of false positives compared to the other nets. This implies that the vessel diameter is less overestimated.

Finally, the distal part shows a different pattern. The DSC value falls below 0.8 for the 2D nets and the difference between the left and right branch is accentuated (Fig. 5.(a)). Moreover, the 3D UNet barely recognises 50% of the vessel (Fig. 5.(b)). A relevant aspect is that the 2D scratch has a sensitivity value of more than 0.9, outperforming the other 2D nets by up to 10% (Fig. 5.(c)).

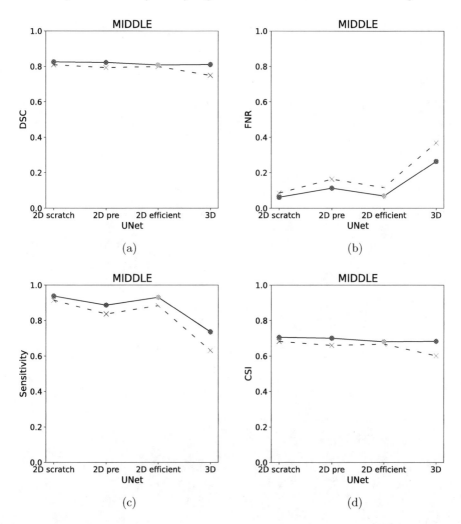

Fig. 4. Parameter values in middle region in y-axis, network type in x-axis. (a) Dice Similarity coefficient (DSC) parameter. (b) False Negative Rate (FNR) parameter. (c) Sensitivity parameter. (d) Critical Success Index (CSI) parameter. Dot represents right coronary tree (RCT). Cross represents left coronary tree (LCT).

Figure 6 presents a summary of all the parameters measured for the three artery regions. It becomes apparent that all four methods are equivalent when the proximal segments are reconstructed with preference to the 2D from scratch or the 2D efficient. Nevertheless, as we move away from the aorta, in the middle region, the 3D network indicators become worse. For the distal segments, this is clear and the 3D network becomes useless.

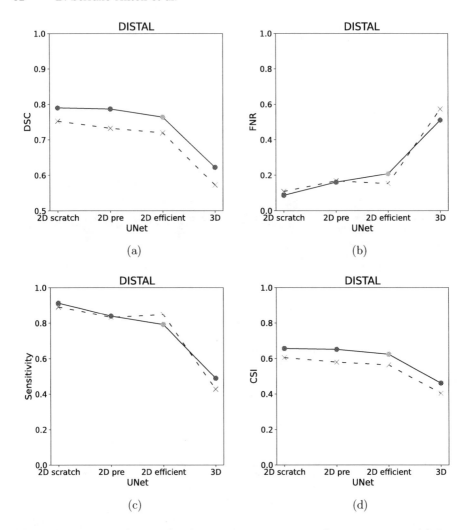

Fig. 5. Parameter values in distal region in y-axis, network type in x-axis. (a) Dice Similarity coefficient (DSC) parameter. (b) False Negative Rate (FNR) parameter. (c) Sensitivity parameter. (d) Critical Success Index (CSI) parameter. Dot represents right coronary tree (RCT). Cross represents left coronary tree (LCT).

4   Discussion and Conclusion

The task to be addressed in this study is the comparison of the performance of the different UNet for the proximal, middle and distal regions.

What has been observed is a similar pattern for the proximal and middle regions. In this case, the performance of the 2D networks is similar, regardless of the pre-training. This indicates that the number of patients used to train the 2D UNet from scratch is sufficient to obtain a complete and detailed coronary

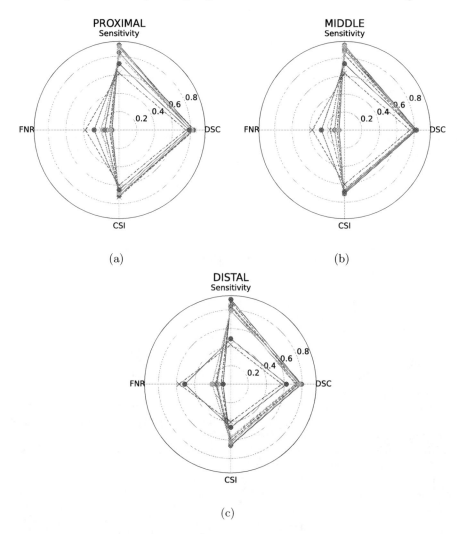

Fig. 6. Radial plots for each of the three regions. (a) Proximal region. (b) Middle region. (c) Distal region. The parameters represented are: dice similarity coefficient (DSC), Sensitivity, false negative rate (FNR) and critical success index (CSI). In purple 2D from scratch UNet, in blue 2D pre-trained UNet, in orange 2D pre-trained efficient UNet and in green 3D UNet. (Color figure online)

tree. In the middle region, however, the differences between RCT and LCT start to become more pronounced, as the latter is more complex and therefore yields worse results.

The most important change in behaviour is observed for the distal region, as the 3D network is not able to reproduce the vessels, as it did in the proximal and middle zones, leading to an increase in false negatives. This region is characterised by narrow arteries and low brightness due to a lack of contrast during the image acquisition. This scarcity of information makes the 3D network fail. Moreover, it is in the distal region where the 2D UNet from scratch obtains better results than the other nets. This may indicate that pre-trained nets also have limiting factors when adjusting their parameters.

In conclusion, the 2D networks perform better than the 3D in all regions, being more evident in the distal region, where the 3D generate incomplete geometries. However, the 2D networks tend to overestimate the vessel calibre. Our study opens the possibility to consider a more complex mechanism for segmentation that uses different tools depending on the region considered. In this way, global performance indicators will improve.

References

1. Cheung, W.K., et al.: A computationally efficient approach to segmentation of the aorta and coronary arteries using deep learning. IEEE Access **9**, 108873–108888 (2021)
2. Gu, L., Cai, X.C.: Fusing 2D and 3D convolutional neural networks for the segmentation of aorta and coronary arteries from CT images. Artif. Intell. Med. **121**, 102189 (2021). https://doi.org/10.1016/j.artmed.2021.102189
3. Hajhosseiny, R., et al.: Clinical comparison of sub-mm high-resolution non-contrast coronary CMR angiography against coronary CT angiography in patients with low-intermediate risk of coronary artery disease: a single center trial. J. Cardiovasc. Magn. Reson. **23**, 1–14 (2021)
4. Kurata, A., et al.: On-site computed tomography-derived fractional flow reserve using a machine-learning algorithm–clinical effectiveness in a retrospective multi-center cohort. Circ. J. **83**(7), 1563–1571 (2019)
5. Otero-Cacho, A., et al.: Validation of a new model of non-invasive functional assessment of coronary lesions by computer tomography fractional flow reserve. REC: CardioClinics (2023)
6. Pan, L.S., Li, C.W., Su, S.F., Tay, S.Y., Tran, Q.V., Chan, W.P.: Coronary artery segmentation under class imbalance using a u-net based architecture on computed tomography angiography images. Sci. Rep. 11(1) (2021). https://doi.org/10.1038/s41598-021-93889-z
7. Panayides, A.S., et al.: Ai in medical imaging informatics: current challenges and future directions. IEEE J. Biomed. Health Inform. **24**(7), 1837–1857 (2020)
8. Serrano-Antón, B., et al.: Coronary artery segmentation based on transfer learning and UNet architecture on computed tomography coronary angiography images. IEEE Access **11**, 75484–75496 (2023). https://doi.org/10.1109/ACCESS.2023.3293090
9. Yang, L., et al.: Serial coronary CT angiography-derived fractional flow reserve and plaque progression can predict long-term outcomes of coronary artery disease. Eur. Radiol. **31**(9), 7110–7120 (2021). https://doi.org/10.1007/s00330-021-07726-y
10. Yasue, H., Matsuyama, K., Matsuyama, K., Okumura, K., Morikami, Y., Ogawa, H.: Responses of angiographically normal human coronary arteries to intracoronary injection of acetylcholine by age and segment. possible role of early coronary atherosclerosis. Circulation **81**(2), 482–490 (1990)

Unsupervised Learning of Cortical Surface Registration Using Spherical Harmonics

Seungeun Lee[1], Sunghwa Ryu[2], Seunghwan Lee[1], and Ilwoo Lyu[1,3(✉)] ⓘ

[1] Department of Computer Science and Engineering, UNIST, Ulsan, South Korea
ilwoolyu@unist.ac.kr
[2] Department of Bio and Brain Engineering, KAIST, Daejeon, South Korea
[3] Graduate School of Artificial Intelligence, UNIST, Ulsan, South Korea

Abstract. We present novel learning-based spherical registration using the spherical harmonics. Our goal is to achieve a continuous and smooth warp field that can effectively facilitate precise cortical surface registration. Conventional spherical registration typically involve sequential procedures for rigid and non-rigid alignments, which can potentially introduce substantial warp distortion. By contrast, the proposed method aims at joint optimization of both types of alignments. Inspired by a recent study that represents a rotation by 6D parameters as a continuous form in the Euclidean domain, we extend the idea to encode and regularize a velocity field. Specifically, a local velocity is represented by a single rotation with 6D parameters that can vary smoothly over the unit sphere via spherical harmonic decomposition, yielding smooth, spatially varying rotations. To this end, our method can lead to a significant reduction in warp distortion. We also incorporate a spherical convolutional neural network to achieve fast registration in an unsupervised manner. In the experiments, we compare our method with popular spherical registration methods on a publicly available human brain dataset. We show that the proposed method can significantly reduce warp distortion without sacrificing registration accuracy.

Keywords: Cortical surface registration · Spherical registration · Spherical harmonics · Unsupervised learning

1 Introduction

Neuroimaging data analysis is one of the most challenging tasks in the field due to the presence of high structural variability across individuals. In structural data analysis, neuroimaging data are often understood on 2-manifolds, which requires non-rigid surface registration [6]. Since a bijective spherical mapping is readily available for cortical surfaces [6], spherical registration is a popular choice to establish a shape correspondence. Here, the goals of spherical registration can

---

S. Lee and S. Ryu—Contributed equally to this work.

C. Wachinger et al. (Eds.): ShapeMI 2023, LNCS 14350, pp. 65–74, 2023.
https://doi.org/10.1007/978-3-031-46914-5_6

be categorized (1) to align the input geometry as closely as possible to the target geometry and (2) to reduce the warp distortion. While most methods focus on (1), the importance of warp distortion is often overlooked, which can mislead statistical shape analysis even with high registration accuracy [13,17,22,25]. In previous attempts, classical non-learning-based spherical registration methods [7,8,13,14,17,22,24] provided a well-established spherical correspondence by directly optimizing the energy function. Due to their high registration accuracy, they hence have been widely used in structural data analysis. Nevertheless, a common drawback of these methods is costly optimization process.

Despite the success of spherical convolutional neural networks (CNNs) in neuroimaging studies [1,2,12,15,16,18–20,23,27], only few have explored CNN-based spherical registration. An early attempt by [4] adapted a volumetric method for spherical registration by flattening out the spherical space to the Euclidean domain. Despite the computational gain, their method suffers from unbalanced sampling of spherical coordinates and boundary discontinuity. [26] extended [24] by incorporating a spherical CNN. However, their convolution can be flipped at the pole, the resulting registration is inconsistent depending on a pole choice. [21] built upon the work of [17] by enhancing the flexibility of spherical warp through the incorporation of a broader range of predefined rotations. Nevertheless, their warp remains constrained to a discrete representation. Furthermore, the existing methods employ rigid alignment only once prior to optimization, which may not be optimal for non-rigid alignment. This can consequently yield substantial warp distortion as reported in [13].

In this paper, we propose a learning-based surface registration method while reducing warp distortion. The main approach of this work follows our previous study [13] that jointly optimize rigid and non-rigid alignments using the spherical harmonics. Unfortunately, spherical warp in [13] is achieved via local displacement rather than vector field integral, which can restrict the amount of spherical warp. Moreover, parameter dependency in the displacement encoding of [13] poses challenges for its extension with deep learning approaches. In this work, we consider local velocity rather than a simple displacement for spherical warp. Here, local velocity can be interpreted as a rotation, which can be efficiently encoded via spherical harmonic decomposition. Consequently, the proposed method enables flexible warp trajectories and deep learning integration. The main contributions of this paper include (1) a new encoding scheme of the warp field, (2) a significant reduction in warp distortion without losing registration accuracy, and (3) validation in registration accuracy, warp distortion, and runtime. Figure 1 shows an overview of the proposed method.

2   Method

2.1   Problem Statement

Given a moving geometric feature M and a target feature F on $\mathbb{S}^2$, the goal is to seek a warp field $\Phi : \mathbb{S}^2 \rightarrow \mathbb{S}^2$ such that $F(\theta, \phi) = M(\Phi(\theta, \phi))$ for $(\theta, \phi) \in$

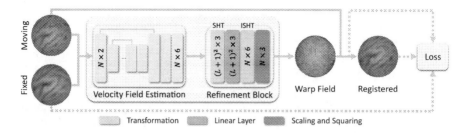

Fig. 1. A schematic overview of the proposed method. The size of the output channels is shown in each box. N is the number of vertices of the spherical tessellation, and L is the harmonic bandwidth. The velocity field is initially estimated by a spherical CNN and further refined by imposing smoothness and allowing joint optimization of rigid and non-rigid alignments through the spherical harmonic transform (SHT) and its inverse (ISHT). The dotted arrows indicate the training phase only.

$[0, \pi] \times [-\pi, \pi]$. In general, spherical registration minimizes the following energy function:

$$\int_{\mathbb{S}^2} \mathcal{L}_{sim}(F(\mathbf{x}), M \circ \Phi(\mathbf{x})) d\mathbf{x} + \alpha \int_{\mathbb{S}^2} \mathcal{L}_{reg}(\Phi(\mathbf{x})) d\mathbf{x}, \tag{1}$$

where $\mathcal{L}_{sim}(\cdot, \cdot) \in \mathbb{R}$ is a similarity term and $\mathcal{L}_{reg}(\cdot) \in \mathbb{R}$ is a regularization term to control the smoothness of Φ for some non-negative weighting factor $\alpha \in \mathbb{R}$. A diffeomorphic mapping can be modeled by introducing a smooth velocity field on the tangent space in a continuous time domain $t = [0, 1]$ as a stationary ordinary differential equation (ODE):

$$\frac{d\Phi(\mathbf{x}, t)}{dt} = v(\Phi(\mathbf{x}, t)), \tag{2}$$

where $v \in T_{\mathbf{x}}\mathbb{S}^2$ is a stationary velocity with the initial condition $\Phi(\mathbf{x}, 0) = \mathbf{x}$. We will describe the computation of the velocity field and its incorporation into a spherical CNN in the remainder of this paper.

2.2 Velocity Encoding

We encode a velocity vector as a spatially varying function using a 6D continuous representation of the rotation space $SO(3)$ [28] to handle vector as independent set of scalar components. More formally, for a velocity v at $\mathbf{x}$, we can find a function that generates a rotation matrix $R_{\mathbf{x}} : \mathbb{R}^6 \to SO(3)$:

$$\Phi_v(\mathbf{x}) = R_{\mathbf{x}}(\{r^{(i)}\}_{i=1,\cdots,6}) \cdot \mathbf{x}, \tag{3}$$

such that

$$R_{\mathbf{x}} \cdot \mathbf{x} = \exp_{\mathbf{x}}(v), \tag{4}$$

where $\{r^{(i)}\}_{i=1,\cdots,6}$ is a set of spatially varying irreducible parameters with respect to $\mathbf{x}$.

2.3   Velocity Field

In our encoding scheme, a smooth velocity field can be interpreted as a smooth change of the 6 parameters, and their order independence allows smoothing over each parameter in its own parametric space. However, the complexity of parameter-wise smoothing is subject to the spherical tessellation, which may also affect the quality of the smoothing depending on the tessellation regularity. To address such an issue, we propose spherical harmonic decomposition that encodes our velocity field and encourages its smoothness, inspired by [13]. Here, we restate our original problem of computing a smooth vector field by finding a proper set of learnable harmonic coefficients $\{c\} \subseteq \mathbb{R}$. Once $\{c\}$ is determined, the ith parameter $r^{(i)}(\theta, \phi)$, $i = 1, \cdots, 6$ can be reconstructed by a linear combination of the irreducible real harmonic basis functions $Y : \mathbb{S}^2 \to \mathbb{R}$:

$$r^{(i)}(\theta, \phi) = c_{00}^{(i)} Y_{00}(\theta, \phi) + \sum_{l=1}^{L} \sum_{m=-l}^{l} c_{lm}^{(i)} Y_{lm}(\theta, \phi), \qquad (5)$$

where r is truncated up to degree of L. This strategy has three major advantages. (1) The velocity field is differentiable because the spherical harmonics consist of trigonometric functions that guarantee C^∞ continuity for $L > 0$. As high-frequency components are truncated by L, the velocity field becomes smoother as L decreases, in which the smoothness can be easily tunable. (2) As discussed in [13], both rigid ($l = 0$, the left term of Eq. (5)) and non-rigid ($l > 0$, the right term of Eq. (5)) alignment components are explicitly encoded and thus can be optimized simultaneously. (3) Smoothing over the proposed encoding is independent of the spherical tessellation.

2.4   Warp Trajectory

Since our velocity field consists of rotation matrices, the warp trajectories on $\mathbb{S}^2$ can be computed easily as matrix multiplications. The trajectories can be efficiently traced in a scaling and squaring fashion on $\mathbb{S}^2$ as numerical integration with k-step recursion [24]:

$$\Phi_{v/2^{(k-1)}}(\mathbf{x}) = \Phi_{v/2^k} \circ \Phi_{v/2^k}(\mathbf{x}). \qquad (6)$$

2.5   Architecture

Velocity Field Estimation. We couple M and F to estimate a set of the initial vertex-wise parameters $\hat{r}^{(i)}(\theta, \phi), i = 1, \cdots, 6$. We integrate a spherical CNN into the proposed framework. Although the framework is flexible in choosing spherical CNNs, we use [9] as our backbone architecture since its spectral convolution is free of tessellation strategies and some resources are reusable. It is noteworthy that the estimated parameters by the spherical CNN may not exhibit spatial smoothness, which requires further refinement. Hereafter, we denote the spherical tessellation by $\mathcal{S} \subset \mathbb{S}^2$.

Refinement Block. To control the smoothness of $\hat{r}$ over $\mathbb{S}^2$ and enable joint optimization of rigid and non-rigid alignments, $\hat{r}$ are converted into the spectral signals using the spherical harmonic transform. This yields the ith harmonic coefficient for degree of l and order of m through the inner product:

$$\hat{c}_{lm}^{(i)} = \int_{\mathbb{S}^2} \hat{r}^{(i)}(\theta, \phi) Y_{lm}(\theta, \phi) d\theta d\phi. \tag{7}$$

For a sufficiently dense spherical tessellation, the above integral can be efficiently approximated by a Riemann sum, as reported in [5,9]. Each spectral component is further refined by its own linear model:

$$c_{lm}^{(i)} = \hat{c}_{lm}^{(i)} \cdot w_{lm}^{(i)} + b_{lm}^{(i)}, \tag{8}$$

where $w_{lm}^{(i)}$ and $b_{lm}^{(i)}$ are learnable parameters. The refined harmonic coefficients are recovered back to vertex-wise rotations by Eq. (5) followed by Eq. (3). We finally compute the warp field by Eq. (6).

Loss Function. To compute the similarity between the registered and fixed surfaces, we consider an L^2 similarity metric as widely used in classical methods [7,13,24]:

$$\mathcal{L}_{sim}(F, M \circ \Phi) = \frac{1}{|\mathcal{S}|} \sum_{\mathbf{x} \in \mathcal{S}} (F(\mathbf{x}) - M \circ \Phi(\mathbf{x}))^2. \tag{9}$$

In the proposed encoding, the smoothness of the velocity field is controlled after the harmonic truncation. For further regularization, we penalize the amount of warp after registration [13]. For a spherical location $\mathbf{x} \in \mathcal{S}$ with its neighborhoods $\mathcal{N}_{\mathbf{x}} \subset \mathcal{S}$, their arc length changes after the spherical warp are measured by

$$\mathcal{L}_{reg}(\Phi) = \frac{1}{2} \sum_{\mathbf{x} \in \mathcal{S}} \sum_{\mathbf{y} \in \mathcal{N}_{\mathbf{x}}} (\arccos(\mathbf{x}^\mathsf{T}\mathbf{y}) - \arccos(\Phi(\mathbf{x})^\mathsf{T}\Phi(\mathbf{y})))^2. \tag{10}$$

In this way, the regularity of the field can be controlled by tuning α. Strong regularization (*i.e.*, large α) forces the field to stay as isometric as possible.

Implementation Details. We use PyTorch for the backend processing and customized CUDA kernels for spherical re-tessellation. For the hyperparameters, we choose $k = 6$, $L = 40$, and $\alpha = 25^{-1}$ for smoothing and refinement block, and 64 input channels, harmonic bandwidth of 80, and depth of 4 for initial velocity estimation (1.19M learnable parameters in total). We use the Adam optimizer [10] at initial learning rate of 0.001 with decay by a factor of 0.5 if no improvement is made in four consecutive epochs.

3   Experimental Setup

3.1   Imaging Data

We used the Mindboggle public dataset [11] with 101 subjects becasue the manual labels of 32 regions of interest (ROIs)[1] are available. The cortical surfaces were reconstructed via the standard FreeSurfer package [6]. For M, we used *average convexity* [6] for registration metric and performance evaluation. For F, we used the *Buckner40* template provided by the official FreeSurfer package [6]. We note that the proposed method is flexible in choosing other geometric features. The left hemisphere was used.

3.2   Baseline Methods

We compared publicly available spherical registration methods: FreeSurfer [7], Spherical Demons (SD) [24], Multimodal Surface Matching (MSM) [17], Hierarchical Spherical Deformation (HSD) [13], and Deep-Discrete Spherical Registration (DDR) [21]. For a fair comparison, we carefully tuned the baseline methods to match the mean squared error (MSE) around 0.31, while keeping their warp field as smooth as possible. In FreeSurfer [7], we disabled the default mean curvature alignment to align with *average convexity* only. We also set the distance term to be 6 for smooth warp. In SD [24], we set 3 iterations for the velocity field smoothness for all multi-resolution stages (icosahedral subdivision: 4, 5, 6, 7). In MSM [17], we set the weight for smoothing to be 0.1 for all multi-resolution stages (icosahedral subdivision: 4, 4, 5, 6). We also used cross-correlation as a similarity metric as suggested for its best performance. In HSD [13], we set $L = 0, 5, 10, 15$ for the respective optimization stages (icosahedral subdivision: 4, 5, 6, 6) by fixing the regularization weight of 169. In DDR [21], we followed the same training configurations as reported in the paper. It is noteworthy that all the baseline methods commonly reported optimization over *average convexity* in their original work. For DDR and our method, we performed 5-fold cross-validation strategy: 60% for training, 20% for validation, and 20% for test. For each fold, we circulated the partitions to achieve full validation. In the test phase, we used the learned model parameters at the peak performance in the validation set. For the inference of unseen data, the inferred warp field is applied to the original mesh. All the experiments were conducted on an Intel Xeon 6248R and an NVIDIA GeForce RTX 3090.

3.3   Evaluation Metrics

For the similarity evaluation, we used the normalized cross-correlation (NCC) between F and $M \circ \Phi$ and Dice of ROIs. As our template has no manual labels, we transferred the manual labels of individuals to a target subject via the established correspondence by each method and then computed the majority for $\mathbf{x} \in \mathbb{S}^2$ to

---

[1] A full list can be found at https://mindboggle.readthedocs.io/en/latest/labels.html.

Table 1. Benchmark in registration accuracy, distortion, CPU runtime, and learnable parameters. The runtime is measured on the original mesh (average: 141K vertices) for the whole process, including model initialization, rigid alignment, remeshing, and model inference. Inference (sec.): 9.02 (DDR) and 8.62 (ours). *Bold*: best. *Blue*: $q < 0.05$.

Method	Accuracy			Log Area			CPU	Param
	MSE	NCC	Dice	Mean	Median	Max	(sec.)	(N)
FreeSurfer	0.313	0.898	**0.873**	0.322	0.267	4.808	439.83	–
SD	0.309	0.898	0.873	0.320	0.250	2.319	42.28	–
MSM	0.311	0.891	0.863	0.573	0.404	7.043	503.23	–
HSD	0.305	0.898	0.872	0.306	0.234	3.340	206.82	–
DDR	0.305	0.885	0.871	0.344	0.251	3.699	15.91	21.26M
Ours	0.307	**0.899**	0.871	**0.289**	**0.225**	**2.277**	**15.48**	1.19M

measure the Dice. For the warp distortion, we measured the absolute logarithm of the area ratio (*Log Area*) given by $|\log_2(\Delta(\mathbf{x})/\Delta(\Phi(\mathbf{x})))|$, where $\Delta(\cdot)$ is a vertex area [17].

4 Results

Table 1 summarizes the overall performance in registration accuracy and warp distortion. We performed a paired t-test against each baseline and reported statistical significance by the false discovery rate (FDR) [3] at $q = 0.05$. There is no performance degradation in registration accuracy for our method. For warp distortion, our method achieves the best performance in most cases. Clearly, a comparable NCC/Dice can be maintained in our method while reducing the warp distortion (see Table 1 and Figs. 2 and 3). This is because the proposed method incorporates diffeomorphic trajectories for the feature alignment and optimizes rigid and non-rigid alignments simultaneously for the distortion reduction. It is noteworthy that although the reported measures can be adjusted by balancing the similarity and regularization terms, there always exists a trade-off between the two terms. In the runtime benchmark, our method offers faster registration than the baseline methods on a single CPU thread (see Table 1).

5 Discussion

Our preliminary findings indicate that our framework can reduce warp distortion without compromising registration accuracy. However, further research is necessary, particularly regarding methodological validation. Firstly, an exhaustive analysis of individual hyperparameters will provide valuable insights into fine-tuning our approach. Secondly, the performance of our method can be evaluated on other high-resolution geometric features. A multi-resolution approach could further enhance our framework [21, 26]. Furthermore, as our method allows

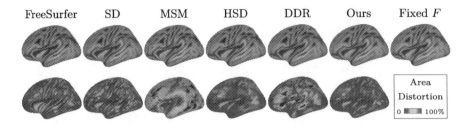

Fig. 2. Average feature and distortion maps across 101 subjects. *Top*: mean *average convexity*. *Bottom*: area distortion by the exponent of the *Log Area*. All the methods show comparable registration accuracy, while our method overall achieves the lowest area distortion. The inflated surface is used for better visualization.

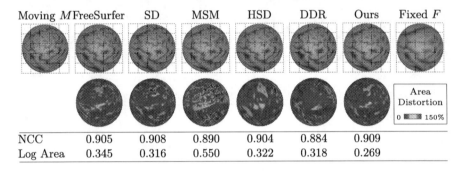

	FreeSurfer	SD	MSM	HSD	DDR	Ours
NCC	0.905	0.908	0.890	0.904	0.884	0.909
Log Area	0.345	0.316	0.550	0.322	0.318	0.269

Fig. 3. The lateral frontal lobe of an example subject. *Top*: warped feature. All the methods align M well with F. *Bottom*: area distortion by the exponent of the *Log Area*. Our method reduces warp distortion without sacrificing registration accuracy.

for flexibility in designing similarity measures, exploring the incorporation of multi-modal geometric features [17] could be an interesting direction for future research.

6   Conclusion

We presented a novel learning-based spherical registration method using the spherical harmonics. We decomposed local velocity into 6D independent parameters and encoded the velocity field as a smooth spherical function via spherical harmonic decomposition. This approach enabled the simultaneous optimization of both rigid and non-rigid alignments, resulting in reduced warp distortion. We also integrated a spherical CNN for fast spherical registration. The experimental results on a human brain dataset demonstrated significant reduction in warp distortion without sacrificing registration accuracy.

Acknowledgments. This work was supported in part by the NRF under Grant RS-2023-00251298 and RS-2023-00266120, in part by the IITP under Grant 2020-0-01336, the Artificial Intelligence Graduate School Program, UNIST.

References

1. Barbaroux, H., Feng, X., Yang, J., Laine, A.F., Angelini, E.D.: Encoding human cortex using spherical CNNs-a study on Alzheimer's disease classification. In: 2020 IEEE 17th International Symposium on Biomedical Imaging (ISBI), pp. 1322–1325. IEEE (2020)
2. Bayrak, R.G., Lyu, I., Chang, C.: Learning subject-specific functional parcellations from cortical surface measures. In: International Workshop on PRedictive Intelligence In MEdicine, pp. 172–180. Springer (2022). https://doi.org/10.1007/978-3-031-16919-9_16
3. Benjamini, Y., Hochberg, Y.: Controlling the false discovery rate: a practical and powerful approach to multiple testing. J. Roy. Stat. Soc.: Ser. B (Methodol.) $57(1)$, 289–300 (1995)
4. Cheng, J., Dalca, A.V., Fischl, B., Zöllei, L., Initiative, A.D.N., et al.: Cortical surface registration using unsupervised learning. Neuroimage 221, 117161 (2020)
5. Chung, M.K.: Heat kernel smoothing on unit sphere. In: 3rd IEEE International Symposium on Biomedical Imaging: Nano to Macro, 2006, pp. 992–995. IEEE (2006)
6. Fischl, B.: Freesurfer. Neuroimage $62(2)$, 774–781 (2012)
7. Fischl, B., Sereno, M.I., Tootell, R.B., Dale, A.M.: High-resolution intersubject averaging and a coordinate system for the cortical surface. Hum. Brain Mapp. $8(4)$, 272–284 (1999)
8. Glaunès, J., Vaillant, M., Miller, M.I.: Landmark matching via large deformation diffeomorphisms on the sphere. J. Math. Imaging Vis. $20(1)$, 179–200 (2004)
9. Ha, S., Lyu, I.: SPHARM-Net: Spherical harmonics-based convolution for cortical parcellation. IEEE Trans. Med. Imaging $41(10)$, 2739–2751 (2022)
10. Kingma, D.P., Ba, J.: Adam: a method for stochastic optimization. In: International Conference on Learning Representations (ICLR) (2015)
11. Klein, A., Tourville, J.: 101 labeled brain images and a consistent human cortical labeling protocol. Front. Neurosci. 6, 171 (2012)
12. Lyu, I., et al.: Labeling lateral prefrontal sulci using spherical data augmentation and context-aware training. Neuroimage 229, 117758 (2021)
13. Lyu, I., Kang, H., Woodward, N.D., Styner, M.A., Landman, B.A.: Hierarchical spherical deformation for cortical surface registration. Med. Image Anal. 57, 72–88 (2019)
14. Lyu, I., et al.: Robust estimation of group-wise cortical correspondence with an application to macaque and human neuroimaging studies. Front. Neurosci. 9, 210 (2015)
15. Ngo, G.H., Khosla, M., Jamison, K., Kuceyeski, A., Sabuncu, M.R.: From connectomic to task-evoked fingerprints: individualized prediction of task contrasts from resting-state functional connectivity. In: Martel, A.L., et al. (eds.) MICCAI 2020. LNCS, vol. 12267, pp. 62–71. Springer, Cham (2020). https://doi.org/10.1007/978-3-030-59728-3_7
16. Parvathaneni, P., et al.: Cortical Surface Parcellation Using Spherical Convolutional Neural Networks. In: Shen, D., et al. (eds.) MICCAI 2019. LNCS, vol. 11766, pp. 501–509. Springer, Cham (2019). https://doi.org/10.1007/978-3-030-32248-9_56
17. Robinson, E.C., et al.: MSM: a new flexible framework for multimodal surface matching. Neuroimage 100, 414–426 (2014)

18. Sedlar, S., Alimi, A., Papadopoulo, T., Deriche, R., Deslauriers-Gauthier, S.: A spherical convolutional neural network for white matter structure imaging via dMRI. In: de Bruijne, M., et al. (eds.) MICCAI 2021. LNCS, vol. 12903, pp. 529–539. Springer, Cham (2021). https://doi.org/10.1007/978-3-030-87199-4_50
19. Seong, S.B., Pae, C., Park, H.J.: Geometric convolutional neural network for analyzing surface-based neuroimaging data. Front. Neuroinform. **12**, 42 (2018)
20. Sinzinger, F.L., Moreno, R.: Reinforcement learning based tractography with so (3) equivariant agents. In: Geometric Deep Learning in Medical Image Analysis (2022)
21. Suliman, M.A., Williams, L.Z., Fawaz, A., Robinson, E.C.: A deep-discrete learning framework for spherical surface registration. In: International Conference on Medical Image Computing and Computer-Assisted Intervention, pp. 119–129. Springer, Cham (2022). https://doi.org/10.1007/978-3-031-16446-0_12
22. Van Essen, D.C.: A population-average, landmark-and surface-based (PALS) atlas of human cerebral cortex. Neuroimage **28**(3), 635–662 (2005)
23. Willbrand, E.H., et al.: Uncovering a tripartite landmark in posterior cingulate cortex. Sci. Adv. **8**(36), eabn9516 (2022)
24. Yeo, B.T., Sabuncu, M.R., Vercauteren, T., Ayache, N., Fischl, B., Golland, P.: Spherical demons: fast diffeomorphic landmark-free surface registration. IEEE Trans. Med. Imaging **29**(3), 650–668 (2009)
25. Yu, C., et al.: Validation of group-wise registration for surface-based functional MRI analysis. In: Proceedings of SPIE-the International Society for Optical Engineering, vol. 11596. NIH Public Access (2021)
26. Zhao, F., et al.: S3Reg: superfast spherical surface registration based on deep learning. IEEE Trans. Med. Imaging **40**(8), 1964–1976 (2021)
27. Zhao, F., et al.: Spherical U-net on cortical surfaces: methods and applications. In: Chung, A.C.S., Gee, J.C., Yushkevich, P.A., Bao, S. (eds.) IPMI 2019. LNCS, vol. 11492, pp. 855–866. Springer, Cham (2019). https://doi.org/10.1007/978-3-030-20351-1_67
28. Zhou, Y., Barnes, C., Lu, J., Yang, J., Li, H.: On the continuity of rotation representations in neural networks. In: Proceedings of the IEEE/CVF Conference on Computer Vision and Pattern Recognition, pp. 5745–5753 (2019)

Unsupervised Correspondence with Combined Geometric Learning and Imaging for Radiotherapy Applications

Edward G. A. Henderson[1]([✉]) [ID], Marcel van Herk[1,2] [ID], Andrew F. Green[3] [ID], and Eliana M. Vasquez Osorio[1,2] [ID]

[1] The University of Manchester, Oxford Rd, Manchester M13 9PL, UK
edward.henderson@postgrad.manchester.ac.uk
[2] Radiotherapy Related Research, The Christie NHS Foundation Trust, Manchester M20 4BX, UK
[3] European Bioinformatics Institute, EMBL-EBI, Cambridge, UK

Abstract. The aim of this study was to develop a model to accurately identify corresponding points between organ segmentations of different patients for radiotherapy applications. A model for simultaneous correspondence and interpolation estimation in 3D shapes was trained with head and neck organ segmentations from planning CT scans. We then extended the original model to incorporate imaging information using two approaches: 1) extracting features directly from image patches, and 2) including the mean square error between patches as part of the loss function. The correspondence and interpolation performance were evaluated using the geodesic error, chamfer distance and conformal distortion metrics, as well as distances between anatomical landmarks. Each of the models produced significantly better correspondences than the baseline non-rigid registration approach. The original model performed similarly to the model with direct inclusion of image features. The best performing model configuration incorporated imaging information as part of the loss function which produced more anatomically plausible correspondences. We will use the best performing model to identify corresponding anatomical points on organs to improve spatial normalisation, an important step in outcome modelling, or as an initialisation for anatomically informed registrations. All our code is publicly available at https://github.com/rrr-uom-projects/Unsup-RT-Corr-Net.

Keywords: correspondence · un-supervised learning · geometric learning · image registration · radiotherapy

1 Introduction

Radiotherapy is used in the treatment of $\sim 80\%$ of Head and Neck (HN) cancer patients [19]. Treatments are planned on a patient's computed tomography (CT)

scan, where the tumour and the organs-at-risk are segmented. These segmentations are also used to establish dose-effect relationships which are ultimately used to improve radiotherapy practice. Modern techniques which allow the investigation of sub-volume dose effects rely on spatial normalisation to map the dose distributions between patients [15]. Examples for these associations in HN radiotherapy include radiation dose to the base of the brainstem and late dysphagia (problems swallowing) [20], dose to the masseter muscle and trismus (limited jaw movement) [2]. In these examples, the authors used intensity-based non-rigid image registration (NRR) to indirectly establish the correspondence of the anatomy between different patients. Improved spatial normalisation, using point-wise correspondences rather than NRR algorithms, would reduce uncertainties in outcome modelling applications. However, manually annotating pair-wise correspondences is a complex and time consuming task, rendering its practice unfeasible.

Another promising use of correspondences in radiotherapy applications is in the initialisation of structure-guided image registration methods. Currently, spline-based registration relies on estimating correspondence based on distance criteria [21] and more advanced finite-element based models rely on a set of boundary conditions, e.g. based on structure curvature [4]. These structure-based registrations are particularly useful for cases with dramatic changes, such as registration of images before/after an intervention [22] or of images separated by a long time period (e.g. paediatric follow-up or re-irradiation settings). A model that can quickly and accurately identify corresponding points on sets of anatomical structures would be particularly effective for incorporation into other non-rigid image registration frameworks.

The aim of this study was to find a solution to automatically identify corresponding anatomical points on organs for radiotherapy applications. In this study, we took an established model for simultaneous correspondence and interpolation estimation in everyday 3D shapes, *Neuromorph* [5], and retrained it on biomedical data, specifically HN organ segmentations from planning CT scans. It has previously been shown that the performance of geometric learning models for tasks involving radiotherapy organ shapes can be dramatically improved by incorporating the associated CT scan imaging [6], an approach not attempted in previous correspondence literature [9,13,17]. Therefore we extended *Neuromorph* in two ways in an attempt to optimise its performance for this application: 1) by directly complementing geometrical features with learned image features, and 2) by adding a novel imaging loss function component. The performance of our resultant correspondence models were compared to a NRR algorithm currently used for outcome modelling.

2   Materials and Method

2.1   Dataset

An open-access dataset of 34 head and neck CT scans with segmentations of the brainstem, spinal cord, mandible, parotid and submandibular glands was used for this study [14]. The segmentations are highly consistent and followed international guidelines, having been produced by an expert and then audited by three observers and a specialist oncologist with at least four years of experience.

2.2   Pre-processing

The CT scans had a $\sim 2.5 \times 1 \times 1$ mm voxel spacing and were truncated at the apex of the lungs to ensure consistency in the length of the cervical section of the spinal cord. The marching cubes algorithm was used to generate 3D triangular meshes for each organ, which were then smoothed with ten iterations of Taubin smoothing. The meshes were simplified using quadric decimation to 3000 triangles for each of the organs apart from the submandibular glands which were simplified to 2000 triangles because of their smaller volume. The organ meshes were then optimised by iteratively splitting the longest and collapsing the shortest edges. The CT scans were rigidly aligned to a single reference patient using *SimpleITK* 2.0.2. The computed transformations were applied directly to the mesh vertices to align the organ shapes thereby avoiding interpolation artefacts.

2.3   Model

The source model used in this study, *Neuromorph*, was originally presented by Eisenberger et al. [5]. *Neuromorph* is a geometric learning model which, when given two 3D triangular meshes, predicts corresponding points and a smooth interpolation between the two in a single forward pass. The model performs unsupervised learning which is crucial for our applications because of the scarcity of high quality 3D data labelled with point-to-point correspondences.

Figure 1 shows a schematic of the original model and one of our modifications to add imaging features. The *Neuromorph* model is formed of two components: a Siamese feature extracting network and an interpolator. The feature-extracting portion consists of two networks with shared features which receive two meshes as input. The encoded shape features are matched using matrix multiplication to produce a correspondence matrix between the input meshes. The correspondence matrix is used to produce a vector which contains the offset between source vertices and their corresponding counterparts in the target mesh. This offset vector provides part of the input to the interpolator, along with the original source vertices and a time-step encoding to provide the number of intermediate steps along which to interpolate the deformation. The interpolator outputs a deformation vector for these time-steps for each vertex in the source mesh.

The feature extractor and interpolator have identical graph neural network architectures, consisting of repeating residual EdgeConv layers [23]. The primary intuition behind success of the *Neuromorph* architecture is that correspondence and interpolation are interdependent tasks that complement each other when optimised in an end-to-end fashion. *Neuromorph* uses a three-component loss function for unsupervised learning. These are: a registration loss, to quantify the overlap of the target and source meshes; an "as-rigid-as-possible" (ARAP) loss, to penalise overly elastic deformations; and a geodesic distance preservation loss, to regularise the predicted pair-wise correspondences. For all models in this study, the weight of the ARAP loss component was increased by a factor of ten compared to the originally proposed value to reduce the elasticity of predicted deformations. For full implementation details of the original *Neuromorph* model, refer to [5].

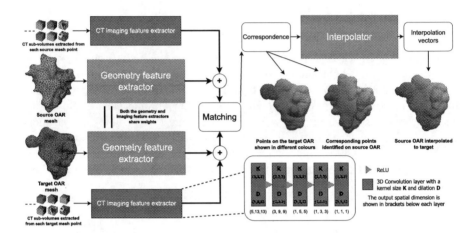

Fig. 1. A schematic of the *Neuromorph* architecture [5] with our first extension to leverage the CT imaging. Additional CNN blocks (red) encode a $7 \times 19 \times 19$ CT sub-volume for each mesh point into an imaging feature vector to complement the geometrical features used to predict point correspondences. An example of the correspondence and interpolation predictions for a pair of parotid glands from different patients is shown with the different colours showing corresponding points. (Color figure online)

2.4 Incorporating Imaging Information

The original *Neuromorph* model predicts point-to-point correspondences solely on the geometric structure of the input meshes. Since the meshes used in this study are organ shapes derived from CT scans, we have additional imaging information which was leveraged using two different approaches.

Complementing Geometrical Features with Image Features. We followed a similar approach as in our previous work to encode image patches for each point on the mesh using a 3D convolutional neural network (CNN) [6]. Figure 1 shows further details of this methodology extension. Cubic 3D image patches of side length $\approx 19\,\mathrm{mm}$ ($7 \times 19 \times 19$ voxel sub-volumes) were extracted from the CT scan for each vertex on the triangular mesh of each organ. This patch size was chosen so that image information $10\,\mathrm{mm}$ outside the organ is within view, including surrounding structures such as bones and air cavities. Figure 2 shows an example slice of a parotid gland contour and demonstrates the field-of-view which these image patches cover. The image patches were normalised from

Hounsfield Units (HU) onto the range $[0, 1]$ using contrast windowing with settings used to visualise soft tissue (W 350HU, L 40HU) [7]. The patches were then encoded using a custom CNN architecture into imaging feature vectors (Fig. 1). These imaging feature vectors were concatenated with the feature vectors created by the geometric feature extractors of the original *Neuromorph* model. Feature matching and correspondence prediction was then performed as before, but now utilising both geometric and imaging information. The imaging and geometric feature extractors of the extended model were optimised simultaneously during training.

Imaging as a Component of the Loss Function. For the second approach, we added a new loss term to calculate the mean-squared error of $7 \times 19 \times 19$ image patches for which the associated vertices are identified as corresponding. The l_{imaging} loss component was calculated as

$$l_{\text{imaging}} = \lambda_{\text{imaging}} \times \|\Pi Y_{\text{CT\_patches}} - X_{\text{CT\_patches}}\|^2 \tag{1}$$

where Π is the predicted correspondence matrix, $Y_{\text{CT\_patches}}$ and $X_{\text{CT\_patches}}$ are the CT image patches of the related target and source mesh points respectively and λ_{imaging} was set to 1000 to balance the contribution with the other components. This hyperparameter value was chosen in preliminary testing from a range spanning $1 \rightarrow 100,000$.

By incorporating the imaging information as a loss component, the model does not require any additional input or modification from the original architecture. However, the rationale of including such an imaging loss was to encourage the model to learn more anatomically feasible correspondences at training time based on the underlying CT scan.

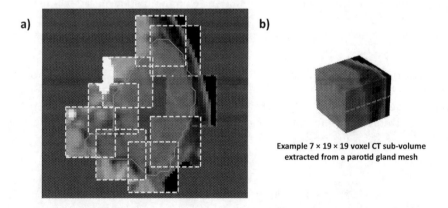

Fig. 2. a) A cross sectional view of a parotid gland mesh showing the field-of-view of the $7 \times 19 \times 19$ CT sub-volumes. Only $\sim 10\%$ of the sub-volume patches in this cross section are shown for clarity of the visualisation. b) A visualisation of one of the 3D CT sub-volumes from the lateral aspect of the parotid.

2.5   Comparison with Non-rigid Image Registration

We compared the performance of the correspondence models with an established NRR algorithm which is a standard approach for aligning images and anatomical structures for radiotherapy applications [2,11,20]. For this comparison, the CT scans were first rigidly registered to a single reference patient, as before, then *NiftyReg* was used to non-rigidly register each pair of patients [10]. The registration performed was a cubic B-spline using normalised mutual information loss with specific parameters: -ln 5 -lp 4 -be 0.001 -smooR 1 -smooF 1 -jl 0.0001. The computed non-rigid transformations were applied to the organ masks which were then meshed as in Sect. 2.2. Corresponding points between the pairwise registered organs were assigned using the nearest neighbours.

2.6   Evaluation Metrics

We implemented each of the three metrics used by Eisenberger et al. [5]:

The Geodesic Error: measures the consistency of shapes for sets of corresponding points [18]. It is defined as the differences between the geodesic distances of pairs of points on the target and the predicted corresponding pairs of points on the source mesh. This metric quantifies the discrepancies in the geodesic distances, resulting from the predicted correspondences, normalised by the square root area of the mesh.

The Chamfer Distance: measures the accuracy of the predicted interpolation. It is defined as the distance between each predicted point on the source mesh to the nearest point on the target [3]. While the chamfer distance is a good measure of the overlap of the predicted shapes, a sufficiently elastic (and anatomically unrealistic) registration can achieve a near perfect (zero) chamfer distance.

The Conformal Distortion: provides insight into the realism of the deformations produced [8]. This metric quantifies the amount of distortion each triangle on the mesh experiences through interpolation. The conformal distortion is a good indicator of the anatomical feasibility of a deformation, with a higher conformal distortion metric value suggesting a more unrealistic registration.

Anatomical Landmark Error We additionally evaluated the correspondence of organ sub-regions using anatomical landmarks identified in the original CT scans. Figure 3 shows the landmarks used in this study which were manually identified in each of the 34 CT scans by a single observer. When the identified landmark was not on the segmentation, the closest point on each mesh was found. The Euclidean distance between the landmark on the target organ and the predicted corresponding landmark point was then found, which we call the landmark error.

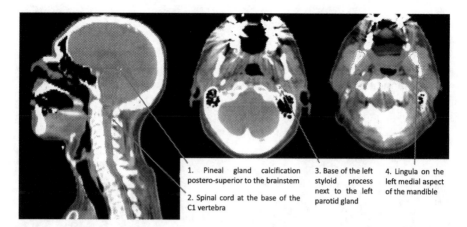

1. Pineal gland calcification postero-superior to the brainstem

2. Spinal cord at the base of the C1 vertebra

3. Base of the left styloid process next to the left parotid gland

4. Lingula on the left medial aspect of the mandible

Fig. 3. CT scan slices showing the locations of the anatomical landmarks used for clinical validation.

2.7 Implementation Details

All models are implemented using *PyTorch* 1.13.0 and *PyG* 2.2.0. *Open3D* 0.13.0 and *PyVista* 0.38.6 were used to perform mesh smoothing and visualisation. All training was performed using a 24 GB NVidia GeForce RTX 3090 and AMD Ryzen 9 3950X 16-Core Processor. The base *Neuromorph* model contained $389,507$ parameters and the extended model with imaging features contained $686,467$ parameters. Models were trained for 75 epochs with the Adam optimiser (learning rate of 0.0001) and used a maximum of 4.8 GB GPU memory.

2.8 Experiments

For this study we evaluated the original model, *Neuromorph*, and two proposed extensions against a NRR baseline. Each model was trained with data from all organs, but with the restriction that only pairs of the same organ were presented to the model, e.g., a pair of left parotid glands, followed by a pair of mandibles, etc. For each configuration we performed a five-fold cross-validation, dividing the data into folds to train five different model parameter sets. Trivial self-pairs were excluded when computing the evaluation metrics in Sect. 2.6, resulting in $7 \times 24^2 = 4032$ pairs for training, $7 \times 3^2 = 63$ pairs for validation and $7 \times 7 \times 6 = 294$ pairs for testing each parameter set. The metric results in the testing fold for all five parameter sets are reported.

A Wilcoxon signed-rank hypothesis test was used to compare the performance of each of the model configurations to the NRR baseline for the anatomical landmark error. The geodesic error and chamfer distances were also calculated for the non-rigidly registered organs, but the conformal distortion could not be computed for the baseline approach since this metric requires a vertex-wise interpolation sequence.

3   Results

Figures 4 and 5 show an example set of correspondence predictions for every organ between a single pair of patients. Identical colours on the organs identify corresponding points. 2D images, either axial or sagittal slices or maximum intensity projections of the CT scans are shown annotated with the organ contour to aid visualisation. The predictions shown were produced by a single model that included the imaging loss during training.

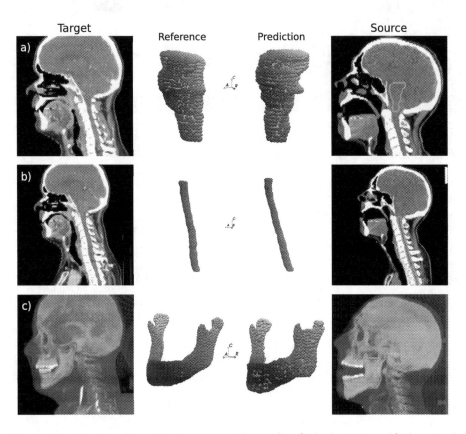

Fig. 4. Visualisation of predicted correspondences for a) the brainstem, b) the spinal cord and c) the mandible between a single pair of patients. The target patient scan and contour is presented in the first column, followed by the reference/target mesh, then the predicted correspondence on the source mesh, and finally, the scan and contour of the source patient. Sagittal slices or a maximum intensity projection (mandible) of the CT scans are shown to improve visualisation clarity.

Figure 6 shows cumulative distributions of the geodesic error, chamfer distance and conformal distortion of all model configurations and organs. The original (*Neuromorph*) model and model with imaging features perform similarly

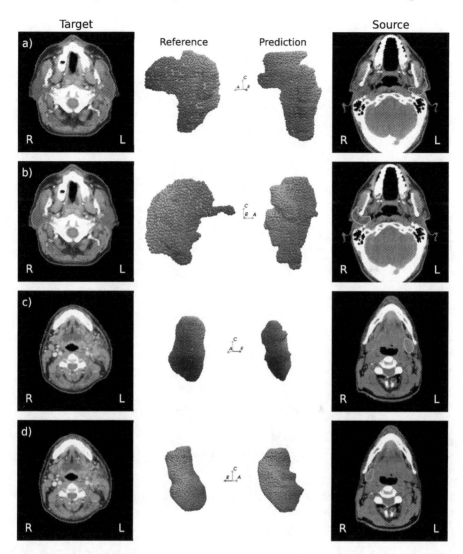

Fig. 5. Visualisation of predicted correspondences for a) the left parotid gland, b) the right parotid gland, c) the left submandibular gland and d) the right submandibular gland between a single pair of patients. The target patient scan and contour is presented in the first column, followed by the reference/target mesh, then the predicted correspondence on the source mesh, and finally, the scan and contour of the source patient. Axial slices of the CT scans are shown.

across most metrics. The original performs better on the geodesic error and conformal distortion for the spinal cord. However, the imaging features model produces less distortion for the parotid and submandibular glands. The model which includes the imaging as an additional loss component performed similarly

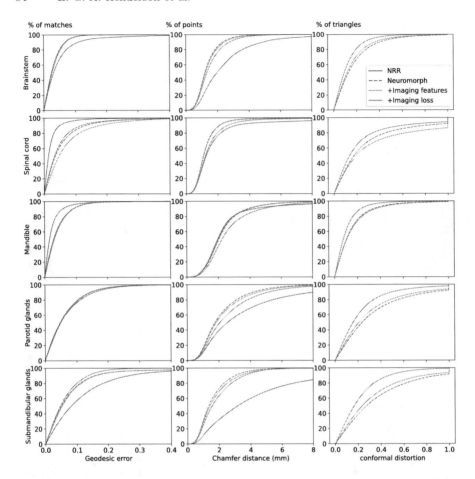

Fig. 6. Cumulative distributions of the geodesic error, chamfer distance and conformal distortion metrics for each of the model configurations and the NRR baseline. Closer to zero is better for all metrics.

to the original for the geodesic error, apart from the submandibular glands for which it outperforms the original. The imaging loss model has a slightly poorer chamfer distance results compared to the original, but greatly improved conformal distortion results, especially for the submandibular glands.

For the geodesic error, the NRR baseline performs better than the correspondence models in the spinal cord and mandible, similarly for the parotid glands and worse for the brainstem and submandibular glands. The correspondence models outperform the NRR baseline for the chamfer distance for all organs apart from the mandible.

Table 1 shows the landmark error distances for each of the model configurations and the NRR baseline. All of the models showed a significant improvement over the baseline for all anatomical landmarks. All correspondence methods per-

Table 1. Landmark error distances for each model configuration. The level of significance of improvement of each model over the NRR baseline according to the Wilcoxon signed-rank test is shown as: * - p value < 0.05, ⁑ - p value < 0.005, † - p value < 0.0005, ‡ - p value < 0.00005.

Model configuration	Median landmark error (IQR) mm			
	Pineal gland	Spinal cord at C1	Styloid process	Mandible lingula
Baseline (NRR)	4.6 (4.4)	3.9 (3.6)	8.4 (7.0)	7.8 (10.8)
Neuromorph	3.7 (2.5) †	2.3 (2.0) ‡	5.6 (5.3) *	3.1 (2.1) ‡
+ Imaging features	3.7 (2.7) †	2.2 (1.9) ‡	6.2 (5.8) *	3.0 (2.4) ‡
+ Imaging loss	3.8 (2.5) ⁑	2.5 (2.2) ‡	6.3 (5.5) *	3.1 (2.1) ‡
Distance from landmark to organ	3.6 (2.4)	2.1 (1.6)	5.4 (5.0)	2.6 (1.8)

form similarly in terms of landmark distance, but subtle differences could exist that are hidden by observer variation. The median distance from the landmark to the organs is shown in the final row and this serves as an indication on the landmark variability and hence a reasonable upper bound of the correspondence accuracy identifiable with this measure.

Figure 7 shows an additional example of correspondences produced by the model including imaging loss. This particular case is interesting as it demonstrates how the model handles the difficult scenario of missing correspondences. One of the patients has an accessory parotid, an anterior extension of the parotid present in > 30% of the population [16], and the other does not. The model was able to robustly handle this case in both directions, i.e. with either patient as the reference.

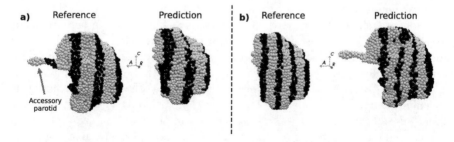

Fig. 7. An example of the model including imaging loss robustly handling a case with missing correspondences for a pair of parotid glands. In a) the reference parotid has an anterior extension (accessory) which is not reproduced on the predicted correspondences. In b) the lack of accessory in the reference does not impact majority of the predicted correspondences shown by the black stripes aligning between the two. Black and white has been used here to show corresponding points instead of the full colormap for clarity.

4   Discussion

In this study we showed that an established neural network for predicting correspondence and smooth interpolation of 3D shapes can be applied to HN organ segmentations from CT scans. We additionally evaluated two methodological extensions to leverage the CT imaging information.

The correspondence models were compared to an intensity-based NRR algorithm regularly used for radiotherapy outcome modelling. The NRR produced better correspondences for the spinal cord and mandible in terms of the geodesic error showing the effectiveness of the image registration method for the more straightforward task of aligning the skeleton and anatomy enclosed by bone. However, the original *Neuromorph* model and extensions all produced significantly lower landmark errors for every organ than the NRR baseline as well as producing better chamfer distance results for soft tissue organs. This promising result demonstrates the potential of such correspondence methods to reduce uncertainties in radiotherapy outcome modelling. Further work is required to quantify the uncertainty reduction and its impact for this purpose.

An intensity-based non-rigid registration algorithm was used as a comparison baseline for the learning-based correspondence models. A mesh-based registration such as coherent point drift could have alternatively been applied for a more direct comparison [12]. However, intensity-based image registration algorithms are the current standard for aligning images and structures for radiotherapy applications, particularly for outcome modelling, and therefore provide a more relevant comparison for this study [2,11,20].

The performance of the original *Neuromorph* model was slightly improved by incorporating imaging information, not as explicit imaging features, but rather by introducing an additional imaging term to the loss function. This configuration does not require imaging at inference time, instead the imaging is used solely when training to additionally encourage the model to match points with similar appearances within the local neighbourhood. Our underlying assumption here was that corresponding locations look similar between patients. This configuration was shown to be particularly effective at regularising the predicted correspondences with substantially reduced conformal distortion results. This indicates that the inclusion of an imaging loss term produced more anatomically feasible and robust deformations. The improvement in conformal distortion metric, whilst hardly affecting performance in terms of the other metrics, makes this particular configuration appealing for future exploration as a starting point for an anatomically informed non-rigid registration method.

The original *Neuromorph* model was also extended to receive imaging input directly and predict correspondences based on geometric and imaging features, but this extension did not improve performance. We believe that this is primarily due to the highly consistent data used for model training. The segmentations were as close to the consensus guidelines as possible which is unlikely in clinical practice. This meant the contours will deviate only slightly from "true" organ boundaries. We envisage providing the imaging information as input to the model to be of greater use in scenarios where the segmentations are more variable, and could be inconsistent with the underlying anatomy. This is an interesting avenue for future work.

While *Neuromorph* is an established model for everyday 3D shapes, we believe this is the first time it has been shown to be effective in biomedical applications. Additionally, while there are other learning based correspondence methods [1,9], this is the first to combine geometric learning and leverage imaging, providing a slight improvement on the original model in terms of anatomical feasibility.

The additional imaging loss component described in Sect. 2.4 utilises the mean squared error for simplicity. This metric is only appropriate when quantifying the similarity of mono-modal scans which are intensity calibrated, such as CT scans used for radiotherapy planning. If the underlying imaging was a cone-beam CT or MRI, an alternative measure such as mutual information or correlation ratio could be used.

Our model was primarily developed with outcome modelling in mind, which relies on inter-patient analysis. Inter-patient correspondence is a more complex task than identifying intra-patient correspondence since there is greater variability in the anatomy. Consequently, we believe that extension to the intra-patient tasks should be straightforward.

5   Conclusion

We have shown that an established model, originally developed for generic 3D shapes can be adapted for applications in biomedical imaging. Specifically, this model could be used to identify corresponding points on 3D organs to improve spatial normalisation in outcome modelling applications, potentially reducing the associated uncertainties and facilitating the development of better radiotherapy treatments. Further, we envision that in the future, such a correspondence tool, which also provides a smooth interpolation, could be deployed at the heart of an effective, anatomically informed non-rigid registration method.

References

1. Attaiki, S., Ovsjanikov, M.: NCP: Neural Correspondence Prior for Effective Unsupervised Shape Matching. ArXiv e-prints (2023). 10.48550/arXiv. 2301.05839
2. Beasley, W., et al.: Image-based data mining to probe dosimetric correlates of radiation-induced trismus. Int. J. Radiat. Oncol. Biol. Phys. **102**(4), 1330–1338 (2018). https://doi.org/10.1016/j.ijrobp.2018.05.054
3. Butt, M.A., Maragos, P.: Optimum design of chamfer distance transforms. IEEE Trans. Image Process. **7**(10), 1477–1484 (1998). https://doi.org/10.1109/83.718487
4. Cazoulat, G., Owen, D., Matuszak, M.M., Balter, J.M., Brock, K.K.: Biomechanical deformable image registration of longitudinal lung CT images using vessel information. Phys. Med. Biol. **61**(13), 4826–4839 (2016). https://doi.org/10.1088/0031-9155/61/13/4826

5. Eisenberger, M., et al.: NeuroMorph: unsupervised shape interpolation and correspondence in one go. In: 2021 IEEE/CVF Conference on Computer Vision and Pattern Recognition (CVPR). IEEE (2021). https://doi.org/10.1109/cvpr46437.2021.00739

6. Henderson, E.G.A., Green, A.F., van Herk, M., Vasquez Osorio, E.M.: Automatic identification of segmentation errors for radiotherapy using geometric learning. In: Wang, L., Dou, Q., Fletcher, P.T., Speidel, S., Li, S. (eds.) Medical Image Computing and Computer Assisted Intervention – MICCAI 2022. MICCAI 2022. LNCS, vol. 13435. Springer, Cham (2022). https://doi.org/10.1007/978-3-031-16443-9_31

7. Hoang, J.K., Glastonbury, C.M., Chen, L.F., Salvatore, J.K., Eastwood, J.D.: CT mucosal window settings: a novel approach to evaluating early t-stage head and neck carcinoma. Am. J. Roentgenol. **195**(4), 1002–1006 (2010). https://doi.org/10.2214/ajr.09.4149

8. Hormann, K., Greiner, G.: Mips: An Efficient Global Parametrization Method. Erlangen-Nuernberg Univ (Germany) Computer Graphics Group, Tech. rep. (2000)

9. Klatzow, J., Dalmasso, G., Martínez-Abadías, N., Sharpe, J., Uhlmann, V.: μMatch: 3D shape correspondence for biological image data. Front. Comput. Sci. **4**, 777615 (2022). https://doi.org/10.3389/fcomp.2022.777615

10. Modat, M., et al.: Fast free-form deformation using graphics processing units. Comput. Methods Programs Biomed. **98**(3), 278–284 (2010). https://doi.org/10.1016/j.cmpb.2009.09.002

11. Monti, S., et al.: Voxel-based analysis unveils regional dose differences associated with radiation-induced morbidity in head and neck cancer patients. Sci. Rep. **7**(1), 7220 (2017). https://doi.org/10.1038/s41598-017-07586-x

12. Myronenko, A., Song, X.: Point set registration: coherent point drift. IEEE Trans. Pattern Anal. Mach. Intell. **32**(12), 2262–2275 (2010). https://doi.org/10.1109/tpami.2010.46

13. Nie, K., Pei, Y., Sun, D., Xu, T.: Deep Supervoxel mapping learning for dense correspondence of cone-beam computed tomography. In: Yu, S., et al. Pattern Recognition and Computer Vision. PRCV 2022. LNCS, vol. 13535. Springer, Cham (2022). https://doi.org/10.1007/978-3-031-18910-4_34

14. Nikolov, S., et al.: Deep learning to achieve clinically applicable segmentation of head and neck anatomy for radiotherapy. ArXiv e-prints (2018). https://doi.org/10.48550/arXiv.1809.04430

15. Palma, G., Monti, S., Cella, L.: Voxel-based analysis in radiation oncology: a methodological cookbook. Physica Med. **69**, 192–204 (2020). https://doi.org/10.1016/j.ejmp.2019.12.013

16. Rosa, M.A., et al.: The accessory parotid gland and its clinical significance. J. Craniofac. Surg. **31**(3), 856–860 (2020). https://doi.org/10.1097/scs.0000000000006092

17. Shi, J., Wan, P., Chen, F.: An unsupervised non-rigid registration network for fast medical shape alignment. In: 2021 43rd Annual International Conference of the IEEE Engineering in Medicine & Biology Society (EMBC). IEEE (2021). https://doi.org/10.1109/embc46164.2021.9631030

18. Shilane, P., Min, P., Kazhdan, M., Funkhouser, T.: The Princeton shape benchmark. In: Proceedings Shape Modeling Applications. IEEE (2004). https://doi.org/10.1109/smi.2004.1314504

19. Strojan, P., et al.: Treatment of late sequelae after radiotherapy for head and neck cancer. Cancer Treat. Rev. **59**, 79–92 (2017). https://doi.org/10.1016/j.ctrv.2017.07.003

20. Vásquez Osorio, E., et al.: Dysphagia at 1 year is associated with mean dose to the inferior section of the brainstem. Int. J. Radiat. Oncol. Biol. Phys. **17**, 0360-3016 (2023). https://doi.org/10.1016/j.ijrobp.2023.06.004

21. Vásquez Osorio, E.M., Hoogeman, M.S., Bondar, L., Levendag, P.C., Heijmen, B.J.M.: A novel flexible framework with automatic feature correspondence optimization for nonrigid registration in radiotherapy. Med. Phys. **36**(7), 2848–2859 (2009). https://doi.org/10.1118/1.3134242

22. Vásquez Osorio, E.M., Kolkman-Deurloo, I.K.K., Schuring-Pereira, M., Zolnay, A., Heijmen, B.J.M., Hoogeman, M.S.: Improving anatomical mapping of complexly deformed anatomy for external beam radiotherapy and brachytherapy dose accumulation in cervical cancer. Med. Phys. **42**(1), 206–220 (2014). https://doi.org/10.1118/1.4903300

23. Wang, Y., Sun, Y., Liu, Z., Sarma, S.E., Bronstein, M.M., Solomon, J.M.: Dynamic graph CNN for learning on point clouds. ACM Trans. Graph. **38**(5), 1–12 (2019). https://doi.org/10.1145/3326362

ADASSM: Adversarial Data Augmentation in Statistical Shape Models from Images

Mokshagna Sai Teja Karanam[1,2]([⊠]), Tushar Kataria[1,2], Krithika Iyer[1,2], and Shireen Y. Elhabian[1,2]([⊠])

[1] Kahlert School of Computing, University Of Utah, Salt Lake City, USA
[2] Scientific Computing and Imaging Institute, University of Utah, Salt Lake City, USA
{mkaranam,tushar.kataria}@sci.utah.edu, krithika.iyer@utah.edu,
shireen@sci.utah.edu

Abstract. Statistical shape models (SSM) have been well-established as an excellent tool for identifying variations in the morphology of anatomy across the underlying population. Shape models use consistent shape representation across all the samples in a given cohort, which helps to compare shapes and identify the variations that can detect pathologies and help in formulating treatment plans. In medical imaging, computing these shape representations from CT/MRI scans requires time-intensive preprocessing operations, including but not limited to anatomy segmentation annotations, registration, and texture denoising. Deep learning models have demonstrated exceptional capabilities in learning shape representations directly from volumetric images, giving rise to highly effective and efficient Image-to-SSM networks. Nevertheless, these models are data-hungry and due to the limited availability of medical data, deep learning models tend to overfit. Offline data augmentation techniques, that use kernel density estimation based (KDE) methods for generating shape-augmented samples, have successfully aided Image-to-SSM networks in achieving comparable accuracy to traditional SSM methods. However, these augmentation methods focus on shape augmentation, whereas deep learning models exhibit image-based texture bias resulting in sub-optimal models. This paper introduces a novel strategy for on-the-fly data augmentation for the Image-to-SSM framework by leveraging data-dependent noise generation or texture augmentation. The proposed framework is trained as an adversary to the Image-to-SSM network, augmenting diverse and challenging noisy samples. Our approach achieves improved accuracy by encouraging the model to focus on the underlying geometry rather than relying solely on pixel values.

Keywords: Statistical Shape Model · Data Augmentation · Adversarial Training

C. Wachinger et al. (Eds.): ShapeMI 2023, LNCS 14350, pp. 90–104, 2023.
https://doi.org/10.1007/978-3-031-46914-5_8

1   Introduction

Statistical shape modeling (SSM) is widely used in the fields of medical image analysis and biological sciences for studying anatomical structures and conducting morphological analysis. It enables shape analysis by facilitating the understanding of the geometrical properties of shapes that are statistically consistent across a population. SSM has diverse applications in neuroscience [17,27], cardiology [8], orthopedics [5,18], and radiology [9,15].

Optimization-based SSM [10] methods typically involve anatomy segmentation, data preprocessing (e.g., image resampling, denoising, rigid registration), and optimizing population-level shape representation i.e., correspondence points (or particles), all of which require substantial expertise-driven workflow, involving intensive preprocessing that can be time-consuming. Deep learning approaches for SSM, train networks to learn the functional mapping from unsegmented images to statistical representations of anatomical structures [2–4,6,8,22]. This shift towards deep learning-based methods offers a more efficient and automated approach to SSM, bypassing the need for extensive manual preprocessing and leveraging the power of neural networks to learn directly from raw imagery data [4,23]. However, deep learning models are notorious for requiring enormous quantities of data to achieve acceptable performance [21], necessitating the use of data augmentation to supplement the available training data [8].

In the field of deep learning for medical image analysis, data augmentation plays a crucial role [1,11,20]. Nevertheless, unlike computer vision applications, acquiring a substantial number of segmented medical images is difficult due to privacy concerns, the substantial human effort and expertise required, and the intensive preprocessing involved [14]. Off-the-shelf data augmentation methods may not generate augmented samples that promote invariances and improve the task-specific generalizability of the model [16]. Therefore, having a large amount of data supplemented with challenging task-specific variations would be extremely beneficial for training deep neural networks and improving model performance. In the field of medical imaging, attempts have been made to employ task-driven automatic data augmentation techniques [12] for image segmentation and classification [11,14,24]. For regression tasks, various strategies have been proposed for handling data augmentation that includes, data-dependent shape augmentation for Image-to-SSM networks [2,6,7] and a mix-up [26] based augmentation by interpolating input samples using the similarity of labels [25].

DeepSSM [6,7], and other variants [2–4], learn to map unsegmented images to shape models, exhibiting comparable performance to traditional SSM [10] methods, as well as in downstream tasks such as atrial fibrillation recurrence [8,18]. DeepSSM relies heavily on offline shape-based data augmentation via kernel density estimation (KDE) in the linear principal component analysis (PCA) subspace [6,7]. The DeepSSM shape augmentation approach entails using generative modeling to sample shapes from probability distribution estimated via KDE. Existing offline methods have three deficiencies: (1) the generation of augmented samples is independent of the task (shape modeling) at hand, (2) the augmentation process focuses on shape augmentation rather than incorporating noise/texture augmentation, neglecting the inherent texture bias often present in deep learning models [19], which can lead to sub-optimal models in shape analysis, and (3) they require extensive offline data compilation, which is time intensive and resource consuming.

We draw inspiration from adversarial domain adaptation [13] and adversarial data augmentation for classification tasks [14] and adapted these ideas to regression tasks with application to Image-to-SSM networks. The regression task poses a greater challenge as it is not straightforward to generate challenging adversarial samples for the learning task at hand due to the absence of label-separating hyperplanes. Consequently, we focus our methodology on generating noise-augmented samples as an alternative to KDE based shape augmentation [6,7]. The proposed method implicitly drives the model to attend to the shape of the underlying object of interest instead of explicit shape augmentation [6,7]. We also demonstrate that data- and task-dependent noise augmentation is better than off-the-shelf noise augmentation with varied variance levels. The proposed augmentation approach is generic enough to be used for any Image-to-SSM network, but here we focus on DeepSSM [7] to showcase the efficacy of the on-the-fly noise augmentation vs offline shape and noise augmentation.

As we focus on image noise augmentation for this work, the shape representation of the augmented images should not be affected. As a result, we employ a contrastive loss to inform the deep learning model that noisy and their corresponding original images should be projected to the same latent representation. This contrastive loss acts as a regularizer to both the augmentation framework and the Image-to-SSM network.

The contributions of this paper are as follows:-

- A computationally efficient, automated, on-the-fly adversarial data augmentation method for regression tasks with better generalization.
- A contrastive loss based regularization that enables enhanced noise generation that is more task- and data-dependent.
- Extensive experiments with Image-to-SSM and downstream tasks on left atrium and femur datasets show the efficacy of the proposed approach.

2   Methodology

This section explains the details of the proposed method and the regularization losses. The block diagram for the proposed approach is shown in Fig. 1.

2.1   Adversarial Data Augmentation Block

The proposed framework for augmentation aims to enhance the performance of the Image-to-SSM network by generating data-dependent noise. This architecture can be especially useful in the context of SSM tasks, where the input data size is typically limited. Due to the paucity of data samples, deep learning

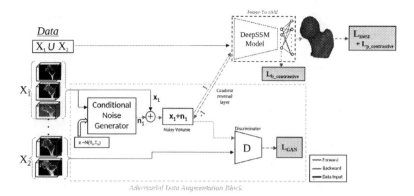

Fig. 1. ADASSM Block Diagram. Conditional Noise Generator uses the input volumes ($\mathbf{x_1}$) to generate noise ($\mathbf{n_1}$), which is added back to the input volume. **Discriminator** used the remaining volumes ($\mathbf{X_2}$) for GAN loss($\mathbf{L}_{GAN}$) to ensure the generated data is in distribution. The noise generator and DeepSSM are connected by a **gradient reversal layer**, which sets up the second adversarial training paradigm. Both the noisy and original volumes are used to train the Image-to-SSM framework (DeepSSM). $\mathbf{L}_{RMSE}$ is the correspondences RMSE loss, $\mathbf{L}_{p\_contrastive}$ & $\mathbf{L}_{b\_contrastive}$ are correspondence and bottleneck contrastive loss.

models may suffer from overfitting, leading to poor generalization performance on unseen data. To address this, we integrate an adversarial [13,14] data augmentation approach for regression tasks with applications in shape modeling. The generator produces adversarial data samples that are difficult for the shape model to project into statistical shape representation.

Conditional Noise Generator: Conditional noise generator receives an image $\mathbf{x_1}$ from input set X_1 and a randomly sampled Gaussian vector, $\mathbf{z}$ as inputs. It then generates noise vector $\mathbf{n_1}$, which is subsequently added to the original input volume $\hat{\mathbf{x}_1} = \mathbf{x_1} + \mathbf{n_1}$, resulting in the generation of a noisy augmented sample.

$$\hat{\mathbf{x}_1} = \mathcal{R} * G(\mathbf{z}, \mathbf{x_1}) \oplus \mathbf{x_1} \tag{1}$$

Here, $\mathcal{R}$ is a hyperparameter for noise perturbation range, G denotes the conditional generator, and $\oplus$ represents voxelwise addition. To achieve controlled augmentations and minimize excessive perturbations, a regularization loss based on total variation (TV) is incorporated [14]. TV loss also enables the generation of noise variations that exhibit smooth transitions, which is crucial for capturing the inherent features of real-world images.

$$\mathbf{L}_{TV} = ||G(\mathbf{z}, \mathbf{x_1})||_2 \tag{2}$$

Discriminator: To further regularize the noise generation a discriminator is used. This discriminator uses the remaining input samples from set $\mathbf{X_2}$, as reference distribution, and noisy samples are treated as samples of input distributions.

The objective of the discriminator is to assist the generator in producing realistic noise while ensuring that the noisy augmented sample comes from the same distribution as the original data. The generative adversarial network (GAN) aims to strike a balance between meaningful data-specific augmentations and excessive perturbations. The loss of the block is given in the equation below:-

$$\mathbf{L}_{\text{GAN}} = \min_{G} \max_{D} \mathbf{E}_{X \in \Omega} \left[\log D(\mathbf{x_2}) \right] + \mathbf{E}_{X \in \Omega} [\log(1 - D(G(\mathbf{z}, \mathbf{x_1}) \oplus \mathbf{x_1}))] \ + \beta \mathbf{L}_{TV} \tag{3}$$

where β is hyperparameter.

2.2   Adversary to Image-To-SSM Network

The noisy augmented sample is fed into the Image-To-SSM network to obtain the predicted shape representation($\hat{\mathbf{y}}$), which is compared to the original shape representation via an RMSE loss.

$$\mathbf{L}(\hat{\mathbf{y}}, \mathbf{y}) = RMSE(\mathbf{y}, \text{DeepSSM}(\hat{\mathbf{x}}_1)) \tag{4}$$

The Image-to-SSM network and GAN framework are put in an adversarial relationship with the help of a gradient reversal layer, as shown in the block diagram in Fig. 1. The objective function in Eq. 5 aims to minimize the Image-To-SSM network error while maximizing the conditional generator G perturbations, setting up a second adversarial objective:

$$\mathbf{L}_{\text{RMSE}} = \mathbf{E}_{X,Y \in \Omega} [\min_{M} \max_{G} \mathbf{L}(\hat{\mathbf{y}}, \mathbf{y})] \tag{5}$$

The framework above allows the augmentation model to search along the adversarial direction, leading to the generation of challenging noise augmentations that facilitate the learning of more robust shape features.

2.3   Image-To-SSM Network:

The DeepSSM Model employs a deterministic encoder and a deterministic linear decoder. The reconstructed correspondences are obtained as the output of the Image-to-SSM network, providing both shape representation as well as low dimensional latent features for each input volume.

2.4   Shape Regularization Loss:

Noise augmentation affects only the texture of the input volume and does not affect the underlying shape. We hypothesize that the shape representation (after DeepSSM) of the augmented noisy volume ($\hat{\mathbf{x}}_1$) and original volume ($\mathbf{x}_1$), should be closer in the shape space. We use a contrastive loss in Eq. 6 as an additional regularizer to ensure that both the Image-to-SSM network and GAN account for this.

$$\mathbf{L}_{contrastive} = -\log(\frac{\exp\left(sim(DeepSSM(\mathbf{x_1}), DeepSSM(\hat{\mathbf{x_1}}))\right)}{\sum^{N} \exp(sim(DeepSSM(\mathbf{x_1}), DeepSSM(\hat{\mathbf{x_1}})))}) \tag{6}$$

We propose to use contrastive loss at two different latent representations as shown in Fig. 1:

1. *Correspondences* ($\mathbf{L}_{\text{p\_contrastive}}$, PC), which are the predicted SSM representation by the Image-to-SSM network. This loss ensures that the augmentation does not effect the final statistical shape representation of the augmented image.
2. *Bottleneck* ($\mathbf{L}_{\text{b\_contrastive}}$, BC), which is the low-dimensional space representation obtained from the Image-to-SSM network. This loss helps the model learn the same latent representation despite noise augmentation.

These regularization losses encourage both the generator and shape model to focus more on shape-related information during the learning process and factor out texture variations. The overall objective function for the proposed model is:

$$\mathbf{L} = \alpha \mathbf{L}_{GAN} + \mathbf{L}_{RMSE} + \lambda_1 \mathbf{L}_{\text{b\_contrastive}} + \lambda_2 \mathbf{L}_{\text{p\_contrastive}} \tag{7}$$

where $\alpha, \lambda_1, \lambda_2$ are hyperparameters.

3 Results

We use the same Image-to-SSM (DeepSSM [6]) model architecture across all experiments to ensure that variations in model performance can only be attributed to different augmentation techniques. As a baseline for comparison, we train an Image-to-SSM architecture without any augmentations (*NoAug*). Additionally, we train another model using KDE [6] augmentation (*KDE* [6]). We also compute two other baselines with off-the-shelf Gaussian noise augmentation with different variances($\sigma = 1$ and 10).

3.1 Metrics

Root Mean Squared Error (RMSE): To measure the error, we calculate the average relative mean squared error (RMSE) between the predicted 3D correspondences and this is achieved by computing the RMSE for the x, y, and z coordinates and averaging them as shown in (7)

$$RMSE = \frac{1}{3}(RMSE_x + RMSE_y + RMSE_z) \tag{8}$$

For N 3D correspondences, $RMSE_x = \sqrt{\frac{||C_x - C'_x||_2^2}{N}}$. The same calculation is applied to $RMSE_y$ and $RMSE_z$ for the respective coordinates. Additionally, we calculate the RMSE error for each correspondence point as, $RMSE_i = \sqrt{\frac{||C_x^i - C'^i_x||_2^2 + ||C_y^i - C'^i_y||_2^2 + ||C_z^i - C'^i_z||_2^2}{3}}$

The per-point RMSE helps us assess the accuracy of DeepSSM in modeling various local anatomical features. For all experiments, the shape representations were calculated on same test data (held-out data) using the trained DeepSSM model and were only used for inference.

Surface-to-Surface Distance (mm): The surface-to-surface distance is measured between the ground truth mesh and the mesh reconstructed from the predicted correspondences by DeepSSM. This distance provides a more precise measure of how well the correspondences adhere to the shape and indicates their suitability for anatomy segmentation.

Furthermore, we validate the effectiveness of DeepSSM with the learned shape representations by utilizing its correspondences for various downstream analysis applications. The specific downstream applications vary for each dataset and are described in separate subsections below.

3.2   Femur

Data Description and Processing: The femur dataset consists of 49 CT images of the femur bone, of which 42 are considered healthy with no morphological abnormalities. DeepSSM [6,7] requires generating point distribution models (or correspondences) for the training images. We use ShapeWorks [10] to optimize a shape model with 1024 correspondences. Along with the training and validation data, we also randomly selected 7 controls and 2 CAM-FAI scans for testing the DeepSSM Model. To meet GPU memory requirements, each image is downsampled by a factor of 2 from $260 \times 184 \times 235$ ($0.5\,\mathrm{mm}$ isotropic voxel spacing) to isotropic voxel spacing of 1mm with dimensions of $130 \times 92 \times 117$. The training images are divided into training and validation sets with 80% -20% split.

Training Specifics: Empirically we set $\alpha = 1$ and $\beta = 0.1$. The default parameters and configuration are used to get the result for KDE [6] augmentation for femur dataset. The training process involves optimizing the loss on correspondences for 1500 epochs, employing a data augmentation framework based on validation loss. For all ADASSM experiments, involving the proposed regularization losses, a learning rate of 5e-5 is utilized while ADASSM itself is trained with a learning rate of 1e-5. The generator and discriminator learning rates are set to 5e-3, except for ADASSM+BC+PC, which employs a learning rate of 1e-3. A batch size of 4 is used for training all the proposed models and baseline models. In the generator, $\mathcal{R}$ is set to 500 for ADASSM experiments.

Evaluation and Analysis: Figure 2 visualizes the RMSE results alongside the surface-to-surface distance. Augmenting the DeepSSM model with Gaussian noise with different variances without any adversarial training improves the RMSE when compared to the KDE augmentation [6]. However, an increase in the surface distance of the predicted correspondences indicates misalignment with the ground truth femur bone segmentation. This result shows that standard noise augmentation can provide better results for Image-to-SSM networks.

We can observe that for the proposed ADASSM (data-dependent noise augmentation), RMSE results are better compared to both KDE [6] and Gaussian noise. The addition of contrastive regularization losses further improves RMSE error and surface distance, which proves that data-dependent noise-augmented

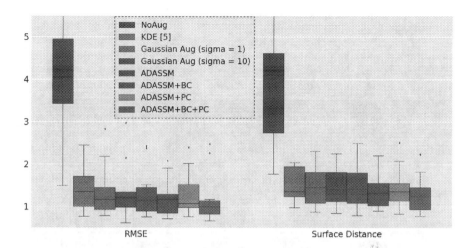

Fig. 2. Femur Test Results. The *RMSE* is the Euclidean distance between the ground truth and predicted correspondences for all the test samples. The Surface-to-Surface distance (mm) is computed by comparing the reconstructed mesh using the ground truth correspondences and the predicted correspondences for all the test samples. λ_1 for ADASSM+BC is 0.5, λ_2 for ADASSM+PC is 0.1 and λ_1, λ_2 for ADASSM+BC+PC is 0.5. Y-axis is the magnitude of the errors displayed.

samples are better compared to Gaussian augmented samples and shape augmentation [6].

Visualizations of surface-to-surface distance for the best, median, and worst cases for the test set are shown in Fig. 3. Upon careful examination of these visualizations, we can observe that the proposed models demonstrate a remarkable reduction in errors in critical regions such as the greater trochanter, growth plate, femoral neck, and epiphyseal lines in the best-case scenario. In the median case, a detailed analysis of both views reveals that in view (1), the KDE baseline [6] exhibits some errors around the trochanter region that are substantially reduced by the proposed models. In view (2), the error around the trochanter region is completely reduced with the ADASSM+BC+PC model. In the worst-case scenario, view (2) displays the majority of errors in the KDE baseline [6], but these errors are significantly diminished, particularly in the lower trochanter region, with the employment of the ADASSM variants.

Downstream Task - Group Differences: To evaluate the effectiveness of the learned shape representations using the proposed models, we conducted a downstream analysis. The experiments were aimed at evaluating whether the models can accurately capture group differences [18] in medically relevant regions. To achieve this, we formed two groups: a control group and a pathology group consisting of CAM-FAI cases. We calculated the mean differences (μ_{normal} and $\mu_{cam-FAI}$) between these groups and visualized the differences on a mesh and compared the group differences obtained using the predicted correspondences

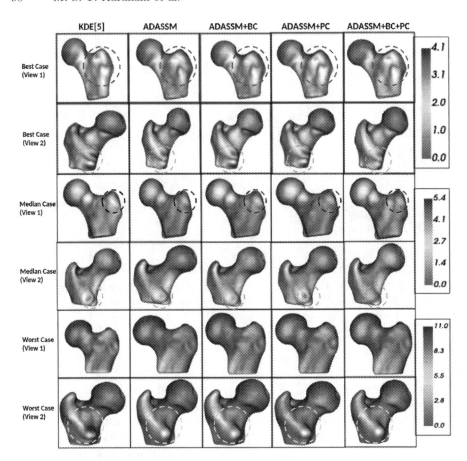

Fig. 3. Femur Surface-to-Surface distance (mm) visualization: Shape reconstruction error is displayed as a heatmap on the ground truth reconstructed meshes. The results for different proposed models are presented, showcasing their performance in the best, median, and worst-case scenarios.

from the proposed models and the ShapeWorks PDM model. We utilized the entire dataset, including both training and testing samples, for these group differences, and the results are presented in Fig. 4.

Each group difference illustrates the transition from the mean shape of the pathological group to that of the control group, overlaid on the mean pathological scan. Interestingly, we observed that the group differences between the state-of-the-art PDM model and ADASSM+PC were quite similar compared to baseline KDE [6]. In some cases, the proposed models exhibited similar differences in medically relevant regions of the femur, whereas in other areas, the models identified additional variations. These findings suggest that the established correspondences can be employed to characterize CAM deformity effectively.

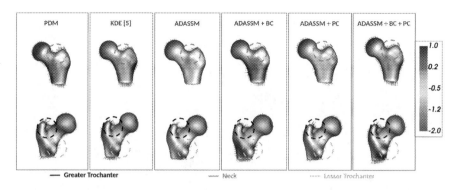

Fig. 4. Femur shape group difference between CAM-FAI and controls. The difference $\mu_{cam} - \mu_{normal}$ and is projected on μ_{cam} with both training and test samples, the arrows denote the direction of the correspondence (particle) movement, and the heatmap showcases the normalized magnitude.

3.3 Left Atrium

Data Description and Processing: The left atrium dataset consists of 176 late gadolinium enhancement (LGE) MRI images from patients that have been diagnosed with atrial fibrillation (AF). These scans are acquired after the first ablation. 80% -20% split is used to split the data where 146 volumes are used to train the model and 30 scans which are used to test the DeepSSM network. To generate point distribution models (or correspondences) for training images, we utilize ShapeWorks [10] to optimize a shape model with 1024 correspondences. For training purposes, the MRIs are downsampled from $235 \times 138 \times 175$ (0.625mm isotropic voxel spacing) to $117 \times 69 \times 87$ (1.25mm voxel spacing) by a factor of 2.

Training Specifics: The results for KDE augmentation [6] on the left atrium dataset are obtained using the default parameters and configuration. With the proposed data augmentation framework, DeepSSM model is trained for 1000 epochs. For the various ADASSM models with the aforementioned regularization losses, a learning rate of 1e-4 is utilized, while ADASSM itself is trained with a learning rate of 5e-3. A batch size of 6 is employed during the training of both the proposed and baseline models. In all ADASSM experiments, $\mathcal{R}$ in the generator is set to 100. Following the publication of the manuscript, we plan to make the training models, implementation code, and relevant hyperparameters publicly available.

Evaluation and Analysis: Figure 5 displays the RMSE results alongside the surface-to-surface distance. We can make the following observations from the bar graph:- 1) When augmenting the DeepSSM model with Gaussian noise of varying variances without any adversarial training, we find that it does not improve performance, which may be because the original dataset already has more intensity variation when compared to CT volumes. 2) By enhancing the data- and

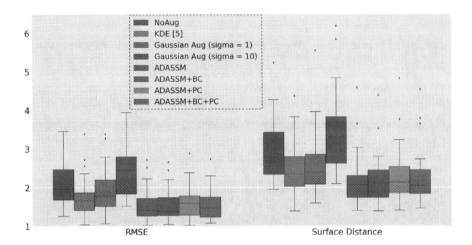

Fig. 5. Left Atrium Test Set Results. The *RMSE* is the Euclidean distance between the ground truth and predicted correspondences for all the test samples. The Surface-to-Surface distance (mm) is computed by comparing the reconstructed mesh using the ground truth correspondences and the predicted correspondences for all the test samples. λ_1 for ADASSM+BC is 0.001, λ_2 for ADASSM+PC is 0.05 and λ_1, λ_2 for ADASSM+BC+PC is 0.05. Y-axis is the magnitude of the errors displayed.

task-dependency of the noise and integrating the proposed data augmentation framework with various regularization losses, the methodology surpasses the baseline Gaussian noise and KDE shape augmentation framework [6]. In the left atrium, the performance disparity between the ADASSM variants is more pronounced compared to the femur. This can be attributed to the significant variations observed in the left atrium, where the proposed method excels in effectively regulating the Image-to-SSM task.

In Fig. 6, visualizations of the surface-to-surface distance are presented for the best, median, and worst cases in the test set. For best-case and median-case views, the proposed model plainly outperforms other methods, while worst-case views are comparable.

Downstream Task - AF Recurrence Prediction: The shape of the left atrium can provide insights into the recurrence of AF [18]. The dataset has binary outcome labels indicating whether patients experienced AF recurrence after ablation. The goal is to estimate the probability of AF recurrence based on the learned shape representations. We use PCA projections of the shape representations as features for a Multi-Layer Perceptron (MLP) for classification. The results are summarized in Table 1. Compared with the traditional SSM [10] and KDE, we observe similar performance. All ADASSM variants perform on par with the PDM, with the ADASSM+PC model outperforming the baselines by capturing better shape descriptors for the left atrium than the PCA scores learned in other models. Due to the fact that the classification model is based on

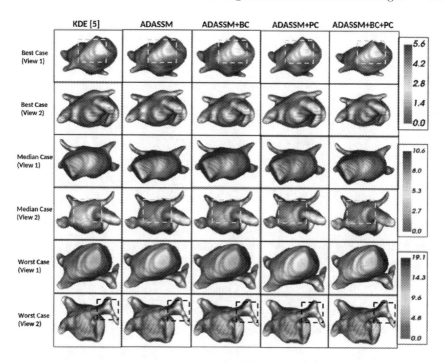

Fig. 6. Left Atrium Surface-to-Surface distance (mm) visualization: Shape reconstruction error is displayed as a heatmap on the ground truth reconstructed meshes. The results for the best, median, and worst-case scenarios are shown for various proposed models. We can observe that the proposed methods are outperforming the baseline results in best and median-case.

the PCA scores of correspondences, ADASSM+PC has the highest accuracy, as the contrastive loss will bring the correspondence's latent space representation closer. But if we train a classifier with non-linear features (other than PCA), ADASSM+BC+PC might result in the best accuracy.

3.4   Training Time

The proposed augmentation method not only improved model performance for both the left atrium and femur datasets but also significantly reduces the training time by approximately 60% compared to the baseline method [6] as shown in Table 2.

Table 1. AF Recurrence Prediction: Accuracy of AF recurrence that uses PCA scores as shape descriptors from different models.

Model	Accuracy
ShapeWorks [10]	53.99% ± 6.20
KDE [6]	52.66% ± 1.77
Gaussian(sigma=1)	50.66% ± 6.22
Gaussian(sigma=10)	51.99% ± 7.11
ADASSM	51.33% ± 2.66
ADASSM+BC	55.99% ± 15.11
ADASSM+PC	67.99% ± 2.67
ADASSM+BC+PC	51.33% ± 2.66

Table 2. Resources Required: Comparison for time taken to run *Augmentation* & *Model* training required on Nvidia RTX 5000 system.

Dataset	KDE[5]	ADASSM
Left Atrium	690.35+122.35 min	336.15 min
Femur	725.35+ 93.5 min	399.6 min

4   Conclusion and Future Work

In this study, we introduced a novel methodology by proposing an adversarial data augmentation framework for generic regression tasks with applicability to Image-to-SSM networks. Using data-dependent noise augmentation, the proposed method seeks to discover effective shape representations for three-dimensional volumes. By generating challenging augmentations during model training, the proposed method eliminates the need for offline data augmentation, effectively training a more accurate Image-to-SSM network. The proposed noise augmentation framework outperforms the shape augmentation framework [8] and standard noise augmentation, demonstrating that data-dependent noise aids the model by implicitly attending to shape. Through downstream task analysis, we confirmed that the proposed method effectively taught models robust shape descriptors that capture pertinent pathology information. In addition, compared to existing frameworks for shape augmentation, the proposed methodology is not only more robust but also faster. The limitation of the proposed framework is that it trains only on data-dependent intensity/noise augmentations and does not take shape augmentation into account. We plan to extend this framework to data-dependent shape augmentation as well.

Acknowledgements. We thank all research members of Dr.Elhabian's lab and the ShapeWorks team for their assistance in discussions and suggestions that helped us improve this work. The National Institutes of Health supported this work under grant numbers NIBIB-U24EB029011, NIAMS-R01AR076120, and NIBIB-R01EB016701.

The content is solely the authors' responsibility and does not necessarily represent the official views of the National Institutes of Health.

References

1. Abdollahi, B., Tomita, N., Hassanpour, S.: Data augmentation in training deep learning models for medical image analysis. In: Nanni, L., Brahnam, S., Brattin, R., Ghidoni, S., Jain, L.C. (eds.) Deep Learners and Deep Learner Descriptors for Medical Applications. ISRL, vol. 186, pp. 167–180. Springer, Cham (2020). https://doi.org/10.1007/978-3-030-42750-4_6
2. Adams, J., Bhalodia, R., Elhabian, S.: Uncertain-DeepSSM: from images to probabilistic shape models. In: Reuter, M., Wachinger, C., Lombaert, H., Paniagua, B., Goksel, O., Rekik, I. (eds.) ShapeMI 2020. LNCS, vol. 12474, pp. 57–72. Springer, Cham (2020). https://doi.org/10.1007/978-3-030-61056-2_5
3. Adams, J., Elhabian, S.: From images to probabilistic anatomical shapes: a deep variational bottleneck approach. In: Wang, L., Dou, Q., Fletcher, P.T., Speidel, S., Li, S. (eds.) Medical Image Computing and Computer Assisted Intervention – MICCAI 2022. MICCAI 2022. LNCS, vol. 13432. Springer, Cham (2022). https://doi.org/10.1007/978-3-031-16434-7_46
4. Adams, J., Elhabian, S.: Fully bayesian vib-deepssm. arXiv preprint arXiv:2305.05797 (2023)
5. Bhalodia, R., Dvoracek, L.A., Ayyash, A.M., Kavan, L., Whitaker, R., Goldstein, J.A.: Quantifying the severity of metopic craniosynostosis: a pilot study application of machine learning in craniofacial surgery. J. Craniofac. Surg. **31**(3), 697 (2020)
6. Bhalodia, R., Elhabian, S., Adams, J., Tao, W., Kavan, L., Whitaker, R.: DeepSSM: A blueprint for image-to-shape deep learning models. arXiv preprint arXiv:2110.07152 (2021)
7. Bhalodia, R., Elhabian, S.Y., Kavan, L., Whitaker, R.T.: DeepSSM: a deep learning framework for statistical shape modeling from raw images. In: Reuter, M., Wachinger, C., Lombaert, H., Paniagua, B., Lüthi, M., Egger, B. (eds.) ShapeMI 2018. LNCS, vol. 11167, pp. 244–257. Springer, Cham (2018). https://doi.org/10.1007/978-3-030-04747-4_23
8. Bhalodia, R., et al.: Deep learning for end-to-end atrial fibrillation recurrence estimation. In: 2018 Computing in Cardiology Conference (CinC). vol. 45, pp. 1–4. IEEE (2018)
9. Bharath, K., Kurtek, S., Rao, A., Baladandayuthapani, V.: Radiologic image-based statistical shape analysis of brain tumours. J. R. Stat. Soc. Ser. C, Appl. Stat. **67**(5), 1357 (2018)
10. Cates, J., Elhabian, S., Whitaker, R.: ShapeWorks: particle-based shape correspondence and visualization software. In: Statistical Shape and Deformation Analysis, pp. 257–298. Elsevier (2017)
11. Chlap, P., Min, H., Vandenberg, N., Dowling, J., Holloway, L., Haworth, A.: A review of medical image data augmentation techniques for deep learning applications. J. Med. Imaging Radiat. Oncol. **65**(5), 545–563 (2021)
12. Cubuk, E.D., Zoph, B., Mane, D., Vasudevan, V., Le, Q.V.: AutoAugment: Learning augmentation policies from data. arXiv preprint arXiv:1805.09501 (2018)
13. Ganin, Y., Ustinova, E., Ajakan, H., Germain, P., Larochelle, H., Laviolette, F., Marchand, M., Lempitsky, V.: Domain-adversarial training of neural networks. J. Mach. Learn. Res. **17**(1), 2096–2030 (2016)

14. Gao, Y., Tang, Z., Zhou, M., Metaxas, D.: Enabling data diversity: efficient automatic augmentation via regularized adversarial training. In: Feragen, A., Sommer, S., Schnabel, J., Nielsen, M. (eds.) IPMI 2021. LNCS, vol. 12729, pp. 85–97. Springer, Cham (2021). https://doi.org/10.1007/978-3-030-78191-0_7

15. Gardner, G., Morris, A., Higuchi, K., MacLeod, R., Cates, J.: A point-correspondence approach to describing the distribution of image features on anatomical surfaces, with application to atrial fibrillation. In: 2013 IEEE 10th International Symposium on Biomedical Imaging, pp. 226–229. IEEE (2013)

16. Geiping, J., Goldblum, M., Somepalli, G., Shwartz-Ziv, R., Goldstein, T., Wilson, A.G.: How much data are augmentations worth? An investigation into scaling laws, invariance, and implicit regularization. arXiv preprint arXiv:2210.06441 (2022)

17. Gerig, G., Styner, M., Jones, D., Weinberger, D., Lieberman, J.: Shape analysis of brain ventricles using SPHARM. In: Proceedings IEEE Workshop on Mathematical Methods in Biomedical Image Analysis (MMBIA 2001), pp. 171–178. IEEE (2001)

18. Harris, M.D., Datar, M., Whitaker, R.T., Jurrus, E.R., Peters, C.L., Anderson, A.E.: Statistical shape modeling of cam femoroacetabular impingement. J. Orthop. Res. **31**(10), 1620–1626 (2013)

19. Hermann, K., Chen, T., Kornblith, S.: The origins and prevalence of texture bias in convolutional neural networks. Adv. Neural. Inf. Process. Syst. **33**, 19000–19015 (2020)

20. Hussain, Z., Gimenez, F., Yi, D., Rubin, D.: Differential data augmentation techniques for medical imaging classification tasks. In: AMIA Annual Symposium Proceedings. vol. 2017, p. 979. American Medical Informatics Association (2017)

21. Krizhevsky, A., Sutskever, I., Hinton, G.E.: ImageNet classification with deep convolutional neural networks. Commun. ACM **60**(6), 84–90 (2017)

22. Tóthová, K., et al.: Uncertainty quantification in CNN-based surface prediction using shape priors. In: Reuter, M., Wachinger, C., Lombaert, H., Paniagua, B., Lüthi, M., Egger, B. (eds.) ShapeMI 2018. LNCS, vol. 11167, pp. 300–310. Springer, Cham (2018). https://doi.org/10.1007/978-3-030-04747-4_28

23. Xu, H., Elhabian, S.Y.: Image2SSM: Reimagining statistical shape models from images with radial basis functions. arXiv preprint arXiv:2305.11946 (2023)

24. Xu, J., Li, M., Zhu, Z.: Automatic data augmentation for 3D medical image segmentation. In: Martel, A.L., et al. (eds.) MICCAI 2020. LNCS, vol. 12261, pp. 378–387. Springer, Cham (2020). https://doi.org/10.1007/978-3-030-59710-8_37

25. Yao, H., Wang, Y., Zhang, L., Zou, J.Y., Finn, C.: C-mixup: improving generalization in regression. Adv. Neural. Inf. Process. Syst. **35**, 3361–3376 (2022)

26. Zhang, H., Cisse, M., Dauphin, Y.N., Lopez-Paz, D.: mixup: Beyond empirical risk minimization. arXiv preprint arXiv:1710.09412 (2017)

27. Zhao, Z., Taylor, W.D., Styner, M., Steffens, D.C., Krishnan, K.R.R., MacFall, J.R.: Hippocampus shape analysis and late-life depression. PLoS ONE **3**(3), e1837 (2008)

Body Fat Estimation from Surface Meshes Using Graph Neural Networks

Tamara T. Mueller[1,2($\boxtimes$)], Siyu Zhou[1], Sophie Starck[1], Friederike Jungmann[2],
Alexander Ziller[1,2], Orhun Aksoy[1], Danylo Movchan[1], Rickmer Braren[2],
Georgios Kaissis[1,2,4], and Daniel Rueckert[1,3]

[1] Institute for AI in Medicine and Healthcare, Faculty of Informatics,
Technical University of Munich, Munich, Germany
tamara.mueller@tum.de
[2] Department of Diagnostic and Interventional Radiology, Faculty of Medicine,
Technical University of Munich, Munich, Germany
[3] Department of Computing, Imperial College London, London, UK
[4] Institute for Machine Learning in Biomedical Imaging, Helmholtz-Zentrum Munich,
Munich, Germany

Abstract. Body fat volume and distribution can be a strong indication for a person's overall health and the risk for developing diseases like type 2 diabetes and cardiovascular diseases. Frequently used measures for fat estimation are the body mass index (BMI), waist circumference, or the waist-hip-ratio. However, those are rather imprecise measures that do not allow for a discrimination between different types of fat or between fat and muscle tissue. The estimation of visceral (VAT) and abdominal subcutaneous (ASAT) adipose tissue volume has shown to be a more accurate measure for named risk factors. In this work, we show that triangulated body surface meshes can be used to accurately predict VAT and ASAT volumes using graph neural networks. Our methods achieve high performance while reducing training time and required resources compared to state-of-the-art convolutional neural networks in this area. We furthermore envision this method to be applicable to cheaper and easily accessible medical surface scans instead of expensive medical images.

1 Introduction

The estimation of body composition measures refers to the qualification and quantification of different tissue types in the body as well as the estimation of their distribution throughout the body. These measures can function as risk factors of individuals and be an indicator for health and mortality risk [1,12]. One component of body composition analysis is the estimation of fatty tissue volume in the body. The strong correlation between body composition and disease risk has lead to a routine examination of measures indicating body composition in medical exams. The body mass index (BMI), for example, measures the ratio

---

T. T. Mueller and S. Zhou—These authors contributed equally to this work.

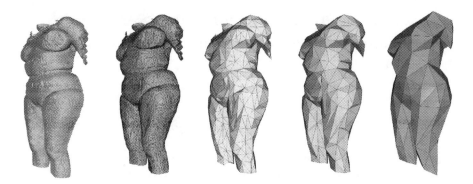

Fig. 1. Visualisation of body surface meshes at different decimation rates; The most left mesh shows the original mesh, then left to right are visualisations of decimated meshes with ten thousand, one thousand, five hundred and two hundred faces.

between a person's weight and height and has been shown to be an indicator for developing cardiovascular diseases, type 2 diabetes, as well as overall mortality [3,12,28,32]. Additionally, the waist circumference and waist-hip-ratio can be used as an indication for body fat distribution [6,25,42,48]. These metrics are easy, fast, and cheap to assess. However, they have strong limitations. They are imprecise as they do not allow for a more accurate assessment of the distribution of body fat or to differentiate between weight that stems from muscle or fat tissue. Understanding the specific differences between different types of fatty tissue and their impact on health risks is crucial for accurately assessing an individual's risk factors and enabling personalised medical care. Towards this goal, several works have investigated methods to identify variations of fat distribution in the body and the quantification of fatty tissues [29,54].

Body fat can be divided into different types of fat. Two commonly investigated types are *visceral fat* (VAT), which surrounds the abdominal organs, and *abdominal subcutaneous fat* (ASAT), which is located beneath the skin. Studies have shown that especially visceral fat can have a negative impact on a person's health [8,40,47]. Therefore, a separate analysis of VAT and ASAT is an important step towards gaining accurate insights into body composition. Several works have investigated a precise estimation of VAT and ASAT volumes from medical images, like magnetic resonance (MR) [29] and computed tomography (CT) images [23], dual-energy X-ray absorptiometry (DXA) assessment [41], or ultrasound imaging [7]. Deep learning techniques have shown promising results in analysing these medical images in order to estimate body composition values [23,29,43,53].

In this work, we perform VAT and ASAT volume prediction from full body triangulated surface meshes using graph neural networks (GNNs). We show that GNNs allow to utilise the full 3D data at hand, thereby achieving better results than state-of-the-art convolutional neural networks (CNNs) on 2D silhouettes, while requiring significantly less training time and therefore resources. Both ours and related work, such as [29], use data extracted from MR images. However,

MR imaging is a very expensive technique, which is highly unequally distributed around the globe. The access to MR scanners in lower income countries is much more limited [18]. Furthermore, the acquisition of MR images is time consuming and very unlikely to be used for routine exams. Given the light computational weight and fast nature of our method, we envision it to be applied to data acquired from much simpler surface scans in the future and enable an incorporation into routine medical examination.

2   Background and Related Work

In the following, we summarise related works on body fat estimation from medical (and non-medical) images, define triangulated meshes and the concept of graph neural networks and show some of their application to medical data, with a focus on surface meshes.

2.1   Body Fat Estimation from Medical Imaging

Body fat estimation has been part of routine medical assessments for decades through the analysis of simple measurements such as BMI or waist circumference [17]. However, more elaborate ways such as using proxy variables derived from medical images, like dual energy X-ray absorptiometry (DXA), CT or MR images, have achieved more accurate results. Multiple studies have successfully assessed patient body composition based upon DXA [15,22,41]. Hemke et al. [23] and Nowak et al. [43] show successful utilisation of CT images for body composition assessment. Works like [31] use segmentation algorithms to identify fatty tissue in MR scans, from which body composition values can be derived. Tian et al. [50] estimate body composition measures based on 2D photography, not even requiring medical imaging techniques. Many of these approaches focus on predicting specific types of adipose tissue [29,31,36,39]. One idea, that has been followed by several works is the utilisation of silhouettes, a binary 2D projection of the outline of the body extracted from images. Xie et al. [54] use silhouettes generated from DXA whole-body scans to estimate shape variations and Klarqvist et al. [29] use silhouettes derived from MR Images for VAT and ASAT volume estimation using CNNs. The latter use two-dimensional coronal and sagittal silhouettes of the body outline and predict VAT and ASAT volume using convolutional neural networks. The silhouettes are extracted from the full-body magnetic resonance (MR) scans of the UK Biobank dataset [49]. In our work, we propose to switch from full medical images or binary silhouettes to surface meshes for fat volume prediction, which allows to integrate the full potential of the 3D surface into deep learning methods, while using the light-weight and fast method of graph neural networks (GNNs).

2.2   Triangulated Meshes

In this work, we use triangulated surface meshes of the body outline. A mesh structure can be interpreted as a specific 3D representation of a graph. A graph

$G := (V, E)$ is defined by a set of nodes V and a set of edges E, connecting pairs of nodes. The nodes usually contain node features, which can be summarised in a node features matrix $\mathbf{X}$. A triangulated mesh M has the same structure, commonly holding the 3D coordinates of the nodes as node features. All edges form triangular faces that define the surface of the object of interest –in our case: body surfaces. A visualisation of such meshes can be found in Fig. 1.

2.3   Graph Neural Networks

Graph neural networks have opened the field of deep learning to non-Euclidean data structures such as graphs and meshes [11]. Since their introduction by [20] and [46], they have been utilised in various domains, including medical research [2,14]. Graphs are, for example, frequently used for representations of brain graphs [9], research in drug discovery [10], or bioinformatics [55,56]. One native data structure that benefits from the utilisation of graph neural networks are surface meshes [11]. GNNs on mesh datasets have also advanced research in the medical domain such as brain morphology estimation [5], which can be used for Alzheimer's disease classification, or for the predicting of soft tissue deformation in image-guided neurosurgery [45].

In general, GNNs follow a so-called message passing scheme, where node features are aggregated among neighbourhoods, following the underlying graph structure [13,24,27,30]. This way, after each iteration, a new embedding for the node features is learned. In this work, we use Graph SAGE [21] convolutions, which were designed for applications on large graphs. The mean aggregator architecture for a node $v \in \mathcal{V}$ at step k is defined as follows:

$$h_v^k = \sigma \left(\mathbf{W} \cdot \text{MEAN}(\{h_v^{k-1}\} \cup \{h_u^{k-1}, \forall u \in \mathcal{N}_v\}) \right). \tag{1}$$

$\mathcal{N}_v$ is the neighbourhood of node v, $\mathbf{W}$ is a learnable weight matrix, and MEAN the mean aggregator, which combines the node features of v at the previous step and the node features of v's neighbours.

3   Methods

We construct three different model architectures: (a) a graph neural network, (b) a simple convolutional neural network (CNN), and (c) a DenseNet and compare their performance. All models are trained using the Adam optimiser [26] and Shrinkage loss [38] and all results reported are cross-validated based on a 5-fold data split. We use a Quadro RTX 8 000 GPU for our experiments and all models predict both targets –VAT and ASAT– with the same network, following the approach from [29].

GNN Architecutre. We perform a whole-graph regression task on the input meshes. The model architecture consists of a three-layer GNN with SAGE graph convolutions [21] and batch normalisation layers, followed by a max aggregation

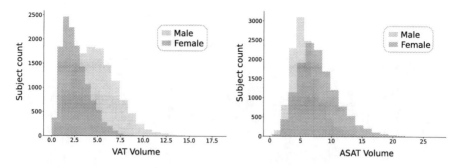

Fig. 2. Distribution of VAT (left) and ASAT (right) volume of male and female subjects in the cohort. Male subjects tend to have more VAT volume, whereas female subjects tend to have more ASAT volume.

and a three-layer multi-layer perceptron (MLP). Hyperparameters such as learning rate and GNN layers are selected by manual tuning. All GNNs are trained for 150 epochs.

CNN Architecture. In order to compare our results to the work by Klarqvist et al. [29], we also train a DenseNet and a simpler CNN on the silhouette data. DenseNet is a CNN which is more densely connected, where each layer takes all previous outputs as an input. For our DenseNet implementation, we follow the architecture in [29]. We additionally construct a simpler CNN architecture that consists of three 2D convolutions, followed by a three-layer MLP, matching the design of the graph neural networks. Both convolutional networks are trained for 20 epochs on a 2D input image, that consist of a sagittal and a coronal view of the binary silhouette masks of the MR images, following the pipeline in [29].

4   Experiments and Results

We use a subset of the UK Biobank dataset [49], which is a large-scale medical database. It contains a variety of imaging data, genetics, and life-style information from almost 65 000 subjects and was acquired in the United Kingdom. In this work, we use the neck-to-knee magnetic resonance images of a subset of 25 298 subjects, for which the labels are available (12 210 male and 13 088 female). The mean age of this cohort is 62.95 years. The VAT and ASAT distributions of male and female subjects are visualised in Fig. 2. We can see that female subjects tend to have a higher ASAT volume, whereas male subjects tend to have more VAT. As labels, we used the reported VAT and ASAT volumes in the UK Biobank (field IDs: 22407 and 22408) (Fig. 3).

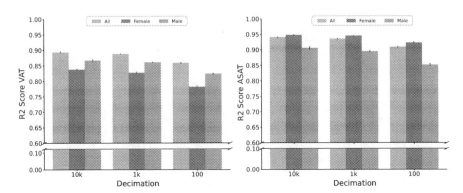

Fig. 3. R2 scores of VAT (left) and ASAT (right) predictions for male, female, and all subjects at different decimation rates of the input meshes.

4.1 Data Processing

The experiments in this work are performed on triangulated body surface meshes that are extracted from the neck-to-knee MR images from the UK Biobank [44]. These were acquired in stations and merged through stitching [33]. In order to extract the surface meshes, we first perform an algorithmic whole-body segmentation by a succession of morphological operations on the stitched MR scans. We then convert these segmentations into surface meshes using the marching cubes algorithm [37] and the open3d library [57]. In order to investigate how much the surface meshes can be simplified, we decimate them into meshes consisting of different numbers of faces. We use meshes with 10 000, 5 000, 1 000, 500, 200, and 100 faces. The number of nodes is always half the number of faces, following Euler's formula for triangular meshes [16]. Subsequently, the meshes are registered into a common coordinate system, using the iterative closest point algorithm [4]. As a reference subject, the most average subject in the dataset was selected based on height, weight, and age. The resulting decimated and registered surface meshes are then used for graph learning. Figure 1 shows an example of a body surface mesh at different decimation rates

4.2 Results

Table 1 summarises the results of the GNNs and CNNs for ASAT and VAT volume prediction. We report the 5-fold cross-validation results on the test set of the best performing models, evaluated on the validation loss. We compare the results of our graph neural networks (GNNs) with the results achieved by the DenseNet from [29] and the results of a simpler CNN (which we call *CNN* in the tables). We furthermore report the training times of all models, measured by the full training process for 150 and 20 epochs for GNNs and CNNs, respectively. All GNNs are trained on the body surface meshes, whereas the CNNs are trained on the silhouettes, following the approach proposed in [29]. We evaluate

Table 1. Results for **VAT** and **ASAT** volume estimation; We report the R2 scores on the test set with standard deviations based on 5-fold cross validation, as well as the training times of the full training in minutes.

Tissue	Model	Decim.	Test R2	Time (min)
VAT	GNN (ours)	100	0.858 ± 0.001	**8.36**
		200	0.872 ± 0.001	8.63
		500	0.882 ± 0.001	9.01
		1k	0.888 ± 0.001	10.11
		5k	**0.893 ± 0.002**	22.36
		10k	0.893 ± 0.003	37.75
	CNN (ours)	–	0.874 ± 0.001	16.20
	DenseNet	–	0.878 ± 0.004	95.79
ASAT	GNN (ours)	100	0.909 ± 0.001	**8.36**
		200	0.921 ± 0.002	8.63
		500	0.931 ± 0.001	9.01
		1k	0.935 ± 0.002	10.11
		5k	0.938 ± 0.000	22.36
		10k	**0.941 ± 0.002**	37.75
	CNN (ours)	–	0.921 ± 0.002	16.20
	DenseNet	–	0.934 ± 0.002	95.79

the GNNs on body surface meshes at different decimation rates of ten thousand, five thousand, one thousand, 500, 200, and 100 faces per mesh (see Fig. 1 for a visualisation of some of these decimated meshes). The best test performances are highlighted in bold, so are the shortest training times. We can see that the simpler CNN architecture almost matches performance of the DenseNet proposed by [29], while requiring less training time. The GNNs outperform the CNN and the DenseNet, when the utilised meshes are not heavily decimated. But even highly decimated surface meshes with one hundred faces, only result in minor performance loss while requiring less than ten times less training time compared to the DenseNet. We envision the utilisation of the surface meshes and graph neural networks to allow for more efficient model training and the utilisation of the full 3D structure of the body, while keeping resource requirements low.

Male and female subjects show different distributions in VAT and ASAT volume. While male subjects tend to have more VAT, females tend to have more ASAT. Figure 2 shows the distributions of the fat volumes of the two sex groups. We therefore compare the results of our method for female and male subjects separately. Table 2 summarises the results of all GNNs and CNNs for VAT and ASAT volume prediction split by sex. The best performing model for each fat type and sex is highlighted in bold. We can see that the predictions of VAT volume tends to be better on male subjects whereas the prediction of ASAT volume achieves slightly higher scores for the female subject. The GNNs,

Table 2. Results of **VAT** and **ASAT** volume prediction split by subject sex; all reported values are R2 scores on the test set, cross-validated across 5 folds.

Fat tissue	Model	Decimation	Female R2	Male R2
VAT	GNN (ours)	100	0.782 ± 0.004	0.824 ± 0.003
		200	0.804 ± 0.006	0.840 ± 0.003
		500	0.815 ± 0.008	0.854 ± 0.003
		1k	0.827 ± 0.004	0.861 ± 0.001
		5k	0.831 ± 0.006	**0.868 ± 0.002**
		10k	**0.837 ± 0.002**	0.867 ± 0.004
	CNN (ours)	–	0.804 ± 0.003	0.845 ± 0.002
	DenseNet	–	0.811 ± 0.006	0.849 ± 0.006
ASAT	GNN (ours)	100	0.923 ± 0.003	0.852 ± 0.004
		200	0.934 ± 0.001	0.870 ± 0.006
		500	0.940 ± 0.002	0.890 ± 0.002
		1k	0.945 ± 0.001	0.895 ± 0.004
		5k	0.945 ± 0.000	0.903 ± 0.002
		10k	**0.948 ± 0.001**	**0.906 ± 0.005**
	CNN (ours)	–	0.934 ± 0.002	0.870 ± 0.002
	DenseNet	–	0.944 ± 0.001	0.891 ± 0.003

however, seem to show a slightly lower gap in performance between the sex groups. We attribute the difference in performance on the different fatty tissue types to the varying distributions in fat volume between the sex groups.

5   Discussion and Conclusion

In this work, we introduce a graph neural network-based method that enables adipose tissue volume prediction for visceral (VAT) and abdominal subcutaneous (ASAT) fat from triangulated surface meshes. The assessment of fatty tissue has high clinical relevance, since it has been shown to be a strong risk factor for diseases like type 2 diabetes and cardiovascular diseases [28,32]. Especially a separate estimation of the two different fat tissues VAT and ASAT has shown to be a relevant medical assessment, since VAT is known to have a higher correlation with disease development compared to ASAT [8,40,47]. We here use graph neural networks and triangulated surface meshes, extracted from full-body MR scans and show that they achieve accurate VAT and ASAT volume predictions. We investigate how different decimation rates impact model performance and training times. Figure 4 visualises this correlation. The bars in the left figure show the average ASAT volume prediction R2 scores on the test set of the GNNs trained on the differently decimated meshes. The overlaid line plot notes the corresponding training times. We can see that at one thousand faces, we reach an

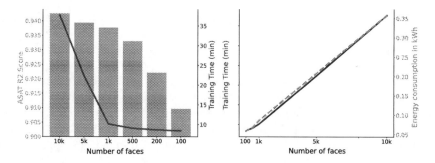

Fig. 4. Relationship between training time and decimation rate of the meshes; The left plot shows the ASAT R2 scores (bars) and the corresponding training time, the right plot shows the linear relation between the training time or the energy consumption in kWh and the number of faces of the meshes.

optimal trade-off between training time and performance. Training the GNN on the meshes with one thousand faces only takes about 10 min and achieves high results of 0.893 R2 on VAT and 0.935 on ASAT volume prediction. On the right in Fig. 4, we visualise the linear relation between the training time and the number of faces in the meshes. Training time also corresponds linearly to energy consumption in kWh. We attribute the comparably high performance of the strongly decimated meshes to the fact that the most outer coordinates/nodes still remain in the meshes, which carry a lot of information about the outline of a body.

The light-weight nature of GNNs allows for the usage of the full 3D data, while significantly reducing resource requirements and run time compared to 3D image-based methods. This shows great promise in the effort of bridging the gap between cheap, fast, but imprecise measures –such as BMI and waist circumference– and time-consuming, costly, but accurate methods such as medical imaging (CT, MR, or DXA).

6   Limitations and Future Work

We see high potential in the utilisation of surface meshes and graph neural networks, given that the full 3D data can be utilised compared to only using binary silhouette projections like in [29]. The low training times as well as the high scores of the GNNs show the successful application to fat volume prediction. We note that we compare the run time of the training loops only. This does not include any pre-processing that is required for both silhouette-based and surface mesh-based approaches. The GNN architecture is based on SAGE graph convolutions [21], because they achieved the best results in our experiments, compared to graph attention networks [51] and graph convolutional networks [27]. A potential improvement of our method would be the utilisation of other mesh-specific convolutions such as adaptive graph convolution pooling [19] or FeaStNet [52]. Another interesting direction to explore is the utilisation of deeper GNNs. Li et al. [34], for example, introduce a method that enables the utilisation of deeper GNNs without over-smoothing –a commonly known problem with GNNs. Over-smoothing refers to the issue that deep GNNs do not achieve high performance because all node embeddings in the graph converge to the same value [35].

Our experiments are performed on surface meshes, that were extracted from MR images. However, we envision this method to work equally well on designated surface scans, without requiring expensive and time-consuming MR scans. We intend to investigate this in future work and apply our method to surface scans, which are for example acquired for dermatological examinations. This would eliminate the need for expensive MR scans and could lead to an embedding of this technique into routine medical examination.

Acknowledgements. TM and SS were supported by the ERC (Deep4MI - 884622). This work has been conducted under the UK Biobank application 87802. SS has furthermore been supported by BMBF and the NextGenerationEU of the European Union.

References

1. Afshin, A., Reitsma, M.B., Murray, C.J.: Health effects of overweight and obesity in 195 countries. N. Engl. J. Med. **377**(15), 1496–1497 (2017)
2. Ahmedt-Aristizabal, D., Armin, M.A., Denman, S., Fookes, C., Petersson, L.: Graph-based deep learning for medical diagnosis and analysis: past, present and future. Sensors **21**(14), 4758 (2021)
3. Anderson, M.R., et al.: Body mass index and risk for intubation or death in SARS-COV-2 infection: a retrospective cohort study. Ann. Intern. Med. **173**(10), 782–790 (2020)
4. Arun, K.S., Huang, T.S., Blostein, S.D.: Least-squares fitting of two 3-D point sets. IEEE Trans. Pattern Anal. Mach. Intell. **PAMI-9**(5), 698–700 (1987). https://doi.org/10.1109/TPAMI.1987.4767965
5. Azcona, E.A., et al.: Interpretation of brain morphology in association to Alzheimer's disease dementia classification using graph convolutional networks on triangulated meshes. In: Reuter, M., Wachinger, C., Lombaert, H., Paniagua, B., Goksel, O., Rekik, I. (eds.) ShapeMI 2020. LNCS, vol. 12474, pp. 95–107. Springer, Cham (2020). https://doi.org/10.1007/978-3-030-61056-2_8
6. Baioumi, A.Y.A.A.: Comparing measures of obesity: waist circumference, waist-hip, and waist-height ratios. In: Nutrition in the Prevention and Treatment of Abdominal Obesity, pp. 29–40. Elsevier (2019)
7. Bazzocchi, A., Filonzi, G., Ponti, F., Albisinni, U., Guglielmi, G., Battista, G.: Ultrasound: which role in body composition? Eur. J. Radiol. **85**(8), 1469–1480 (2016)
8. Bergman, R.N., et al.: Why visceral fat is bad: mechanisms of the metabolic syndrome. Obesity **14**(2S), 16S (2006)
9. Bessadok, A., Mahjoub, M.A., Rekik, I.: Graph neural networks in network neuroscience. IEEE Trans. Pattern Anal. Mach. Intell. **45**(5), 5833–5848 (2022)
10. Bonner, S., et al.: A review of biomedical datasets relating to drug discovery: a knowledge graph perspective. Briefings Bioinform. **23**(6), bbac404 (2022)
11. Bronstein, M.M., Bruna, J., LeCun, Y., Szlam, A., Vandergheynst, P.: Geometric deep learning: going beyond Euclidean data. IEEE Signal Process. Mag. **34**(4), 18–42 (2017)

12. Calle, E.E., Rodriguez, C., Walker-Thurmond, K., Thun, M.J.: Overweight, obesity, and mortality from cancer in a prospectively studied cohort of us adults. N. Engl. J. Med. **348**(17), 1625–1638 (2003)

13. Chiang, W.L., Liu, X., Si, S., Li, Y., Bengio, S., Hsieh, C.J.: Cluster-GCN: an efficient algorithm for training deep and large graph convolutional networks. In: Proceedings of the 25th ACM SIGKDD International Conference on Knowledge Discovery & Data Mining, pp. 257–266 (2019)

14. Ding, K., et al.: Graph convolutional networks for multi-modality medical imaging: Methods, architectures, and clinical applications. arXiv:2202.08916 (2022)

15. Direk, K., et al.: The relationship between DXA-based and anthropometric measures of visceral fat and morbidity in women. BMC Cardiovasc. Disord. **13**, 1–13 (2013)

16. Euler, L.: De summis serierum reciprocarum. Commentarii academiae scientiarum Petropolitanae, pp. 123–134 (1740)

17. Fan, Z., Chiong, R., Hu, Z., Keivanian, F., Chiong, F.: Body fat prediction through feature extraction based on anthropometric and laboratory measurements. PLoS ONE **17**(2), e0263333 (2022)

18. Geethanath, S., Vaughan, J.T., Jr.: Accessible magnetic resonance imaging: a review. J. Magn. Reson. Imaging **49**(7), e65–e77 (2019)

19. Gopinath, Karthik, Desrosiers, Christian, Lombaert, Herve: Adaptive graph convolution pooling for brain surface analysis. In: Chung, Albert C. S.., Gee, James C.., Yushkevich, Paul A.., Bao, Siqi (eds.) IPMI 2019. LNCS, vol. 11492, pp. 86–98. Springer, Cham (2019). https://doi.org/10.1007/978-3-030-20351-1_7

20. Gori, M., Monfardini, G., Scarselli, F.: A new model for learning in graph domains. In: Proceedings of 2005 IEEE International Joint Conference on neural networks. vol. 2(2005), pp. 729–734 (2005)

21. Hamilton, W., Ying, Z., Leskovec, J.: Inductive representation learning on large graphs. In: Advances in Neural Information Processing Systems. vol. 30 (2017)

22. Harty, P.S., et al.: Novel body fat estimation using machine learning and 3-dimensional optical imaging. Eur. J. Clin. Nutr. **74**(5), 842–845 (2020)

23. Hemke, R., Buckless, C.G., Tsao, A., Wang, B., Torriani, M.: Deep learning for automated segmentation of pelvic muscles, fat, and bone from CT studies for body composition assessment. Skeletal Radiol. **49**, 387–395 (2020)

24. Huang, Q., He, H., Singh, A., Lim, S.N., Benson, A.R.: Combining label propagation and simple models out-performs graph neural networks. arXiv preprint arXiv:2010.13993 (2020)

25. Jacobs, E.J., et al.: Waist circumference and all-cause mortality in a large us cohort. Arch. Intern. Med. **170**(15), 1293–1301 (2010)

26. Kingma, D.P., Ba, J.: Adam: A method for stochastic optimization. CoRR abs/1412.6980 (2014)

27. Kipf, T.N., Welling, M.: Semi-supervised classification with graph convolutional networks. arXiv:1609.02907 (2016)

28. Kivimäki, M., et al.: Overweight, obesity, and risk of cardiometabolic multimorbidity: pooled analysis of individual-level data for 120 813 adults from 16 cohort studies from the USA and Europe. Lancet Public Health **2**(6), e277–e285 (2017)

29. Klarqvist, M.D., et al.: Silhouette images enable estimation of body fat distribution and associated cardiometabolic risk. npj Digital Med. **5**(1), 105 (2022)

30. Kong, K., et al.: Flag: Adversarial data augmentation for graph neural networks. arXiv preprint arXiv:2010.09891 (2020)

31. Küstner, T., et al.: Fully automated and standardized segmentation of adipose tissue compartments via deep learning in 3D whole-body MRI of epidemiologic cohort studies. Radiol. Artif. Intell. **2**(6), e200010 (2020)

32. Larsson, S.C., Bäck, M., Rees, J.M., Mason, A.M., Burgess, S.: Body mass index and body composition in relation to 14 cardiovascular conditions in UK biobank: a mendelian randomization study. Eur. Heart J. **41**(2), 221–226 (2020)

33. Lavdas, I., Glocker, B., Rueckert, D., Taylor, S., Aboagye, E., Rockall, A.: Machine learning in whole-body MRI: experiences and challenges from an applied study using multicentre data. Clin. Radiol. **74**(5), 346–356 (2019)

34. Li, G., Muller, M., Thabet, A., Ghanem, B.: DeepGCNs: can GCNs go as deep as CNNs? In: Proceedings of the IEEE/CVF International Conference on Computer Vision, pp. 9267–9276 (2019)

35. Li, Q., Han, Z., Wu, X.M.: Deeper insights into graph convolutional networks for semi-supervised learning. In: Proceedings of the AAAI Conference on Artificial Intelligence. vol. 32 (2018)

36. Linder, N., et al.: Estimation of abdominal subcutaneous fat volume of obese adults from single-slice MRI data - regression coefficients and agreement. Eur. J. Radiol. **130**, 109184 (2020). https://doi.org/10.1016/j.ejrad.2020.109184, https://www.sciencedirect.com/science/article/pii/S0720048X20303739

37. Lorensen, W.E., Cline, H.E.: Marching cubes: a high resolution 3D surface construction algorithm. In: Seminal Graphics: Pioneering Efforts that Shaped the Field, pp. 347–353 (1998)

38. Lu, X., Ma, C., Ni, B., Yang, X., Reid, I., Yang, M.H.: Deep regression tracking with shrinkage loss. In: Proceedings of the European Conference on Computer Vision (ECCV), pp. 353–369 (2018)

39. Lu, Y., et al.: Sex-specific equations to estimate body composition: derivation and validation of diagnostic prediction models using UK biobank. Clin. Nutr. **42**(4), 511–518 (2023). https://doi.org/10.1016/j.clnu.2023.02.005, https://www.sciencedirect.com/science/article/pii/S0261561423000341

40. Matsuzawa, Y., Nakamura, T., Shimomura, I., Kotani, K.: Visceral fat accumulation and cardiovascular disease. Obes. Res. **3**(S5), 645S-647S (1995)

41. Messina, C., et al.: Body composition with dual energy X-ray absorptiometry: from basics to new tools. Quant. Imaging Med. Surg. **10**(8), 1687 (2020)

42. Neeland, I.J., et al.: Visceral and ectopic fat, atherosclerosis, and cardiometabolic disease: a position statement. Lancet Diab. Endocrinol. **7**(9), 715–725 (2019)

43. Nowak, S., et al.: Fully automated segmentation of connective tissue compartments for CT-based body composition analysis: a deep learning approach. Invest. Radiol. **55**(6), 357–366 (2020)

44. Petersen, S.E., et al.: Imaging in population science: cardiovascular magnetic resonance in 100,000 participants of UK biobank-rationale, challenges and approaches. J. Cardiovasc. Magn. Reson. **15**(1), 1–10 (2013)

45. Salehi, Y., Giannacopoulos, D.: PhysGNN: a physics-driven graph neural network based model for predicting soft tissue deformation in image-guided neurosurgery. Adv. Neural. Inf. Process. Syst. **35**, 37282–37296 (2022)

46. Scarselli, F., Gori, M., Tsoi, A.C., Hagenbuchner, M., Monfardini, G.: The graph neural network model. IEEE Trans. Neural Netw. **20**(1), 61–80 (2008)

47. Shuster, A., Patlas, M., Pinthus, J., Mourtzakis, M.: The clinical importance of visceral adiposity: a critical review of methods for visceral adipose tissue analysis. Br. J. Radiol. **85**(1009), 1–10 (2012)

48. Song, X., et al.: Comparison of various surrogate obesity indicators as predictors of cardiovascular mortality in four European populations. Eur. J. Clin. Nutr. **67**(12), 1298–1302 (2013)

49. Sudlow, C., et al.: Uk biobank: an open access resource for identifying the causes of a wide range of complex diseases of middle and old age. PLoS Med. **12**(3), e1001779 (2015)

50. Tian, I.Y., et al.: Predicting 3D body shape and body composition from conventional 2D photography. Med. Phys. **47**(12), 6232–6245 (2020)

51. Veličković, P., Cucurull, G., Casanova, A., Romero, A., Lio, P., Bengio, Y.: Graph attention networks. arXiv preprint arXiv:1710.10903 (2017)

52. Verma, N., Boyer, E., Verbeek, J.: FeaStNet: feature-steered graph convolutions for 3D shape analysis. In: Proceedings of the IEEE Conference on Computer Vision and Pattern Recognition, pp. 2598–2606 (2018)

53. Wang, B., Torriani, M.: Artificial intelligence in the evaluation of body composition. In: Seminars in Musculoskeletal Radiology. vol. 24, pp. 030–037. Thieme Medical Publishers (2020)

54. Xie, B., et al.: Accurate body composition measures from whole-body silhouettes. Med. Phys. **42**(8), 4668–4677 (2015)

55. Yi, H.C., You, Z.H., Huang, D.S., Kwoh, C.K.: Graph representation learning in bioinformatics: trends, methods and applications. Briefings Bioinform. **23**(1), bbab340 (2022)

56. Zhang, X.M., Liang, L., Liu, L., Tang, M.J.: Graph neural networks and their current applications in bioinformatics. Front. Genet. **12**, 690049 (2021)

57. Zhou, Q.Y., Park, J., Koltun, V.: Open3D: A modern library for 3D data processing. arXiv:1801.09847 (2018)

Geometric Learning-Based Transformer Network for Estimation of Segmentation Errors

Sneha Sree[1(✉)], Mohammad Al Fahim[1], Keerthi Ram[2],
and Mohanasankar Sivaprakasam[1,2]

[1] Indian Institute of Technology Madras, Chennai, India
snehacumsali@gmail.com, mohan@ee.iitm.ac.in
[2] Healthcare Technology Innovation Centre, IIT Madras, Chennai, India
keerthi@htic.iitm.ac.in

Abstract. Many segmentation networks have been proposed for 3D volumetric segmentation of tumors and organs at risk. Hospitals and clinical institutions seek to accelerate and minimize specialists' efforts in image segmentation, but in case of errors generated by these networks, clinicians would have to edit the generated segmentation maps manually.

Problem Statement: Given a 3D volume and its putative segmentation map, we propose an approach to identify and measure erroneous regions in the segmentation map. Our method can estimate error at any point or node in a 3D mesh generated from a possibly erroneous volumetric segmentation map, serving as a Quality Assurance tool.

Method: We propose a graph neural network-based transformer based on the Nodeformer architecture to measure and classify the segmentation errors at any point. We have evaluated our network on a high-resolution μCT dataset of the human inner-ear bony labyrinth structure by simulating erroneous 3D segmentation maps. Our network incorporates a convolutional encoder to compute node-centric features from the input μCT data, the Nodeformer to learn the latent graph embeddings, and a Multi-Layer Perceptron (MLP) to compute and classify the node-wise errors.

Results: Our network achieves a mean absolute error of ~ 0.042 over other Graph Neural Networks (GNN) and an accuracy of 79.53% over other GNNs in estimating and classifying the node-wise errors, respectively. We also put forth vertex-normal prediction as a custom pretext task for pre-training the CNN encoder to improve the network's overall performance. Qualitative analysis shows the efficiency of our network in correctly classifying errors and reducing misclassifications.

Keywords: 3D Segmentation error detection · geometric learning

1 Introduction

Medical image segmentation is crucial to isolate and analyze specific structures or regions of interest in a medical image to aid in the diagnosis, treatment planning,

C. Wachinger et al. (Eds.): ShapeMI 2023, LNCS 14350, pp. 118–132, 2023.
https://doi.org/10.1007/978-3-031-46914-5_10

and monitoring of diseases or conditions. Deep learning models have evolved in accuracy, versatility, and deployment-readiness for automatic segmentation of various organs across diverse medical imaging modalities [4,13,14]. Still, automated medical image segmentation needs output review, as models are known to be overconfident, although dealing with natural biological variations and diversity in pathological presentation. There is a need for an automated method of predicting and identifying segmentation errors to aid in improving the segmentation maps in erroneous regions.

Related Works: Many recent works have studied the problem of detecting segmentation errors. Kronman et al. [10] proposed a geometrical segmentation error detection and correction method in which they detect segmentation errors by casting rays from the interior of the initial segmentation map to its outer surface. Altman et al. [2] created an automatic contour quality assurance method that utilizes a knowledge base of historical data. Chen et al. [3] proposed supervised geometric attribute distribution models to identify contour errors accurately. The Reverse Classification Accuracy method [12] identifies failed segmentations to predict the CMRI segmentation metrics, achieving a strong correlation with the predicted metrics and visual quality control scores. Alba et al. [1] utilized a random forest classifier with statistical, pattern, and fractal descriptors to detect segmentation contour failures directly without the need for intermediate regression of segmentation accuracy metrics. Roy et al. [15] presented an approach that directly incorporates a quality measure or prediction confidence within the segmentation framework. This measure is derived from the same model, eliminating the need for a separate model to evaluate quality. By leveraging model uncertainty, their approach avoids the requirement of training an independent classifier for evaluation, which could introduce additional prediction errors.

Graph Neural Networks (GNN) are deep learning algorithms that can extract features from complex graph structures through message-passing. They are particularly suited for processing three-dimensional data and extracting geometric features to capture and analyze the data structure [18].

Henderson et al. [9] proposed a quality assurance tool for identifying segmentation errors in 3D organs-at-risk (OAR) segmentations using a geometric learning method by considering the parotid gland. Their study focuses on the parotid gland in head-neck CT scans.

Inspired by this work [9], we propose a novel segmentation error identification network to predict and classify segmentation errors in the inner ear human bony labyrinth using Nodeformer [20], an advanced Transformer based GNN. We also investigate the effect of pre-training tasks on improving the encoding of node feature vectors for GNNs. The key contributions of our work are: **(1)** We propose a novel 3D segmentation error estimation network based on graph learning, capable of handling graphs with millions of nodes generated from 3D segmentation maps. **(2)** We present VertNormPred, a novel pretext task for pre-training the encoder of our network. It involves predicting the node-wise vertex normals to capture the graph's geometric relationships and surface orientations. **(3)** We quantitatively and qualitatively evaluate our network against other GNN models to estimate and classify node-wise segmentation errors.

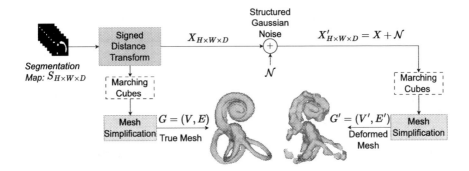

Fig. 1. Simulation of true and perturbed meshes for self-supervised learning of segmentation errors. Each specimen's SDT was perturbed 100 times to produce 100 different deformed segmentations and meshes. The mesh simplification process utilized Taubin smoothing and quadric error decimation techniques to achieve smoother mesh representations.

2   Methods

Formulation: Let S be the input segmentation map of the μCT volume I. Let T be the true segmentation map of I. There exists a deformation on S which operates in the voxel grid to transform S to approach T (limited to nearest neighbors interpolation). The alternative (and finer) domain for mutating S is the surface mesh, computed through a discrete Marching Cubes algorithm [11] f on (per-label) extracted contours.

By defining contours as zero-crossings on a signed Euclidean Distance Transform, we have an additional interim domain of the distance transform, which though residing on the voxel grid, offers some unique properties. For instance, take $X' = SDT(S)$ to be the Signed Distance Transform of S, and likewise, for the true segmentation map, define $X = SDT(T)$. A dense deformation mapping S to T is modeled conveniently as an additive distortion of X with a structured (sparse) 'noise' field: $X' = X + \mathcal{N}$, and recovering T from S becomes estimating and subtracting the noise in X'. Further, the discrete distance transform domain can be interpolated to match the resolution of the surface mesh.

Thus, estimating a per-voxel additive correction on X', conditioned on I would lead to determining the location and magnitude of errors in segmentation. This is mapped to learning from ground truth segmentations T through known random perturbations applied in the form $X' = SDT(T) + \mathcal{N}_{sim}$, leading to a self-supervised learning problem, as shown below.

$$X = SDT(T),\ T = X \leq 0$$
$$X' = X + \mathcal{N}_{sim},\ S = X' \leq 0 \tag{1}$$

Instead of solving this in the SDT domain, we proceed to the mesh domain to setup a per-mesh-vertex estimation of $\mathcal{N}(v)$ conditioned on I, which is equivalent to a corrective field in the interpolated SDT space.

Graph Learning: The surface mesh of a segmentation map S, computed through an operation such as the Discrete Marching Cubes, is representable as a graph $G' = (V', E')$, whose nodes are the mesh vertices, and edges the sides of the triangular faces.

$$G' = (V', E') = f(X') \tag{2}$$

A vertex v_i can be localized in the voxel grid of I to assign an interpolated intensity value. Extending further, a local subvolume in I can be defined around v_i. Finally, v_i is connected to nearby vertices forming a local topological arrangement conditioned on image structure. To capture these relationships jointly in the mesh and image domain, we propose to use graph neural networks.

The learning task is the prediction of node-wise segmentation errors by predicting node-wise Signed Distances (SD) and classifying the node-wise SD into different ranges, given the μCT subvolume centred at each node v, and the entire mesh G'.

The GNN is setup as

$$\hat{\mathcal{N}}(v') := h_\theta(G', I) \quad \forall v' \in V' \tag{3}$$

and optimized as

$$\theta^* = \arg\min \left\| \hat{\mathcal{N}} - \mathcal{N}_{sim} \right\|_2^2 \tag{4}$$

Modeling: We propose a graph learning network based on NodeFormer [20], an advanced Transformer based model designed for efficient node classification on large graphs. NodeFormer incorporates an all-pair message-passing method on adaptive latent structures, enabling information exchange between all nodes by effectively capturing the local and global context. To handle larger graphs, Nodeformer employs the kernelized Gumbel-Softmax operator [20], enabling scalability to millions of nodes.

Our intuition behind the model design was, a CNN encoder can capture contextual details from the μCT data, while the GNN effectively utilizes the local neighborhood of the graph, considering the associated data for each node v. By leveraging the graph's local neighborhood based on data, the GNN can analyze the relationships and connectivity between graph elements, allowing the model to incorporate both the image contextual information from μCT data and the geometric structure of the input. This approach enables the model to exploit the information provided by the local neighborhood of each graph element, enhancing its ability to analyze and process the input data effectively.

2.1 Architecture

We choose a CNN consisting of two 3D Conv layers, each followed by ReLU activation functions as a feature extractor to produce node-wise representations of a $5 \times 5 \times 5$ μCT subvolume centered around each node. The extracted node features are embedded with the perturbed graph's edge connectivity information and passed on to the graph transformer network, consisting of three Nodeformer Conv layers. This takes in the graph-embedded node-wise representations and performs all pair message-passing, updating each node's representation. We consider three Nodeformer Conv layers with eight attention heads, and Batch Normalization and a Leaky ReLU activation function followed each layer. Finally, a Multi-Layer Perceptron (MLP) consisting of three fully connected layers, wherein each layer was followed by a ReLU activation function, Batch Normalization, and a Dropout regularization, processes the updated node-wise representations to produce node-wise SD predictions (using Tanh activation function in the last layer) or classifications (using Softmax activation function in the last layer).

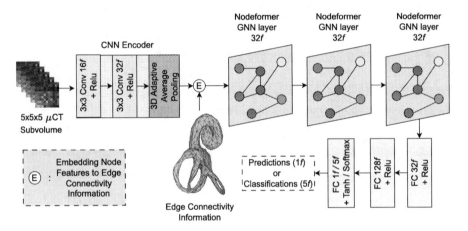

Fig. 2. The proposed graph learning-based transformer network for predicting and classifying node-wise errors. #f represents the number of output channels/nodes. Given the $5 \times 5 \times 5$ μCT subvolume centred at each node and the edge connectivity information of the perturbed mesh, the model predicts the errors at each node.

For classification, predicted node-wise SDs are classified into five classes as shown in Fig. 5, ranging from SDs of -0.16 mm to $+0.16$ mm. Nodes falling into the higher end of the range, exceeding $+0.16$ mm, suggest the occurrence of out-segmentation errors in broad regions. Conversely, nodes with SDs below -0.16 mm indicate in-segmentation errors specifically within narrow regions. These observations highlight the correlation between SDs and the likelihood of realistic segmentation errors in different regions of interest. Figure 2 illustrates the proposed network architecture for node-wise SDs prediction and classification.

2.2   Pre-training Tasks

Towards improving the prediction of node-wise SDs, we incorporated the pre-training transfer learning technique by initializing the model with pre-trained weights obtained from training on different pretext tasks. This approach allows leveraging the knowledge and representations learned during the pretext task to tackle the mainstream tasks [21].

We considered the following three pretext tasks: our custom 1) Vertex Normal Prediction (VertNormPred), 2) μCT volume Reconstruction (ReconCT), and 3) Masked μCT volume Reconstruction (MaskReconCT) tasks. In the VertNorm-Pred task, we train the CNN model shown in Fig. 3(a) to predict the node-wise vertex normal X_{vn} given the $5 \times 5 \times 5$ μCT subvolume centred around a node. While generating the dataset using the marching cubes algorithm, we also obtained the ground truth node-wise vertex normals for each mesh. This task enabled the model to capture geometric relationships and surface orienta-tions. Since neighboring nodes and their orientations influence node-wise SDs [5], understanding surface properties through vertex normal prediction significantly improved the accuracy of the SDs predictions.

In the ReconCT task, we train an encoder-decoder network illustrated in Fig. 3(b) to reconstruct $5 \times 5 \times 5$ μCT subvolumes. This task allowed the CNN encoder to extract essential features from the node-wise μCT data.

In the MaskReconCT [8] task, we focus on reconstructing pixel-wise randomly masked $5 \times 5 \times 5$ μCT subvolumes using an encoder-decoder network shown in Fig. 3(b). We train the model to infer missing regions in the data. By learning

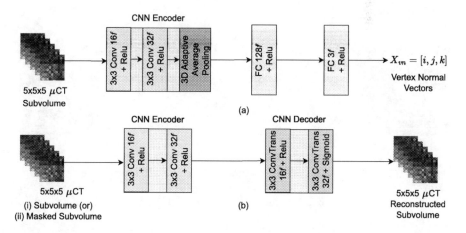

Fig. 3. a) Vertex Normal Prediction (VertNormPred) network predicts the node-wise vertex normals given the $5 \times 5 \times 5$ μCT subvolume centred at each node. b) (i) CT volume Reconstruction (ReconCT) network and (ii) Masked CT volume Reconstruction (MaskReconCT) network reconstructs the $5 \times 5 \times 5$ μCT subvolume given μCT or pixel-wise randomly masked μCT subvolume centred at each node, respectively.

to fill these gaps, the model becomes more adept at estimating SDs, especially when parts of the μCT are incomplete.

We initialized the CNN encoder of our model with the pre-trained weights obtained from the CNN encoder of the models shown in Fig. 3 from these pretext tasks to facilitate node-feature extraction. The pretext tasks: VertNormPred, ReconCT, and MaskReconCT, improved the model in capturing the μCT bony labyrinth structure for the mainstream task of prediction/classification of node-wise SD.

3   Dataset Description

We use the publicly available OpenAIRE's human bony labyrinth dataset [19] to evaluate the method. The dataset consists of clinical Computed Tomography (CT) volumes, co-registered high-resolution micro-CT (μCT) volumes, segmentation maps, and surface models of 23 human bony labyrinths. We used 22 specimens of μCT volumes of 0.06 mm isotropic voxel size and their corresponding segmentation maps.

3.1   Generation of Training Data

We generate the perturbed segmentation maps by perturbing the SDT by addition of noise of the true segmentation map 100 times, ensuring the Hausdorff distances of the perturbed segmentation maps are in the range of (7–65). Figure 4 illustrates the simulation of a perturbed segmentation map obtained from a perturbed SDT.

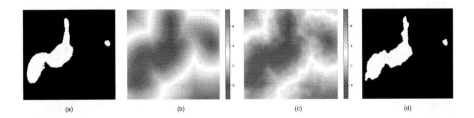

Fig. 4. One of the slices of (a) true segmentation map, (b) distance transform, (c) perturbed distance transform after addition of noise to distance transform and (d) perturbed segmentation map obtained from the perturbed distance transform (c).

We use the marching cubes algorithm to obtain the triangular mesh manifolds of the perturbed segmentation maps. The complex geometry of the human bony labyrinth led to generating a mesh with numerous triangles, resulting in a graph with nodes in the order of 10^5. We use Taubin smoothing [16] and quadratic error decimation techniques to smoothen the mesh. We consider the mesh vertices as nodes (V) of the graph and the sides of the triangular faces of the mesh as edges (E). The simulation of true and deformed mesh is shown in Fig. 1.

To calculate the node-wise SD, we perform bi-linear interpolation between the nodes of the perturbed mesh and the voxels of the ground truth SDT. Note that the generated node-wise errors correspond to the node-wise SDs of the true segmentation. For classification, we split these node-wise SDs into five classes ranging from -0.16 mm to $+0.16$ mm.

4   Experiments and Results

Towards fine-grained prediction of node-wise SDs, we trained and evaluated our model for regression of node-wise SDs. To also identify the errors in different ranges, we trained and evaluated our model for classification to classify the predicted node-wise SDs into different classes.

For all the experiments, we considered the perturbations tied to the 14 volumes in the training set, while maintaining a similar procedure, 2 volumes are

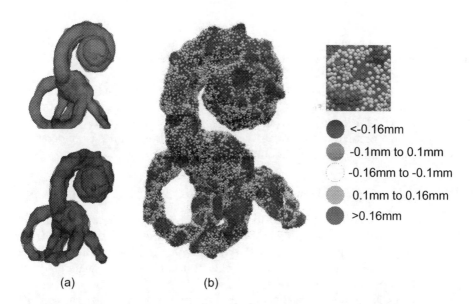

(a)   (b)

Fig. 5. Visualization of the true, perturbed meshes, and the node-wise SD classes. (a) At the top, the true mesh (in blue) is overlaid with the perturbed mesh (in green). At the bottom, the perturbed mesh (in green) is overlaid with the true mesh (in blue). The overlapping region between the true and perturbed meshes reveals where internal and external segmentation errors occur. (b) The node-wise SDs in the perturbed mesh are distributed into five classes indicated by colours varying from red to blue, and the class ranges are shown above. (Color figure online)

Table 1. Comparison of Nodeformer with different pre-trained weights against other models for regression of node-wise SDs.

GNN	Pretraining	MAE ↓	MSE ↓
Spline	ReconCT	0.06994	0.00986
GAT	–	0.06946	0.00913
GAT	VertNormPred	0.0694	0.00968
Spline	MaskReconCT	0.06783	0.00802
GAT	ReconCT	0.06755	0.00884
GAT	MaskReconCT	0.06705	0.00903
Spline	–	0.06032	0.00762
Spline	VertNormPred	0.05728	0.00757
Nodeformer	–	**0.04536**	**0.00475**
Nodeformer	**MaskReconCT**	**0.04397**	**0.00451**
Nodeformer	**ReconCT**	**0.04254**	**0.00444**
Nodeformer	**VertNormPred**	**0.04182**	**0.00429**

designated for the validation set, and 6 volumes are allocated for the testing set. So, with these perturbations, we got 1400 examples for training, 200 for validation, and 600 for the test set.

We have quantitatively and qualitatively evaluated our model against Spline Conv [7] and GAT [17] based GNN models for regression and classification of node-wise SDs. We have also evaluated the models using pre-trained weights from the three pretext tasks. We also performed ablation studies to understand the contribution of each block in our proposed model.

4.1   Implementation Details

We train the network in Fig. 3(a) for the VertNormPred task, where we minimize the *Cosine Similarity loss* between the predicted and ground truth node-wise vertex normals. We train the network in Fig. 3(b) for ReconCT and MaskReconCT tasks, where we minimize the *L1 loss* between the generated and original $5 \times 5 \times 5$ μCT node-wise subvolumes.

For regression of node-wise SDs, we train the models to minimize the *Smooth L1 loss* between the predicted node-wise SDs and the node-wise SDs obtained using interpolation (GT SDs). We used the Mean Absolute Error (MAE) and Mean Square Error (MSE) metrics to quantify the performance of the models trained for regression. For the classification of node-wise SDs, we train the models to minimize the *Cross Entropy loss* between the predicted and GT SD classes. We used the F1 score, Precision, Recall, and Accuracy metrics to quantify the performance of the models trained for classification. We trained all the networks for 100 epochs, using a learning rate of $1e^{-3}$ and a cosine annealing scheduler with a weight decay of $1e^{-3}$. Both the regression and classification models utilized the AdamW optimizer, while the pre-training networks employed the Adadelta optimizer. Models are implemented using PyTorch and PyG [6], and the training process was carried out in a workstation using an i5-1035G4 CPU and NVIDIA 24 GB RTX 3090 GPU.

4.2   Results and Discussion

In Table 1, it can be observed that our model built upon Nodeformer can predict node-wise SDs efficiently, and additionally, using pre-trained weights improved the prediction. Among the evaluated models, our model with a CNN encoder initialized with VertNormPred pre-trained weights yielded the lowest MAE score of 0.04182. This signifies a substantial improvement of ∼30.6% compared to Spline Conv GNN without any pre-trained weights.

Table 2. Comparison of Nodeformer with different pre-trained weights against other models for classification of node-wise SD classes.

GNN	Pretraining	f1 Score ↑	Precision ↑	Recall ↑	Accuracy (%) ↑
Spline	MaskReconCT	0.4872	0.5445	0.5124	66.3
GAT	VertNormPred	0.5186	0.605	0.5398	69.03
GAT	MaskReconCT	0.5024	0.5649	0.5425	71.22
Spline	VertNormPred	0.5367	0.612	0.5487	71.76
GAT	ReconCT	0.5746	0.6289	0.5927	72.17
Spline	ReconCT	0.5871	0.6136	0.614	72.28
GAT	–	0.5623	0.6487	0.567	72.4
Spline	–	0.5582	0.6181	0.5779	71.53
Nodeformer	**MaskReconCT**	**0.5986**	**0.6693**	**0.589**	**74.55**
Nodeformer	**ReconCT**	**0.6695**	**0.72**	**0.7131**	**76.57**
Nodeformer	**–**	**0.6899**	**0.7343**	**0.6693**	**78.82**
Nodeformer	**VertNormPred**	**0.6943**	**0.7384**	**0.6835**	**79.53**

In Table 2, our model, initialized with pre-trained encoder weights of the Vert-NormPred task, gave an overall accuracy of 79.53%. This signifies a substantial improvement of ∼8% in accuracy compared to the Spline Conv GNN without any pre-training task, indicating a significant improvement in the model's ability to identify different ranges of segmentation errors.

Tables 1 and 2 show that our model has benefited from using the pre-trained weights of the VertNormPred task, indicating that the prediction of the node-wise vertex normals during pre-training has helped the encoder of our model in capturing the intricate surface orientations and geometric inter-node relationships in the bony labyrinth structure. This has helped further improve the prediction of node-wise SDs.

From Fig. 6, it is evident that our model, using Nodeformer, outperforms the other models in the classification of node-wise SDs. Our model qualitatively exhibits improved classification of node-wise SD classes, with significantly fewer

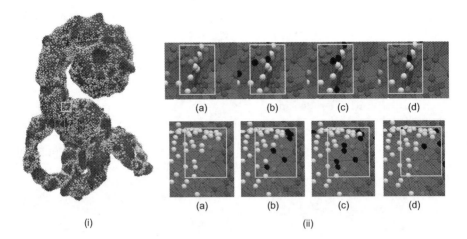

Fig. 6. Visual illustration of the classification of node-wise SDs by all the models. (i) The actual node-wise SD classes in the perturbed mesh. (ii) Two distinct regions within the graph to showcase the predicted node classes compared to the ground truth node classes. The first and second rows show the zoomed-in regions in the pink and yellow boxes in the perturbed mesh (i). (a) GT perturbed node-wise SD classes, (b)–(d) predicted node-wise SD classes by Spline Conv, GAT, and our model, respectively. The black-coloured nodes denote incorrect predictions. The yellow boxes in (ii) (a)–(d) show how well our model can classify the node-wise SD classes with respect to the GT node-wise SD classes with the least number of black nodes. (Color figure online)

black-colored nodes representing incorrect predictions than those obtained using Spline Conv and GAT models. This highlights our model's superior performance and effectiveness in accurately classifying the node classes in the given graph.

In our experiments, we observed that incorporating pre-trained weights from the pretext tasks positively impacted the performance of the models in the regression of node-wise SDs. However, using pre-trained weights for the classification task did not result in much significant improvement.

4.3   Ablation Study

To evaluate the extent to which Nodeformer effectively learns meaningful information from the geometric structure of the segmentation, we performed an ablation study for the classification task that involved removing the GNN component entirely and directly passing on the node-wise representations from the encoder to the MLP decoder. Also, to evaluate the importance of node feature extraction using the CNN encoder and pre-trained weights, we experimented by passing the μCT subvolumes through a linear layer as node feature representations to the Nodeformer instead of passing them through the CNN encoder.

Figure 8 demonstrates the significance of incorporating geometrical structure learning using Nodeformer and the CNN encoder to extract node features

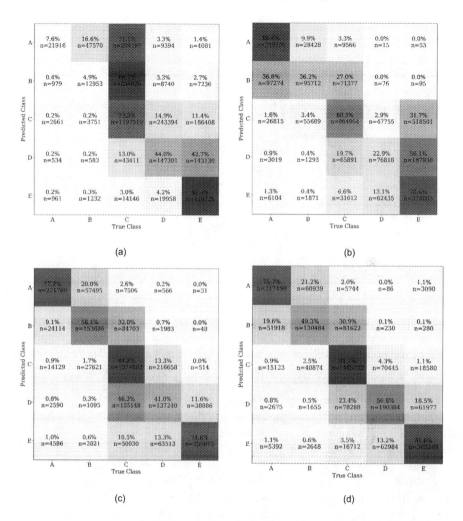

Fig. 7. Confusion Matrices of (a) CNN-MLP: Node features from the CNN encoder are directly given to the MLP for classification, (b) GNN-MLP: Node feature vectors are obtained from a linear layer instead of the CNN encoder are passed on to the Nodeformer and MLP for classification, (c) our complete model, and (d) our model with the CNN encoder initialized with the VertNormPred pre-trained weights. Labels A: (< -0.16 mm), B: (-0.16 mm to -0.1 mm), C: (-0.1 mm to 0.1 mm), D: (0.1 mm to 0.16 mm), E: (>0.16 mm).

in identifying segmentation errors by comparing their performance in classification. Upon removing Nodeformer from our model (CNN-MLP), the classification performance for error identification is notably poor. This emphasizes the importance of Nodeformer in capturing the geometrical information required for error analysis.

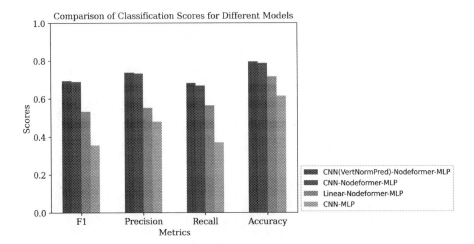

Fig. 8. Comparison of Classification Scores between the different blocks of our model, described in Sect. 4.3. The plot provides a visual representation of the distribution and relative performance of the models based on their classification scores.

Furthermore, using a linear layer to extract node features from μCT subvolumes instead of the CNN encoder also resulted in poor performance, as shown in Fig. 8. This highlights the importance of Conv layers in effectively capturing the node-centred μCT information necessary for accurate error classification.

Regarding using pre-trained weights, our model with the CNN encoder initialized with pre-trained weights from the VertNormPred task gave the best classification performance regarding accuracy, precision, recall, and F1 score, as shown in Figs. 7 and 8.

By comparing Fig. 7(a) and Figs. 7(b)–(d), it can be observed that removing the GNN component (Nodeformer) led to a notable decrease in the model's performance in classifying errors in different ranges. Specifically, it fails to identify internal errors (recall score of 7.6%). Figure 7(b) shows that the Linear-Nodeformer-MLP (GNN-MLP) model can identify internal and external errors but fail to identify the intermediate ones. From Fig. 7(c) and (d), it is clear that the models using Nodeformer for geometrical structure learning with a CNN encoder for node feature extraction were capable of identifying errors in all ranges and using pre-trained weights reduced misclassification in some classes and significantly improved the recall score.

5   Conclusion

Our work introduced a Nodeformer-based graph learning network as a Quality Assurance (QA) tool to evaluate errors in the automatic segmentation of medical images. To our knowledge, this is the first work that addresses segmentation errors in the 3D data of the human inner-ear bony labyrinth structure. The complexity of the inner-ear human bony labyrinth structure gave rise to graphs with nodes in the order of 10^5. Our network, built upon Nodeformer, can scale up to millions of nodes and easily handle human inner-ear bony labyrinth graphs. To boost the performance of our network, we also proposed a custom Vertex Normal Prediction pretext task for pre-training the CNN encoder of our network. We have evaluated our network against other GNN models with pre-trained weights from different pretext tasks for regression and classification of node-wise segmentation errors. We have qualitatively shown how well our model can correctly classify segmentation errors and reduce misclassifications. We have also conducted an ablation study to show the strengths of individual modules of our network, along with loading the pre-trained weights from the Vertex Normal Prediction pretext task, for classification. This study motivates further research into developing and advancing QA techniques and tools for measuring, classifying, and correcting segmentation errors.

References

1. Alba, X., Lekadir, K., Pereanez, M., Medrano-Gracia, P., Young, A.A., Frangi, A.F.: Automatic initialization and quality control of large-scale cardiac MRI segmentations. Med. Image Anal. **43**, 129–141 (2018)
2. Altman, M., et al.: A framework for automated contour quality assurance in radiation therapy including adaptive techniques. Phys. Med. Biol. **60**(13), 5199 (2015)
3. Chen, H.C., et al.: Automated contouring error detection based on supervised geometric attribute distribution models for radiation therapy: a general strategy. Med. Phys. **42**(2), 1048–1059 (2015)
4. Chen, Y., et al.: Fully automated multiorgan segmentation in abdominal magnetic resonance imaging with deep neural networks. Med. Phys. **47**(10), 4971–4982 (2020)
5. Chen, Z., Zhang, H.: Learning implicit fields for generative shape modeling. In: Proceedings of the IEEE/CVF Conference on Computer Vision and Pattern Recognition, pp. 5939–5948 (2019)
6. Fey, M., Lenssen, J.E.: Fast graph representation learning with PyTorch Geometric. In: ICLR Workshop on Representation Learning on Graphs and Manifolds (2019)
7. Fey, M., Lenssen, J.E., Weichert, F., Müller, H.: SplineCNN: fast geometric deep learning with continuous b-spline kernels. In: Proceedings of the IEEE Conference on Computer Vision and Pattern Recognition, pp. 869–877 (2018)
8. He, K., Chen, X., Xie, S., Li, Y., Dollár, P., Girshick, R.: Masked autoencoders are scalable vision learners. In: Proceedings of the IEEE/CVF Conference on Computer Vision and Pattern Recognition, pp. 16000–16009 (2022)
9. Henderson, E.G., Green, A.F., van Herk, M., Vasquez Osorio, E.M.: Automatic identification of segmentation errors for radiotherapy using geometric learning. In: Wang, L., Dou, Q., Fletcher, P.T., Speidel, S., Li, S. (eds.) MICCAI 2022. LNCS, vol. 13435, pp. 319–329. Springer, Cham (2022). https://doi.org/10.1007/978-3-031-16443-9_31
10. Kronman, A., Joskowicz, L.: A geometric method for the detection and correction of segmentation leaks of anatomical structures in volumetric medical images. Int. J. Comput. Assist. Radiol. Surg. **11**, 369–380 (2016)
11. Lorensen, W.E., Cline, H.E.: Marching cubes: a high resolution 3D surface construction algorithm. In: Seminal Graphics: Pioneering Efforts that Shaped the Field, pp. 347–353 (1998)

12. Robinson, R., et al.: Automated quality control in image segmentation: application to the UK biobank cardiovascular magnetic resonance imaging study. J. Cardiovasc. Magn. Reson. **21**(1), 1–14 (2019)
13. Ronneberger, O., Fischer, P., Brox, T.: U-net: convolutional networks for biomedical image segmentation. In: Navab, N., Hornegger, J., Wells, W.M., Frangi, A.F. (eds.) MICCAI 2015. LNCS, vol. 9351, pp. 234–241. Springer, Cham (2015). https://doi.org/10.1007/978-3-319-24574-4_28
14. Roth, H.R., et al.: Deep learning and its application to medical image segmentation. Med. Imaging Technol. **36**(2), 63–71 (2018)
15. Roy, A.G., Conjeti, S., Navab, N., Wachinger, C., Initiative, A.D.N., et al.: Bayesian quicknat: model uncertainty in deep whole-brain segmentation for structure-wise quality control. Neuroimage **195**, 11–22 (2019)
16. Taubin, G.: Curve and surface smoothing without shrinkage. In: Proceedings of IEEE International Conference on Computer Vision, pp. 852–857. IEEE (1995)
17. Velickovic, P., Cucurull, G., Casanova, A., Romero, A., Lio, P., Bengio, Y., et al.: Graph attention networks. stat **1050**(20), 10–48550 (2017)
18. Wang, Y., Sun, Y., Liu, Z., Sarma, S.E., Bronstein, M.M., Solomon, J.M.: Dynamic graph CNN for learning on point clouds. ACM Trans. Graph. (ToG) **38**(5), 1–12 (2019)
19. Wimmer, W., Anschuetz, L., Weder, S., Wagner, F., Delingette, H., Caversaccio, M.: Human bony labyrinth dataset: co-registered CT and micro-CT images, surface models and anatomical landmarks. Data Brief **27**, 104782 (2019)
20. Wu, Q., Zhao, W., Li, Z., Wipf, D.P., Yan, J.: Nodeformer: a scalable graph structure learning transformer for node classification. In: Advances in Neural Information Processing Systems, vol. 35, pp. 27387–27401 (2022)
21. Yosinski, J., Clune, J., Bengio, Y., Lipson, H.: How transferable are features in deep neural networks? In: Advances in Neural Information Processing Systems, vol. 27 (2014)

On the Localization of Ultrasound Image Slices Within Point Distribution Models

Lennart Bastian[(✉)], Vincent Bürgin, Ha Young Kim, Alexander Baumann,
Benjamin Busam, Mahdi Saleh, and Nassir Navab

Computer Aided Medical Procedures, Technical University of Munich,
Munich, Germany
lennart.bastian@tum.de

Abstract. Thyroid disorders are most commonly diagnosed using high-resolution Ultrasound (US). Longitudinal nodule tracking is a pivotal diagnostic protocol for monitoring changes in pathological thyroid morphology. This task, however, imposes a substantial cognitive load on clinicians due to the inherent challenge of maintaining a mental 3D reconstruction of the organ. We thus present a framework for automated US image slice localization within a 3D shape representation to ease how such sonographic diagnoses are carried out. Our proposed method learns a common latent embedding space between US image patches and the 3D surface of an individual's thyroid shape, or a statistical aggregation in the form of a statistical shape model (SSM), via contrastive metric learning. Using cross-modality registration and Procrustes analysis, we leverage features from our model to register US slices to a 3D mesh representation of the thyroid shape. We demonstrate that our multi-modal registration framework can localize images on the 3D surface topology of a patient-specific organ and the mean shape of an SSM. Experimental results indicate slice positions can be predicted within an average of 1.2 mm of the ground-truth slice location on the patient-specific 3D anatomy and 4.6 mm on the SSM, exemplifying its usefulness for slice localization during sonographic acquisitions. Code is publically available: https://github.com/vuenc/slice-to-shape.

Keywords: Ultrasound · Multi-modal Registration · Statistical Shape Models

1 Introduction

High-resolution Ultrasound (US) has detected the presence of thyroid nodules in up to 69% of scans in randomly selected individuals [18]. While only 7–15% of cases develop into malignant tumors, periodic screening is an essential prophylactic measure in the early diagnosis and treatment of a debilitating disease. B-mode US has been identified as the primary tool for diagnosing malignant

---

L. Bastian, V. Bürgin and H.Y. Kim—Equal contribution.

© The Author(s), under exclusive license to Springer Nature Switzerland AG 2023
C. Wachinger et al. (Eds.): ShapeMI 2023, LNCS 14350, pp. 133–144, 2023.
https://doi.org/10.1007/978-3-031-46914-5_11

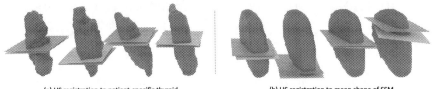

(a) US registration to patient-specific thyroid   (b) US registration to mean shape of SSM

Fig. 1. We propose the task of slice-to-shape registration. Our method successfully localizes ultrasound images to 3D thyroid shapes from the same individual (a) and the mean of a statistical shape model (b). The ground truth image plane is depicted in green, and the prediction in yellow. (Color figure online)

thyroid nodules for its ease of use, lack of harmful ionizing radiation [2], and exceptional soft-tissue resolution [8]. A rise in the prevalence of thyroid cancer over the past decades has been mainly attributed to an increase in early detection with the help of more frequent US screenings [18]. However, US is highly operator-dependent and requires significant training to generate clear and accurate images. Furthermore, US images are abundant with noise and artifacts induced by physical properties such as phase aberrations and attenuation, which can introduce uncertainty and yield inconsistent diagnoses in thyroid nodule classification [8] and thyroid volume [22] estimation. Methods that could improve thyroid scanning and US image quality are therefore highly sought after to increase early detection rates of thyroid cancer worldwide.

Observing a thyroid nodule's evolution is a critical diagnostic protocol [8]. To understand how a nodule changes over time and if it is potentially developing into a malignant tumor, clinicians meticulously match the thyroid morphology to a previous US acquisition. Such a procedure requires significant dexterity, elaborate training, and a cognitive 3D reconstruction while simultaneously conducting a complex medical evaluation. Providing automated support for this procedure could not only alleviate the cognitive burden on clinicians but also reduce costs by enabling less experienced individuals to conduct these scans semi-autonomously. Furthermore, this technological aid could extend potentially life-saving diagnoses to remote communities that lack specialized medical expertise [6,26,34]. A fundamental challenge in automating such a procedure is intraorgan localization.

To address this problem generally, we therefore propose a framework for slice-to-shape registration. Existing 2D–3D registration methods in the medical domain either register image slices to individual 3D volumes [24,30] or aggregated volumes in the form of a medical atlas [33]. While image-atlas-based methods have been demonstrated effective for navigational support, deformable image registration is ill-conditioned and difficult to regularize [13]. Furthermore, registration inaccuracies can yield unrepresentative voxel intensities in the atlas, adversely affecting downstream slice-to-atlas registration. We propose the task of slice-to-shape registration, by registering US slices to a 3D shape representation either directly obtained from an individual's organ segmentation contour, or aggregated shapes in the form of a statistical shape model (SSM) for a more general localization approach which does not require pre-operative acquisitions.

The medical imaging literature has not explored the registration of medical images to 3D organ shapes, particularly for statistical point distributions (see Fig. 1). We thus propose a self-supervised metric learning pipeline to enable matching and registration across US images and a 3D mesh representation of an organ. We leverage unsupervised correspondence estimation to generate a point distribution model (PDM) [4] and use these correspondences to map image patches to a corresponding location on SSM shapes during training. Patch features are extracted using separate deep neural networks, and their cross-modal representations are used to localize a US query slice inside the SSM. Despite having limited supervision and learning on a geometric surface representation, our pipeline for partial thyroid registration successfully localizes US slices from unseen subjects in an SSM.

Our main contributions can be summarized as follows:

- We propose the task of slice-to-shape registration in medical imaging for 3D organ shapes and SSMs.
- Our slice-to-shape correspondence pipeline enables registration of ultrasound slices to 3D thyroid shapes through multi-modal contrastive metric learning.
- We evaluate the capabilities of our model for US slice localization on patient-specific 3D thyroid meshes and SSMs, demonstrating for the first time that 2D US images can be localized within a geometric statistical distribution without prior patient-specific acquisitions.

2   Related Work

2.1   SSMs and Image Atlases in Medical Imaging

Statistical Shape Models (SSMs) and Image Atlases have distinct yet interconnected roles in medical imaging. Since the early 1990s, SSMs have been widely employed for their efficient encapsulation of shape variations and usefulness in enhancing the robustness of segmentation techniques [19]. Both methodologies have found utility in diverse applications such as segmentation [27], registration [5,10], shape classification [16,23], and image augmentation [31,32]. More recently, SSMs have been used as priors to improve the robustness of medical image segmentation in deep neural networks [27], enhance myocardial motion tracking [21], and facilitate the segmentation of the prostate in trans-rectal ultrasound [28]. In these applications, SSMs typically need to be deformably registered with an organ instance in the image space [27]. However, SSMs have also been employed in a broader context of registration tasks, such as in percutaneous ultrasound [7], as a regularization tool in radiation planning [5], or in the correction of cardiac slices [3].

Medical image atlases provide an integrated representation of shape and appearance, offering additional features beneficial for registration. However, building and utilizing an image atlas can be computationally intensive, and the quality heavily depends on accurate image registration, which can incur significant manual labor [13]. Automated methods for deformable registration have been extensively explored [13]. However, less-than-perfect registration results can yield voxels that are not representative of any actual human anatomy, adversely affecting downstream applications. This work focuses on SSMs in the form of point distribution models (PDMs) obtained through unsupervised correspondence estimation [4], which are preferred in some applications as a generalizable and lightweight statistical organ representation [1].

2.2 Multi-modal Registration

Multi-modal registration has been established as a cornerstone for surgical navigation as well as pre-operative and general acquisition planning [20]. Several data modalities have been proposed for navigational support, including multi-template medical atlases [17], or using MRI for US acquisition planning [20]. Deep-learning-based methods have recently proven useful for multi-modal registration due to their robustness to initialization and accuracy [24]. Markova et al. propose to learn dense features from MRI and US modalities, which are combined in a matching module using a confidence threshold and processed with RANSAC to retrieve the pose. Alternatively, learning modality-invariant features can be achieved by sampling triplets and driving together similar latent descriptions of two deep neural networks through a contrastive or triplet loss [11].

While multi-modal slice-to-volume registration is of high interest to the medical community [12,24], the registration of image slices to statistical shape representations has not been explored extensively. Ghanavati et al. first proposed registering US slices to an image atlas generated by deformable registration of CT images [14,15]. More recently, Yeung et al. proposed localizing slices within an atlas of the fetal brain [33]. In contrast to these methods, we propose to localize slices within a PDM. This incurs significant challenges, as PDMs lack image intensity features and, unlike image atlases, only contain sparse and noisy surface samples. The localization of slices to a PDM could overcome the deficits image atlases suffer due to registration inaccuracies, as surface correspondence estimation is generally less ill-posed than the deformable registration of dense volumes. Unlike previous works, we therefore propose the task of *slice-to-shape* registration. We use a triplet-learning loss to construct a common cross-domain embedding space between US images and organ contours, enabling image localization with respect to a 3D statistical geometry.

3   Method

In the following, we propose a deep learning framework to register a US image slice to a 3D shape of the same organ. We first leverage self-supervised contrastive learning to learn a common latent feature space between patches extracted from compounded 3D US data and patches representing local portions of the 3D organ surface. We then sample patches from both modalities during inference, establishing an optimal matching via distance in the latent space. The slice location can then be approximated based on these matches through several iterations of a refinement algorithm, eliminating false-positive matches and narrowing down the region on the surface where the slice is likely located.

We use a discretized signed distance field (SDF) to represent the 3D thyroid surface. The SDF is constructed from one of three possible sources, depending on the task:

- a patient-specific mesh created from the 3D ultrasound segmentation labels (registering a US slice to the 3D SDF of the same patient)
- the SSM's mean shape
- a shape sampled from the SSM distribution

For the SSM, we construct a point distribution model (PDM) to use as a registration target for US patch features. Points are first brought into correspondence using S3M [4] to form a matrix $X \in \mathbf{R}^{n \times d}$, with d point coordinates for each of the n samples. One can then form the mean shape $\bar{X} \in \mathbf{R}^d$ and covariance matrix $S \in \mathbf{R}^{d \times d}$ over the n samples. Since S has rank $n - 1$, the matrix has $n - 1$ eigenvectors v_j with eigenvalues λ_j. Considering the sum

$$s = \bar{X} + \sum_{j=1}^{n-1} \alpha_j \lambda_j v_j, \qquad \alpha_j \sim \mathcal{N}(0, 1) \tag{1}$$

then $s \sim \mathcal{N}\left(\bar{X}, S\right)$, which describes the desired distribution of the SSM. Next, we use the correspondences X to register the US scan with the SDF voxel grid.

We proceed by learning a joint embedding space between the compounded US images and the SDF for each of the three tasks. During inference, an unseen US scan can be localized within the learned embedding space of the SDF by comparing patches in the embedding space of both networks. For inference, we use axis-aligned slices extracted from the compounded volume instead of raw B-mode US slices and make the simplifying assumption that slices have a certain thickness in the longitudinal (z-axis) direction. Figure 2 depicts the proposed method.

3.1   Encoder Training

Two 3D CNN encoders [9] are used to encode cube-shaped patches from each modality into a common embedding space. These are trained with a weighted soft-margin triplet loss [11] to ensure that geometrically corresponding patches are mapped to similar regions in the embedding space. Our triplets consist of anchor patches sampled from the US data and corresponding positive/negative patches sampled from the SDF grid. All patches are sampled near the thyroid surface. Anchor patches from the US data and positive/negative patches from the SDF are then fed into the two respective encoders. Given the embeddings e_0 of a US anchor patch and e_+, e_- of the corresponding positive/negative SDF

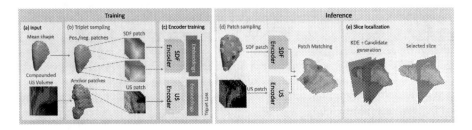

Fig. 2. Our framework for US localization within a PDM. The input comprises US data and a 3D shape model (a). Anchor, positive, and negative patches are sampled from US and SDF modalities (b), after which we learn a joint embedding across modalities with a triplet loss (c). During inference, sampled patches are encoded and matched based on their similarity in the embedding space (d). We employ Kernel Density Estimation (KDE) and an iterative refinement to localize the position of a candidate slice on the 3D shape model (e).

patches, respectively, as well as a hyperparameter α, the weighted soft-margin loss is

$$L(e_0, e_+, e_-) = \log(1 + \exp(\alpha(\|e_0 - e_+\| - \|e_0 - e_-\|))) \qquad (2)$$

To generate an approximately uniform distribution of samples across the organ's surface, we first use farthest point sampling [25] on the vertices of the scan-specific segmentation label mesh (whose coordinates agree with the US image space). A positive SDF patch is then defined as being in the same position as the US anchor patch: For the SSM samples, this requires transferring the coordinates to the semantically corresponding location in the SDF grid, which we achieve by using the learned point-to-point correspondences of S3M [4]. For each positive sample, a negative SDF patch is sampled uniformly from the SDF mesh vertices among a percentile (e.g., 50%) of vertices furthest away from the positive patch center.

3.2   Slice Localization

During inference, we proceed by regressing a US image slice based on similar patches across the two embedding spaces. We first sample patch centers from the two modalities to localize a US image in the SDF representation. SDF patches are sampled via farthest point sampling of the mesh vertices. We sample patches from the US slice near the thyroid surface using the available ground-truth segmentation labels. In practice, these could be obtained by a real-time segmentation network as in [22].

Next, patches are fed through the respective encoders to obtain patch embeddings, to which we apply the Hungarian matching algorithm using the Euclidean distance to establish cross-modality matches between similar patches in the embedding space. From these matches, we then estimate the slice location using the Procrustes algorithm [29].

To reduce the impact of false-positive matches, we run an iterative search, narrowing down the search space based on the matches of the previous iteration. To this end, we employ a kernel density estimation along the longitudinal axis of the thyroid, and several local maxima are taken as candidate locations. For each candidate location, the cross-modality matching algorithm is repeated using SDF patches only sampled around this neighborhood, and a slice candidate is generated via the Procrustes algorithm. The final prediction is the candidate slice with the lowest Procrustes loss.

In detail, the iterative algorithm works as follows: the coordinates of the SDF patches identified as matches are projected to the longitudinal axis (z axis in our coordinate system), and a kernel density estimation (KDE) that fits a Gaussian mixture density to the projected coordinates is computed. The m_{KDE} largest local maxima of the density are found and taken as candidate locations. For each candidate z-coordinate z_j, a restricted mesh is computed that only contains vertices with a z coordinate in the range $[z_j \pm w_{\mathrm{restr}}]$. This restricted mesh is used to sample SDF patches near the mesh surface for this candidate in the next iteration. The process is repeated until m_{step} restriction steps have been performed. The slice transformation with the lowest Procrustes loss is output as the algorithm's prediction. The hyperparameters m_{KDE}, w_{restr} and m_{step} as well as the number of sampled SDF and US patches, are tuned on a subset of the data.

In the above, we restrict ourselves to slices parallel to the axial image plane, a restriction imposed by the kernel density estimation algorithm we require for outlier elimination. However, this is the same direction in which the thyroid is typically scanned during diagnosis and should, therefore, not be limiting in practice.

4   Experiments

Thyroid Dataset. We evaluate our method on a publically available dataset of freehand US scans of healthy thyroids acquired from volunteers aged 24–39 years [22]. Each US sweep is compounded to a 3D resolution of $0.12 \times 0.12 \times 0.12$ mm. We use the right thyroid lobes of a subset of 16 patients for which ground truth segmentation maps are available. Although the US slices were labeled under the supervision of radiologists, the segmentation boundaries contain significant noise, making the correspondence and matching tasks challenging. Figure 1 (a) depicts various right thyroid lobes from the patient dataset. The thyroid dataset has high variance in the segmentation contours of a given sample and in the overall anatomical size and morphology between samples [22].

4.1   Multi-modal Registration

We evaluate the proposed multi-modal registration method by matching US slices to three types of 3D surface representations (c.f. Fig. 1): patient-specific

3D labels, the mean shape of an SSM, and shapes sampled from an SSM, as described in Sect. 3.

All experiments in this section are carried out with a 4-fold cross-validation over the 16 samples. The slice localization accuracy is also tested with two input voxel patch sizes at the 0.12 mm compounding resolution: (32, 32, 32) and (64, 64, 8). The first two dimensions correspond to the x-y axes of the axial plane, while the longitudinal axis (z coordinate) projects into the axial image plane. The longitudinal axis is the direction along which all US sweeps are acquired. The patches protrude 3.84 mm and 0.96 mm into the longitudinal axis, respectively.

The following experiments are conducted to evaluate the proposed method. For US image to patient-SDF registration, we learn to match US slices to the 3D SDF representation of the same US acquisition. We also ablate over two training strategies to train an encoder that enables matching from the US slices to the shape model. We train the 3D SDF encoder with patches from the mean shape or shapes sampled from the SSM, with $\alpha_j \sim \mathcal{N}(0, 0.5)$ according to Eq. 1. The latter strategy could be considered a form of data augmentation.

To evaluate the generalization capability of our model to the slice matching task, we generate 50 slices evenly spaced along the longitudinal axis (parallel to the axial image plane) for each validation thyroid lobe. The models trained to match US image patches to SSM samples are all evaluated for their slice prediction capabilities on the mean shape. We report the following metrics for the proposed matching method: the translational error and absolute rotational error between the predicted and ground truth slice and the percentage of predictions with translational error less than 10% and 15% of the longitudinal-axis length of the mean thyroid lobe. The mean shapes generated from the train set of each cross-validation fold have a length of 39.5 ± 0.655 mm.

5   Results and Discussion

Multi-modal Registration. Table 1 depicts the results of our multi-modal registration model. The model can accurately register US slices to the 3D SDF contours from the same patient anatomy for both input embedding patch sizes. The accuracy obtained for this registration is as low as 1.21 ± 0.08 mm translational error and $2.27°$ rotational error, with 96.47% of slices predicted within 10% of the thyroid lobe length for the (64, 64, 8) patch shape. These results demonstrate that it is possible to match US acquisition slices to a patient-specific topological thyroid representation with an accuracy sufficient for acquisition and surgical planning [8].

The best overall slice matching performance on registration to the mean shape is achieved with a patch shape of (32, 32, 32), yielding an average translational error of 4.60 mm and a rotational error of $2.39°$, with 75.59% of predicted slices located within 15% of the ground truth slice, and 51.16% within 10%.

Both of these slice-matching methods have noteworthy practical ramifications in a clinical setting. For example, if a nodule detected in a patient's thyroid is

Table 1. Results for the US slice registration to either an SDF representation of the patient anatomy or the mean shape from our SSM for two encoder patch sizes. We evaluate the translational (mm) and rotational error (in absolute degrees) of the predicted slice to the ground truth US slice and the percentage of slices within 10% and 15% distance along the z-axis of the mean shape in mm. For all experiments, the registration target is either the SDF of an individual patient (Patient SDF) or the mean shape of an SSM (Mean Shape). During training, we ablate over learning to match features to a Patient SDF, Mean Shape, or SSM samples with $\alpha \sim N(0, 0.5)$ (see Sect. 3).

Patch Dimension	Train Source/Reg. Target	Trans. error (mm)	Rot. error (°)	10% thresh	15% thresh
(32, 32, 32)	Patient SDF	1.82 ± 0.10	2.32 ± 0.15	90.97% ± 1.02	95.71% ± 0.77
	Mean Shape	4.60 ± 0.42	2.39 ± 0.25	51.16% ± 7.28	75.59% ± 4.27
	SSM/Mean Shape	5.08 ± 0.40	2.64 ± 0.48	42.38% ± 9.70	68.75% ± 6.13
(64, 64, 8)	SDF Patient	1.21 ± 0.08	2.27 ± 0.08	96.47% ± 0.46	98.75% ± 0.49
	Mean Shape	4.95 ± 0.41	3.90 ± 1.52	44.38% ± 10.33	71.50% ± 3.51
	SSM/Mean Shape	5.38 ± 0.54	4.12 ± 1.05	40.38% ± 9.04	68.38% ± 5.91

suspected to be malignant, the patient will be recommended for biopsy or follow-up screenings. Furthermore, our methodology generalizes to the more general mean shape representation, albeit with lower registration accuracy. Localizing US slices in this manner could enable acquisition planning with respect to an SSM without additional patient-specific acquisitions. SSMs can be easily shared and deployed in acquisition systems. They can be represented compactly and do not contain possibly identifying information. This could ease the accessibility of such a system for clinics that do not have the resources to curate large medical atlases.

5.1 Limitations and Future Work

To sample image patches corresponding to the SDF's surface, our method requires a segmentation contour of the thyroid lobe during training and inference. In future work, this could be mitigated during inference by learning to segregate features on the boundary of the organ from other parts of the image during training. Furthermore, our US slice to SSM matching considers only the mean shape and samples from the distribution during inference. Using samples from the SSM as data augmentation proves inferior to matching directly to the PDM mean shape. However, future works could explore learning to encode the entire SSM distribution for patch correspondence and general slice localization, as this could increase localization accuracy.

6   Conclusion

This work presents an automated method for US slice localization to aid in surgical and acquisition planning. By formulating the localization problem as a 2D-to-3D registration to a 3D SDF, the proposed method localizes 2D US slices within two different geometric representations of the patient's anatomy. We demonstrate that our unsupervised correspondence method is robust to the heterogenous and noisy thyroid topology across a set of individuals. Furthermore, we propose a pipeline that enables registration of a US slice to not only the surface of the patient anatomy but also a more general statistical representation across a population. Consistent localization of 2D US slices without a previous acquisition could enable several applications, including improved automated robotic scanning, sonographic acquisition planning, or guidance for hands-on US or anatomical training. Perhaps a glimpse into the complex thyroid anatomy in the form of a single US image can yield more insight than previously realized. We are confident this work will advance research in automated thyroid scanning and diagnosis, which has the potential to improve the quality of life of millions suffering from thyroid disorders worldwide.

Acknowledgements and Disclosure. The thyroid dataset used for all experiments is publicly available. Vincent Bürgin is supported by the DAAD program Konrad Zuse Schools of Excellence in Artificial Intelligence, sponsored by the Federal Ministry of Education and Research. The authors declare no conflicts of interest.

References

1. Adams, J., Khan, N., Morris, A., Elhabian, S.: Learning spatiotemporal statistical shape models for non-linear dynamic anatomies. Front. Bioeng. Biotechnol. **11**, 1086234 (2023)
2. Azizi, G., Faust, K., Ogden, L., Been, L., Mayo, M.L., Piper, K., Malchoff, C.: 3-D ultrasound and thyroid cancer diagnosis: a prospective study. Ultrasound Med. Biol. **47**(5) (2021)
3. Banerjee, A., Zacur, E., Choudhury, R.P., Grau, V.: Optimised misalignment correction from cine MR slices using statistical shape model. In: Papież, B.W., Yaqub, M., Jiao, J., Namburete, A.I.L., Noble, J.A. (eds.) MIUA 2021. LNCS, vol. 12722, pp. 201–209. Springer, Cham (2021). https://doi.org/10.1007/978-3-030-80432-9_16
4. Bastian, L., et al.: S3M: scalable statistical shape modeling through unsupervised correspondences. arXiv preprint arXiv:2304.07515 (2023)
5. Berendsen, F.F., Van Der Heide, U.A., Langerak, T.R., Kotte, A.N., Pluim, J.P.: Free-form image registration regularized by a statistical shape model: application to organ segmentation in cervical MR. Comput. Vis. Image Underst. **117**(9), 1119–1127 (2013)
6. Chai, H.H., et al.: Successful use of a 5G-based robot-assisted remote ultrasound system in a care center for disabled patients in rural China. Front. Publ. Health **10**, 915071 (2022)

7. Chan, C.S., Edwards, P.J., Hawkes, D.J.: Integration of ultrasound-based registration with statistical shape models for computer-assisted orthopedic surgery. In: Medical Imaging 2003: Image Processing, vol. 5032, pp. 414–424. SPIE (2003)
8. Cheng, A., Lee, J.W.K., Ngiam, K.Y.: Use of 3D ultrasound to characterise temporal changes in thyroid nodules: an in vitro study. J. Ultrasound (2022)
9. Çiçek, Ö., Abdulkadir, A., Lienkamp, S.S., Brox, T., Ronneberger, O.: 3D U-net: learning dense volumetric segmentation from sparse annotation. In: Ourselin, S., Joskowicz, L., Sabuncu, M.R., Unal, G., Wells, W. (eds.) MICCAI 2016. LNCS, vol. 9901, pp. 424–432. Springer, Cham (2016). https://doi.org/10.1007/978-3-319-46723-8_49
10. Ellingsen, L.M., Chintalapani, G., Taylor, R.H., Prince, J.L.: Robust deformable image registration using prior shape information for atlas to patient registration, **34**(1) (2010)
11. Feng, M., Hu, S., Ang, M.H., Lee, G.H.: 2D3D-matchnet: learning to match keypoints across 2D image and 3D point cloud. In: ICRA 2019 (2019)
12. Ferrante, E., Paragios, N.: Slice-to-volume medical image registration: a survey. Med. Image Anal. **39**, 101–123 (2017)
13. Fu, Y., Lei, Y., Wang, T., Curran, W.J., Liu, T., Yang, X.: Deep learning in medical image registration: a review. Phys. Med. Biol. **65**(20), 20TR01 (2020)
14. Ghanavati, S., Mousavi, P., Fichtinger, G., Abolmaesumi, P.: Phantom validation for ultrasound to statistical shape model registration of human pelvis. In: Medical Imaging 2011: Visualization, Image-Guided Procedures, and Modeling, vol. 7964, pp. 855–862. SPIE (2011)
15. Ghanavati, S., Mousavi, P., Fichtinger, G., Foroughi, P., Abolmaesumi, P.: Multislice to volume registration of ultrasound data to a statistical atlas of human pelvis. In: Medical Imaging 2010: Visualization, Image-Guided Procedures, and Modeling, vol. 7625, pp. 213–222. SPIE (2010)
16. Grassi, L., Väänänen, S.P., Isaksson, H.: Statistical shape and appearance models: development towards improved osteoporosis care. Curr. Osteoporos. Rep. **19**, 676–687 (2021)
17. Guerreiro, F., et al.: Evaluation of a multi-atlas CT synthesis approach for MRI-only radiotherapy treatment planning. Physica Medica (2017)
18. Haugen, B.R., Alexander, E.K., Bible, K.C., Doherty, G.M., Mandel, S.J., et al.: 2015 American thyroid association management guidelines for adult patients with thyroid nodules and differentiated thyroid cancer: the American thyroid association guidelines task force on thyroid nodules and differentiated thyroid cancer. Thyroid **26**(1) (2016)
19. Heimann, T., Meinzer, H.P.: Statistical shape models for 3D medical image segmentation: a review, **13**(4), 543–563 (2009)
20. Hennersperger, C., et al.: Towards MRI-based autonomous robotic US acquisitions: a first feasibility study, **36**(2), 538–548 (2017)
21. Hu, X., Chen, X., Liu, Y., Chen, E.Z., Chen, T., Sun, S.: Deep statistic shape model for myocardium segmentation. arXiv preprint arXiv:2207.10607 (2022)
22. Krönke, M., Eilers, C., Dimova, D., Köhler, M., Buschner, G., et al.: Tracked 3D ultrasound and deep neural network-based thyroid segmentation reduce interobserver variability in thyroid volumetry. PLoS ONE **17**(7) (2022)
23. Lüdke, D., Amiranashvili, T., Ambellan, F., Ezhov, I., Menze, B.H., Zachow, S.: Landmark-free statistical shape modeling via neural flow deformations. In: Wang, L., Dou, Q., Fletcher, P.T., Speidel, S., Li, S. (eds.) MICCAI 2022. LNCS, vol. 13432, pp. 453–463. Springer, Cham (2022). https://doi.org/10.1007/978-3-031-16434-7_44

24. Markova, V., Ronchetti, M., Wein, W., Zettinig, O., Prevost, R.: Global multimodal 2D/3D registration via local descriptors learning. In: Wang, L., Dou, Q., Fletcher, P.T., Speidel, S., Li, S. (eds.) MICCAI 2022. LNCS, vol. 13436, pp. 269–279. Springer, Cham (2022). https://doi.org/10.1007/978-3-031-16446-0_26

25. Moenning, C., Dodgson, N.A.: Fast marching farthest point sampling. Eurographics 2003 - Posters (2003)

26. Naceri, A., et al.: Tactile robotic telemedicine for safe remote diagnostics in times of corona: system design, feasibility and usability study. IEEE Robot. Autom. Lett. **7**(4), 10296–10303 (2022)

27. Raju, A., Miao, S., Jin, D., Lu, L., Huang, J., Harrison, A.P.: Deep implicit statistical shape models for 3D medical image delineation. In: Proceedings of the AAAI Conference on Artificial Intelligence, vol. 36, pp. 2135–2143 (2022)

28. Samei, G., Karimi, D., Kesch, C., Salcudean, S.: Automatic segmentation of the prostate on 3D trans-rectal ultrasound images using statistical shape models and convolutional neural networks. arXiv preprint arXiv:2106.09662 (2021)

29. Schönemann, P.H.: A generalized solution of the orthogonal procrustes problem. Psychometrika **31**(1), 1–10 (1966)

30. Song, X., et al.: Cross-modal attention for MRI and ultrasound volume registration. In: de Bruijne, M., et al. (eds.) MICCAI 2021. LNCS, vol. 12904, pp. 66–75. Springer, Cham (2021). https://doi.org/10.1007/978-3-030-87202-1_7

31. Tang, Z., Chen, K., Pan, M., Wang, M., Song, Z.: An augmentation strategy for medical image processing based on statistical shape model and 3D thin plate spline for deep learning. IEEE Access **7**, 133111–133121 (2019)

32. Uzunova, H., Wilms, M., Forkert, N.D., Handels, H., Ehrhardt, J.: A systematic comparison of generative models for medical images. Int. J. Comput. Assist. Radiol. Surg. **17**(7), 1213–1224 (2022)

33. Yeung, P.H., Aliasi, M., Haak, M., Xie, W., Namburete, A.I.: Adaptive 3D localization of 2D freehand ultrasound brain images. In: Wang, L., Dou, Q., Fletcher, P.T., Speidel, S., Li, S. (eds.) MICCAI 2022. LNCS, vol. 13434, pp. 207–217. Springer, Cham (2022). https://doi.org/10.1007/978-3-031-16440-8_20

34. Zhang, Y.Q., Yin, H.H., He, T., Guo, L.H., Zhao, C.K., Xu, H.X.: Clinical application of a 5G-based telerobotic ultrasound system for thyroid examination on a rural island: a prospective study. Endocrine **76**(3), 620–634 (2022)

FSJP-Net: Foreground and Shape Joint Perception Network for Glomerulus Detection

Qiuchi Han[1], Xiuxiu Hu[2], Pingsheng Chen[2], and Siyu Xia[1]([✉]) [iD]

[1] School of Automation, Southeast University, Nanjing 210096, China
xsy@seu.edu.cn
[2] Department of Pathology, School of Medicine, Southeast University, Nanjing 210096, China

Abstract. Morphological changes in the glomerulus play a vital role in the diagnosis of kidney diseases. However, the detection of the glomerulus in the actual medical situation is challenging due to various factors such as lesions, tissue changes, and staining. These factors raise problems like high foreground-background similarity, blurred contours, and irregular shapes, thus pose difficulties for both physicians and Automatic computer detection. To address these challenges, we propose a foreground-aware feature extraction method, which is used to fully extract foreground information. Furthermore, we design the Foreground and Shape Joint Perception Network (FSJP-Net), a detection network that integrates object foreground information and shapes information, which improves the recall and precision of glomerular detection by fusing the extracted foreground and elliptical shape information from different feature extraction branches. The experiments demonstrate the effectiveness and superiority of our proposed method in detecting various categories of glomeruli.

Keywords: Glomerulus detection · Feature fusion · Foreground perception · Ellipse

1 Introduction

Observation of glomeruli is a fundamental basis for pathological diagnosis of kidney disease. The morphology, size, structure of the glomerulus exhibit distinctive characteristics among different types of kidney diseases [1]. Particularly, The elliptical shape of the glomerulus, a prominent characteristic of the glomerulus, can be altered due to glomerular lesions [2]. For instance, the alterations in the glomerular flatness or the long-to-short axis ratio can indicate glomerular injury or inflammation, whereas diabetic nephropathy may exhibit glomerular deformation or flattening [3]. Therefore, analyzing the elliptical shape of the glomerulus not only enables physicians to discern the type and extent of the lesion, but also facilitate precise diagnosis. Thus, it is a meaningful task to accurately describe the elliptical characteristics of the glomerulus.

© The Author(s), under exclusive license to Springer Nature Switzerland AG 2023
C. Wachinger et al. (Eds.): ShapeMI 2023, LNCS 14350, pp. 145–156, 2023.
https://doi.org/10.1007/978-3-031-46914-5_12

With a considerable number of glomeruli that approximately 1 million to 1.5 million in the adult kidney [4], the number of observable glomeruli in a single kidney section sample even can be a few hundred. Traditional glomerulus observation is based on a manual search and a microscope, which is time-consuming and prone to errors [5]. The use of computer-assisted physicians for observation and diagnosis can greatly improve the efficiency of testing, enable accurate diagnosis, and support treatment decisions. With the utilization of whole slide scanners can form kidney section sample slides into high-resolution images of kidney section samples, facilitating computer-based observations. However, efficient automated glomerular detection methods are still required to effectively observate and precisely analyse the glomerulus.

In recent works, morphological processing methods have been used for glomerular detection. Kato et al. [6] proposed an improved HOG descriptor that enables comprehensive detection of numerous glomeruli in whole kidney section images. Simon et al. [7] employed the Local Binary Patterns (LBP) to extract informative glomerular features, which were subsequently classified using Support Vector Machines (SVM). Ginley et al. [8] utilized Gabor filters to detect the boundaries of the glomerulus. But these methods still have limitations of over-reliance on graphic and pixel features of the images.

With the rapid development of deep learning and computer vision, research efforts of glomerular pathology image have predominantly focused on the following areas. Refined contour segmentation of the glomerulus using segmentation networks, such as UNet [9], AlexNet [10]. The detection and localization of glomerulus using anchor-based object detection networks, including CNN [11], Faster-Rcnn [12], and Yolo [13]. The classification works of glomerulus using methods such as ResNet [14], and GoogleNet [15]. Most of these existing methods are typically pre-trained on datasets consisting of non-medical images, so they cannot fully account for the unique characteristics of medical images and the specific detection environment of the glomerulus.

Yang et al. [16] have noted the contour shape of the glomerulus and tried to detect glomeruli with a bonding circle. However, it should be noted that the majority of glomerular contours in Whole-slide Images (WSIs) exhibit an elliptical shape rather than a standard circle. Therefore, the simplistic use of standard circles to describe glomerular contour characteristics is clearly inadequate in capturing the diverse nature of glomerular shapes. Furthermore, the current study did not fully address the situation of a number of glomeruli with complex morphologies, which are usually caused by lesions, biological tissue changes, and staining variations. These complexities manifest as blurred edges, irregular shapes, and significant similarities between foreground and background in actual medical observations. Consequently, the automatic detection of glomerulus faces significant challenges in accurately identifying such intricate structures.

To address the aforementioned questions, we propose a detection model that fuses the foreground information and the elliptical contours information, enabling the detection of both normal and complex morphology glomeruli. For the issue of unclear glomerular contours and irregular shapes, a multiscale foreground-aware module is optimally designed to leverage foreground cue

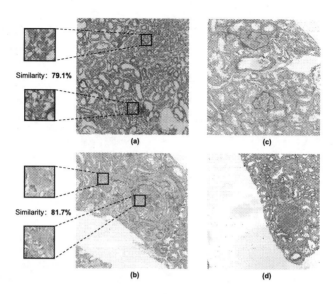

Fig. 1. Major challenges in glomerular detection. (a)(b) high similarity of foreground and background. (c) irregular shapes. (d) blurred edges.

vectors for enhance the discrimination between foreground and background in the feature map, thereby improving the accuracy of glomerular detection. Our contributions can be summarized as follows:

1. We design a foreground-aware feature extraction module for complex morphological glomerulus.
2. We propose FSJP-Net, a novel lightweight anchor-free detection network that fuses object foreground information and contour shape information, effectively integrating the advantages of different feature extraction branches to improve the recall and precision of complex glomerular detection. Experimental results demonstrate its superiority over many commonly used glomerular detection methods.
3. We propose to use ellipse as bounding ellipse for deep learning to automatically detect and describe glomerulus, which reduces the arithmetic power waste caused by accurate edge segmentation, but maximizes the restoration of glomerular structural information and provides a good foundation for subsequent quantitative pathological analysis.

2   Method

2.1   Overview of the Framework

The overview of our proposed FSJP-Net is shown in Fig. 2, which comprises two distinct feature extraction branches, namely the foreground perception branch

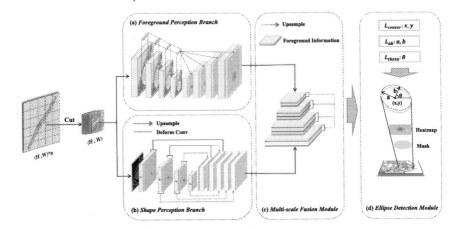

Fig. 2. Overview of FSJP-Net. (a) Foreground perception branch for extracting foreground information. (b) Shape perception branch for extracting shape information. (c) Multiscale fusion module for a fusion of different branch features. (d) Ellipse detection module.

and the shape perception branch. Additionally, we incorporate a multiscale feature fusion module and an ellipse detection module. The pre-processed input image is extracted from complementary feature information through the foreground perception branch and shape perception branch. The obtained feature maps from both branches are then through the multiscale fusion module to produce a novel feature map encompassing both foreground and shape information. All modules, including the ellipse detection module, are jointly trained in an end-to-end manner utilizing the anchor-free network [17].

2.2   Foreground Perception Branch

As shown in Fig. 1(c), it is observed that the contours of certain glomeruli exhibit distortions that cannot be accurately approximated as ellipses. Moreover, Fig. 1(d) depicts glomerulus with highly blurred contours, rendering them devoid of any useful shape information. In order to tackle these challenges, relying on a simple feature extraction method for glomerular images is evidently insufficient. Recognizing the presence of additional distinctive image features within the glomerular contours, we propose a novel method that extracts the foreground information of the glomerulus at multiple scales within the foreground perception branch, thus generating foreground cue vectors. In particular, when faced with inadequate glomerular contour features, the foreground cue vectors are weighted and embedded to enhance the final feature map with foreground cues. This empowers the network to pay greater attention to potential foreground regions, consequently improving the recall of glomerular detection.

As shown in Fig. 3, in order to generate a foreground-aware feature map that contains foreground information. Following data pre-processing, an input image

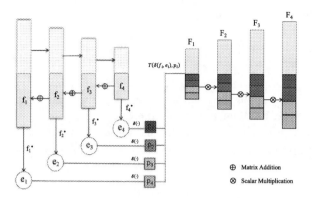

Fig. 3. Details of foreground perception branch.

of size $H \times W$ is first encoded by a structure based on a feature pyramid network (FPN) [18]. The encoding structure performs downsampling with sampling multiples of $\times 4$, $\times 8$, $\times 16$, and $\times 32$, resulting in a 4-layer pyramidal feature map. To incorporate higher-level context, a top-down upsampling process is applied, and the encoded feature map f_i is obtained by merging the upper-level feature map with the corresponding lateral connection. To introduce learnable matrix parameters, a 1x1 convolution is applied to f_i, as represented by Eq. 1.

$$f_i^* = \theta(f_i), i = 1, 2, 3, 4 \tag{1}$$

where $\theta(\cdot)$ is an operation that includes 1×1 convolution, normalization, and ReLU function. $f^* \in R^{d \times H \times W}$ denoted the learnable vector matrix of f_i. A one-dimensional foreground information vector e_i is generated from the foreground content through two 1×1 convolution operations. Both f_i and e_i encompass relevant information, with f_i containing background information regarding non-glomerulus regions, and e_i capturing the foreground information associated with glomerulus.

The similarity of two vectors can be described as the magnitude of the scalar value of the dot product, where a higher scalar value indicates a stronger resemblance or parallel trends. The calculation of the dot product of f_i^* and e_i not only generates a learnable similarity matrix but also weights the foreground information according to the feature that the vector similarity increases with the scalar value of the dot product, as shown in Eq. 2.

$$P_i = \delta(f_i^*, e_i) = \sum_n^H \sum_m^W e_i \cdot f_i^*(n, m), i = 1, 2, 3, 4 \tag{2}$$

where $\delta(\cdot)$ is the calculation of the dot product. Then e_i is dotted with the vectors in the $f_i^* \in R^{H \times W}$ matrix respectively. This process yields the foreground information weighted feature map P_i, which encapsulates relevant foreground information, as shown in Eq. 3.

$$F_i^{(n)} = F_i^{(n-1)} + T\left(\delta\left(f_i^*, e_i\right), P_i\right) = F_i^{(n-1)} + \frac{k\left(f_i\right)}{1 + e^{(-P_i)}}, i = 1, 2, 3, 4 \qquad (3)$$

where $T\left(\cdot\right)$ is the foreground-weighted embedding formula containing the sigmoid function. $k\left(\cdot\right)$ is the encoder with f_i as a parameter [19]. The sigmoid function introduces non-linear units to increase the feature representation of embedded information. The final pyramidal feature map, denoted F_i, contains valuable foreground information and is generated recursively by summing $F_i^{(n-1)}$ and $T\left(\cdot\right)$.

2.3   Shape Perception Branch

The shape perception branch is responsible for extracting image features that encompass contour information. To achieve effective key point feature extraction, we leverage the deep layer aggregation (DLA) method, which has demonstrated promising results [20]. By employing the DLA structure, we conduct feature extraction at various resolutions and perceptual field sizes. Furthermore, a hierarchical feature fusion strategy is applied to aggregate features from different levels for generates 13 distinct levels of feature maps denoted as $\{S_{ij} \mid i, j = 1, 2, 3, 4\}$, as shown in Fig 4.

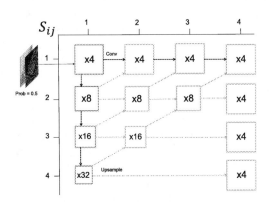

Fig. 4. Details of shape perception branch.

Contour information can be described as a fit to the edges of the image. Edge information extracted from the input image by different edge detection algorithms is summed up with different weights [21], which can superimpose the advantages of different edge extraction algorithms. The edge information are fused into the training process with a certain probability so that the features extracted by the shape perception branch can contain contour information to some extent.

2.4   Multiscale Fusion Module

As shown in Fig. 2(c), the multiscale fusion module performs a deep multiscale fusion of the output results from two feature extraction branches. Foreground information feature maps F_i and shape information feature maps S_{ij} are summed element by element in full dimensionality, thereby the final output feature map Map_{fusion} contains sufficient learnable information.

Algorithm 1. Multiscale Fusion Method

Input: Foreground information features F_i, Shape information features S_{ij};
Output: Multiscale fusion feature Map_{fusion};
1: **for** $i \in [1, 4]$ **do**
2:     $Map_{fusion\ i} = F_i + S_{i,5-i}$;
3: **end for**
4: $Map_{fusion} = \sum_i^4 Map_{fusion\ i}$
5: **return** Map_{fusion}.

2.5   Ellipse Detection Module

The elliptical shape allows the maximum depiction of the glomerular structure while saving arithmetic power. In regressing the bounding ellipse of the glomerulus, the focal loss function and the smooth-L1 loss function were respectively used as L_{center} and $L_{a,b}$ to predict the center (x, y) and the long and short axes a, b of the ellipse.

As shown in Fig. 2(d). Heat maps are introduced to represent the probability of center locations, while heat map loss is used to further refine the center locations. Meanwhile, increasing the learning of the elliptical object mask map helps to improve the regression accuracy of the rotation angle. The mask graph consists of only pixel values 0 and 1, so BCE with Logits Loss, which has high applicability and convergence speed for the bifurcated zero tasks, is chosen as the mask loss, as shown in Eq. 4.

$$L_{mask} = \frac{1}{n} \sum_1^n \left(M_g \cdot log\left(sigmoid\left(M_p\right) + \left(1 - M_g\right) \cdot log\left(1 - sigmoid\left(M_p\right)\right)\right)\right) \tag{4}$$

where M_g represents the label mask, while M_p denotes the predicted value. In addition, the object is more similar to the standard circle while the aspect ratio becomes closer to 1. Consequently, reducing the learning weight assigned to the rotation angle can enhance the detection efficiency to some extent for balancing the relationship between rotation angle and aspect ratio $\frac{a}{b}$. The loss of rotation angle L_{theta} is shown in Eq. 5.

$$L_{theta} = w_t \cdot smooth_{L1}\left(T_p, T_g\right) \tag{5}$$

where w_t is the adaptive weight of the loss of rotation angle, T_p is the predicted value, while T_g is the label value. A Gaussian intersection over union (IoU) loss is also added during training to improve the model performance.

3   Experiments

3.1   Dataset and Implementation Details

The dataset was made with 100 complete WSIs stained with periodic acid-Schiff (PAS) and periodic acid-Silver Methenamine (PASM). The WSIs were first processed using openslide, where we applied a sliding window approach with a step size of 600 and a size of 1200 × 1200. A total of 1690 valid dataset images were selected, ensuring that each image contains at least one glomeruli, of which 250 contained glomeruli with diseased or complex morphology. 1300 images were randomly chosen as the training set while 390 images as the test set. The long and short axes, center coordinates, and deflection angle of the ellipse were used as parameters for the labeling of the data set.

To mitigate the impact of diverse coloring techniques on the color appearance of sample images, we adopt a coloring normalization procedure on both dataset and subsequent test images. Initially, we calculated the color mean and variance of the stained sample images. The target images undergo a grayscale transformation followed by a color reconstitution process, employing the statistics derived from the color mean and variance of stained samples.

Our study is based on the Pytorch framework, using the NVIDIA GeForce RTX 3090 to facilitate training. The dataset adhered to the coco format and underwent 100 epochs of training with the Adam optimizer. The batch size was set to 8 with a learning rate of 1.25×10^{-4}. We set different training weights for accurately estimate the elliptical parameters of the glomerulus. Specifically, the center coordinates, mask, long and short axes, deflection angle, and Gaussian IoU were assigned training weights of 5, 5, 0.1, 0.1, and 15.

3.2   Comparison Experiments

To demonstrate the effectiveness of our proposed FSJP-Net in glomerular detection tasks, we selected several models for comparison. Including YoLov5 [22], which has demonstrated state-of-the-art performance in various object detection tasks, CircleNet [16], which focuses specifically on glomerular shape detection, as well as ElDet [21] and Ellipesenet [20], both specialized in ellipse detection. The evaluation metrics were recall, precision, and mAP with a threshold of 50 to represent as accurately as possible the ability of the network to detect glomeruli.

As shown in Table 1, our method exhibits notable advantages in comparison to conventional object detection methods and glomerulus-specific detection methods, by using the same dataset. Furthermore, as shown in Fig. 5, our method consistently demonstrates superior results in the practical evaluation of glomerular images, including both normal glomeruli and those with intricate morphological features.

Table 1. Results of the comparison experiment.

Methods	Recall	Precision	mAP50
CircleNet	87.75	83.46	86.00
EllipseNet	86.39	84.25	79.76
ElDet	91.42	96.70	90.33
YoLov5	89.06	96.81	92.19
FSJP-Net(Ours)	**94.41**	**97.65**	**94.27**

3.3 Ablation Experiments

To evaluate the effectiveness of the modules in the network, we conducted ablation experiments on the foreground-aware branch, the shape-aware branch, and the multiscale fusion module in FSJP-Net.

Table 2. Ablation experiments of the feature extraction branch.

Methods	Recall	Precision	mAP50
Foreground-only	93.89	91.07	92.79
Shape-only	89.32	97.22	89.31
Joint-Perception	94.41	97.65	94.27

As shown in Table 2, the results are presented for different feature extraction methods. Foreground-only means separately utilized foreground perception branches, while Shape-only means using shape perception branches solely. Additionally, Joint-Perception represents the outcomes of the feature fusion method. Foreground-only exhibits notably higher recall compared to Shape-only with its precision considerably lower. Each branch possesses distinct advantages in that the foreground perception branch fully captures foreground information to enhance recall, while the shape perception branch employs shape information as a constraint to improve accuracy. Consequently, the Joint-Perception results demonstrate a simultaneous enhancement in both recall and precision.

Table 3. Ablation experiments of the feature fusion method.

Methods	Recall	Precision	mAP50
End-add	88.73	89.56	87.24
Multiscale-fusion	94.41	97.65	94.27

Furthermore, we conducted ablation experiments to evaluate the different feature fusion methods. As shown in Table 3, End-add refers to the simple addition of feature maps from both branches at the last layer, while Multiscale-fusion

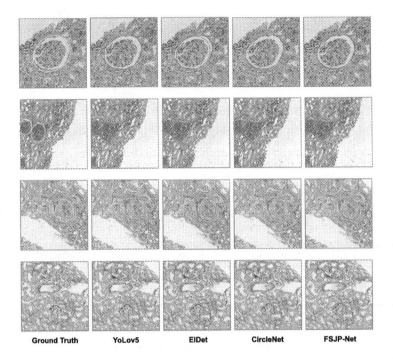

Ground Truth   YoLov5   ElDet   CircleNet   FSJP-Net

Fig. 5. Visualization results of FSJP-Net. The first column represents Ground Truth, while the subsequent three columns correspond to the outcomes obtained from YoLov5, ElDet, and CircleNet. All three methods exhibit missed detection and imprecise object localization. The fifth column displays the outcome of FSJP-Net, which shows superior performance.

denotes the multiscale feature fusion approach. Remarkably, the multiscale fusion approach yields significantly better results compared to the simple feature map summation at the last layer.

3.4   Morphological Analysis

Given the significance of assessing lesions through variations in glomerular volume and internal tissue area, an accurate description of the long and short axes and area of the glomerulus is imperative. Precise edge segmentation is computationally intensive and lacks the ability to quickly capture the long and short axes. Therefore, use a more efficient bounding representation method for detection result to depict the glomerular structure is necessary. Existing bounding representation method of glomerulus detection predominantly comprise rectangular or circular shapes, which inadequately accommodate the majority of elliptical glomeruli. We propose to use bounding ellipse instead of bounding box and bounding circle. To quantify the suitability of the object bounding representation methods in relation to the glomerular structure, we introduce the ratio R_{area}, which compares the glomerular foreground content to the object bounding box

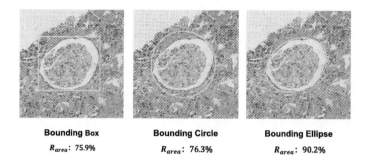

Bounding Box   **Bounding Circle**   **Bounding Ellipse**
R_{area}: 75.9% R_{area}: 76.3% R_{area}: 90.2%

Fig. 6. Fitting of different object frame shapes to glomerular structures.

area. As depicted in Fig. 6, the bounding ellipse exhibits a closer fit to the outer contour of the glomerulus, thus providing a more precise representation of the glomerular structure.

4 Conclusion

In this paper, we propose FSJP-Net, a lightweight anchor-free detection network that combines object foreground perception and shape analysis for glomerular detection of various morphologies, including healthy, diseased and structurally deformed glomeruli. The foreground perception feature extraction module is designed for address the issue of high similarity between glomerular foreground and background. Comparative experiments demonstrate the significant superiority of our proposed network that compared to current mainstream glomerular detection methods. Ablation experiments are conducted to validate the effectiveness of individual modules within our network, indicating the potential for broader applications in other scenarios. Finally, we calculate the ratio between the area of the foreground content and the area of the object bounding box to evaluate the effectiveness of the bounding ellipse in describing the structure of glomeruli.

References

1. Haas, M., et al.: A multicenter study of the predictive value of crescents in IGA nephropathy. J. Am. Soc. Nephrol. **28**(2), 691–701 (2017)
2. Farris, A.B., et al.: Morphometric and visual evaluation of fibrosis in renal biopsies. J. Am. Soc. Nephrol. **22**(1), 176–186 (2011)
3. D'Agati, V.D., Kaskel, F.J., Falk, R.J.: Focal segmental glomerulosclerosis. N. Engl. J. Med. **365**(25), 2398–2411 (2011)
4. Nyengaard, J., Bendtsen, T.: Glomerular number and size in relation to age, kidney weight, and body surface in normal man. Anat. Rec. **232**(2), 194–201 (1992)
5. Puelles, V.G., Hoy, W.E., Hughson, M.D., Diouf, B., Douglas-Denton, R.N., Bertram, J.F.: Glomerular number and size variability and risk for kidney disease. Curr. Opin. Nephrol. Hypertens. **20**(1), 7–15 (2011)

6. Kato, T., et al.: Segmental hog: new descriptor for glomerulus detection in kidney microscopy image. BMC Bioinform. **16**, 1–16 (2015)

7. Simon, O., Yacoub, R., Jain, S., Tomaszewski, J.E., Sarder, P.: Multi-radial LBP features as a tool for rapid glomerular detection and assessment in whole slide histopathology images. Sci. Rep. **8**(1), 2032 (2018)

8. Ginley, B., Tomaszewski, J.E., Yacoub, R., Chen, F., Sarder, P.: Unsupervised labeling of glomerular boundaries using Gabor filters and statistical testing in renal histology. J. Med. Imaging **4**(2), 021102–021102 (2017)

9. Zhang, Y., et al.: U-net-and-a-half: convolutional network for biomedical image segmentation using multiple expert-driven annotations. arXiv preprint arXiv:2108.04658 (2021)

10. Gallego, J., et al.: Glomerulus classification and detection based on convolutional neural networks. J. Imaging **4**(1), 20 (2018)

11. Wilbur, D.C., Smith, M.L., Cornell, L.D., Andryushkin, A., Pettus, J.R.: Automated identification of glomeruli and synchronised review of special stains in renal biopsies by machine learning and slide registration: a cross-institutional study. Histopathology **79**(4), 499–508 (2021)

12. Kawazoe, Y., et al.: Faster R-CNN-based glomerular detection in multistained human whole slide images. J. Imaging **4**(7), 91 (2018)

13. Heckenauer, R., et al.: Real-time detection of glomeruli in renal pathology. In: 2020 IEEE 33rd International Symposium on Computer-Based Medical Systems (CBMS), pp. 350–355. IEEE (2020)

14. Weis, C.A., et al.: Assessment of glomerular morphological patterns by deep learning algorithms. J. Nephrol. **35**(2), 417–427 (2022)

15. Yao, X., Wang, X., Karaca, Y., Xie, J., Wang, S.: Glomerulus classification via an improved Googlenet. IEEE Access **8**, 176916–176923 (2020)

16. Yang, H., et al.: Circlenet: anchor-free detection with circle representation. arXiv preprint arXiv:2006.02474 (2020)

17. Zhou, X., Wang, D., Krähenbühl, P.: Objects as points. arxiv 2019. arXiv preprint arXiv:1904.07850 (2019)

18. Lin, T.Y., Dollár, P., Girshick, R., He, K., Hariharan, B., Belongie, S.: Feature pyramid networks for object detection. In: Proceedings of the IEEE Conference on Computer Vision and Pattern Recognition, pp. 2117–2125 (2017)

19. Zheng, Z., Zhong, Y., Wang, J., Ma, A.: Foreground-aware relation network for geospatial object segmentation in high spatial resolution remote sensing imagery. In: Proceedings of the IEEE/CVF Conference on Computer Vision and Pattern Recognition, pp. 4096–4105 (2020)

20. Chen, J., Zhang, Y., Wang, J., Zhou, X., He, Y., Zhang, T.: EllipseNet: anchor-free ellipse detection for automatic cardiac biometrics in fetal echocardiography. In: de Bruijne, M., et al. (eds.) MICCAI 2021. LNCS, vol. 12907, pp. 218–227. Springer, Cham (2021). https://doi.org/10.1007/978-3-030-87234-2_21

21. Wang, T., Lu, C., Shao, M., Yuan, X., Xia, S.: Eldet: an anchor-free general ellipse object detector. In: Proceedings of the Asian Conference on Computer Vision, pp. 2580–2595 (2022)

22. Zhu, X., Lyu, S., Wang, X., Zhao, Q.: Tph-yolov5: improved yolov5 based on transformer prediction head for object detection on drone-captured scenarios. In: Proceedings of the IEEE/CVF International Conference on Computer Vision, pp. 2778–2788 (2021)

Progressive DeepSSM: Training Methodology for Image-To-Shape Deep Models

Abu Zahid Bin Aziz[1,2(✉)], Jadie Adams[1,2], and Shireen Elhabian[1,2]

[1] Scientific Computing and Imaging Institute, University of Utah,
Salt Lake City, UT, USA
[2] Kahlert School of Computing, University of Utah, Salt Lake City, UT, USA
{zahid.aziz,jadie,shireen}@sci.utah.edu

Abstract. Statistical shape modeling (SSM) is an enabling quantitative tool to study anatomical shapes in various medical applications. However, directly using 3D images in these applications still has a long way to go. Recent deep learning methods have paved the way for reducing the substantial preprocessing steps to construct SSMs directly from unsegmented images. Nevertheless, the performance of these models is not up to the mark. Inspired by multiscale/multiresolution learning, we propose a new training strategy, progressive DeepSSM, to train image-to-shape deep learning models. The training is performed in multiple scales, and each scale utilizes the output from the previous scale. This strategy enables the model to learn coarse shape features in the first scales and gradually learn detailed fine shape features in the later scales. We leverage shape priors via segmentation-guided multi-task learning and employ deep supervision loss to ensure learning at each scale. Experiments show the superiority of models trained by the proposed strategy from both quantitative and qualitative perspectives. This training methodology can be employed to improve the stability and accuracy of any deep learning method for inferring statistical representations of anatomies from medical images and can be adopted by existing deep learning methods to improve model accuracy and training stability.

Keywords: Statistical Shape Modeling · Progressive Learning · Medical Imaging · Deep Supervision

1 Introduction

Statistical shape modeling (SSM) has become vital for quantitative studies of biological and medical data by providing a statistically consistent geometrical description for each shape across a given population. Recent progress in this field has enabled a wide range of clinical and scientific SSM applications, such as bone reconstruction in orthopedics from 2D or 3D medical images [13,18], atrial fibrillation in cardiology [7,15], brain ventricle analysis in neuroscience [8,16,29].

C. Wachinger et al. (Eds.): ShapeMI 2023, LNCS 14350, pp. 157–172, 2023.
https://doi.org/10.1007/978-3-031-46914-5_13

Several shape representations have been introduced and utilized. Among them, deformation-based and correspondence-based models are the most popular [4,10]. While deformation fields can represent shapes directly from images, in this work, we have opted for correspondence-based shape representation as it does not require a reference/atlas. Nevertheless, the proposed training strategy can also be adapted to deformation-based shape representation. Correspondence-based models (also known as point distribution models or PDMs) utilize an ordered set of landmarks or correspondence points placed on the shape surface in a consistent manner across the population. Several algorithms are available for these types of shape representation [10,11,25]. Each algorithm follows a set of time-consuming and labor-intensive preprocessing steps which require domain expertise, including shape segmentation, resampling, smoothing, and alignment. Furthermore, PDM optimization processes and inference on new shapes are computationally expensive and time-consuming.

To ease the burden of manpower and heavy-duty preprocessing, deep learning-based models have been proposed to harness the power of data to learn a functional mapping directly from images to statistical representations of shapes [1–3,5,6]. These works provide a considerable advantage over conventional PDM methods in inference, as they do not require prohibitive, manual preprocessing steps and computationally complex re-optimization. Once a deep network is trained, a PDM can be inferred from a new unsegmented image in seconds. However, in terms of accuracy performance, existing deep learning models have yet to be up to the mark. Here, we propose a training strategy based on progressive learning, deep supervision, and multi-task learning to improve the performance of existing deep learning models.

The proposed methodology draws inspiration from three key concepts: progressive learning that builds on knowledge from prior learned tasks [19], deep supervision that applies loss to intermediate neural network layers [26], and multi-task learning [30] that leverages commonalities and differences across related tasks to improve generalization. The model consists of several progressive blocks, and each block is trained to predict an increasing number of correspondence points, i.e., a shape descriptor or representation at a specific scale. In other words, we predict the correspondence points in a multiscale training process, where each scale leverages the previous scales' output to predict the points. We provide a thorough architecture investigation, exploring the advantages and disadvantages of shared block backbones and the inclusion of an auxiliary segmentation task for improved PDM prediction. Furthermore, we have employed deep supervision to train our models and explored three loss calculation strategies that depend on the intermediate layers where the loss is applied. Finally, we demonstrate that the proposed training strategies significantly improved performance during the training and testing of the existing models. These training strategies can provide effective deep learning-based PDMs for accurate shape representation from images.

2   Related Works

We discuss the related works from three points of view: deep learning-based SSMs, progressive learning, and deep supervision.

Deep Learning-Based SSM Methods: DeepSSM is a state-of-the-art model that can provide statistical shape representations directly from images [5,6]. It uses a principal component analysis (PCA) based data augmentation scheme and has achieved good results on downstream tasks [7]. Probabilistic variants of DeepSSM that add uncertainty quantification have been proposed. For example, Uncertain-DeepSSM focuses on predicting data-dependent and model-dependent uncertainties to overcome the overconfident estimation of the deep learning models [1]. Recently, VIB-DeepSSM and it's fully Bayesian extension have been proposed, which utilize variational information bottleneck to capture the latent representation rather than regressing PCA scores [2]. All of these works supervise the entire dense set of correspondence points compared to the iterative process of incrementally predicting correspondence points of the conventional PDMs. This single-step regression process is error-prone in complex shape regions. The proposed training method can be used in addition to any of these methodologies to achieve more stable training and better performance.

Progressive Learning: Since its introduction in 2017, progressive learning [19] has revolutionized the training process for generative adversarial networks (GAN) and learning applications such as shape representation [23], speech recognition [12, 14], and person re-identification [27]. This incremental training process allows the model to learn a high-level, coarse output representation first, then gradually move on to detailed low-level, fine features. In the context of our task, rather than mapping the feature vectors directly to the final number of correspondence points, progressive learning allows us to map it to a lower number of points first, then gradually increase it to the final number to provide a better shape representation. Here, the mapped points in each scale cover the whole shape.

Deep Supervision: Deep supervision has improved training performances by adding losses in intermediate network layers in a wide range of applications, such as edge detection [21], image segmentation [28], 2D/3D keypoint localization [20], and image classification [20, 26]. We leverage this approach by adding supervision in each level of correspondence point prediction, allowing the model to converge better than existing methods.

3   Methodology

3.1   Datasets

We showcase the proposed training method using two datasets: femur and left atrium.

Femur Dataset: The femur dataset comprises 59 CT scans, with 49 identified as control scans, showcasing healthy subjects without any morphological irregularities in the femur bone. The remaining 10 scans are diagnosed with CAM-FAI, which is a morphological abnormality of the femur characterized by a lack of normal concavity at the femoral head-neck junction [18]. From this pool, we randomly incorporate 42 control images and 8 CAM-FAI images into the training set, reserving the remaining images for testing. Image downsampling is performed at a rate of 50%, resulting in an image size of $130 \times 92 \times 117$ and maintaining a uniform voxel spacing of $1.0\,mm$.

Left Atrium Dataset: The left atrium dataset encompasses 206 late gadolinium enhancement (LGE) MRI images from patients diagnosed with atrial fibrillation (AF), which results in irregular heart rhythm due to abnormal electrical impulses firing in the atrium. Similar to the femur data processing, we downsample these images by 50%, reaching a resolution of $118 \times 69 \times 88$ with a uniform voxel spacing of $1.25\,mm$. We randomly split the instances into 176 images for training, leaving 30 images for the testing phase.

3.2   Training Data

In constructing the multi-scale training data, we first determine the desired number of scales for the progressive training architectures based on the number of correspondence points in the first and last scales. The initial number of points is set at 256 to ensure enough coverage to capture coarse shape features. The maximum number of correspondence points (1024) for specific anatomy is selected empirically, as per the anatomy's size, curvature, and morphological variations. This process is executed using ShapeWorks [9] coarse-to-fine particle splitting strategy until the final correspondence points representation adequately captured the given anatomy's detail. We have selected ShapeWorks to generate PDMs at each scale (256, 512, 1024) because of its ability to generate PDMs with consistent qualitative and quantitative performance [17]. The ground truth PDMs for the test dataset are generated using the pre-optimized shape models of the training data.

Due to the low-sample size that is typical in medical imaging, we have applied model-based data augmentation [1,5] to generate additional realistic training examples. To do so, we applied principal component analysis (PCA) at each level of correspondence point density. A set of M 3D correspondence points for N samples, denoted by $\{y_n\}_{n=1}^{N}$ where $y_n \in \mathbb{R}^{3M}$, is reduced to L dimensional PCA scores $z_n \in \mathbb{R}^{L}$ where L is relatively low (between 15 and 25). These PCA scores can be expressed by a mean vector $\mu \in \mathbb{R}^{3M}$, a diagonal matrix of eigenvalues $\lambda \in \mathbb{R}^{L \times L}$ and matrix of eigenvectors $v \in \mathbb{R}^{3M \times L}$ by the equation: $z_n = v^T (y_n - \mu)$. A distribution is fit to the PCA scores via kernel density estimation (KDE). For the femur data, we use 20 PCA modes (which captured around 99% of the population variability), and for the left atrium, we used 25 PCA modes (which captured around 97% of the variability).To generate a synthetic image, we first draw a random sample from the KDE distribution and then use a technique that involves finding the closest example from a set of input images. This is the same augmentation technique as DeepSSM, and more details can be found there [5].

We have generated 5000 augmented image/correspondence point pairs for the femur dataset and 4000 for the left atrium dataset. The augmented and original images are used for training and validation in an 80:20 ratio.

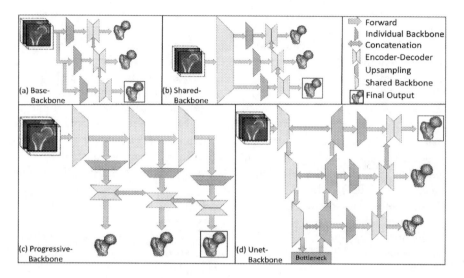

Fig. 1. Proposed model architecture with (a) Base-Backbone, (b) Shared-Backbone, (c) Progressive-Backbone and (d) Unet-Backbone.

3.3 Model Architecture

In this work, we have adopted DeepSSM [6] as the primary building block to underscore the impact of the training strategy. DeepSSM and its various off-shoots use a single deep network to estimate the complete set of correspondence points at the highest resolution/scale directly from unsegmented images. This work aims to demonstrate the efficacy of a multi-scale, progressive learning strategy when used to train these models. Although we demonstrate the proposed work using the original DeepSSM network, the proposed training and loss strategies can be readily applied to other variants. The presented investigation entails four variants of architecture, each contingent on the backbone of every scale.

- **Base-Backbone**: To obtain evidence of the progressive training's improved performance, we need to conduct a proof of concept. Hence, we have experimented with the base progressive architecture (Fig. 1(a)), whereby each scale is predicted by an individual DeepSSM block. From the second scale onwards, every block incorporates latent features from preceding blocks as an auxiliary input. This base architecture, applied initially and subsequently, utilizes a Convolutional Neural Network (CNN) backbone, comprising five convolutional and three max-pooling layers. Following each backbone, an encoder-decoder network is deployed. The encoder is built from three fully-connected layers, with the final layer containing the same number of nodes as the number of PCA modes (L). The decoder comprises a single layer with 3M nodes, where M corresponds to the number of correspondence points for that particular scale. The decoder is initialized with the eigenvalues ($vz_n + \mu$) derived from the principal components serving as weights and the mean shape acting as bias.

– **Shared-Backbone**: Once we have the proof of concept, our goal evolves to explore whether all scales share predictive image features for shape features at different scales, or if each scale requires its own feature extraction. To achieve this, we have implemented an approach of dividing the backbone into two parts: one common for all blocks and the other distinctive for each block. Specifically, we utilize two convolutional and one max-pooling layer from the backbone of the base architecture as the shared backbone, while the remaining three convolution and two map-pooling layers are used individually for each block. The encoder-decoder network is the same as the Base-Backbone architecture. This architecture is shown in Fig. 1(b).

– **Progressive-Backbone**: The previous network utilizes an identical architecture for all scales. However, we are curious to explore the potential benefits of incorporating more layers as the number of correspondence points rises with each scale. Therefore, we have conducted experiments with a progressive backbone, where the number of layers in the shared backbone increases as we progress to later scales (Fig. 1(c)). Despite these changes, the encoder-decoder architecture remains consistent.

– **Unet-Backbone**: Along with weight sharing between the blocks, we want to explore multitasking capabilities and investigate the performance of these models. Specifically, we focus on the task of segmentation, which is a fundamental prerequisite for non deep learning based methods. To achieve this, we have integrated the Progressive-Backbone model into the popular U-net segmentation architecture and pass a fusion of the bottleneck and decoder features to predict each scale of correspondence points (Fig. 1(d)) [24]. Motivation for fusing the bottleneck and decoder features is provided in Sect. 5. The segmentation-guided backbone provides the network with a shape prior, increasing correspondence prediction accuracy. The fused bottleneck and decoder features are passed to a feature extractor consisting of convolution and max-pooling layers to acquire the feature space for each scale. The feature space is connected to an encoder-decoder network, similar to previous architectures, to predict their respective correspondence points. However, in this case, we have not initialized the decoder using PCA, as this imposes linearity on the shape, which may hurt accuracy in the case of complicated shapes.

Each of these architectures uses a filter size of 5 for the convolutional layers and 2 for the max-pooling layers, which are chosen empirically. To ensure optimal performance, we have applied batch normalization after each convolution operation, followed by parametric ReLU (PReLU) activation.

3.4   Loss Function

Our model employs mean squared error (MSE) loss for the predicted correspondence points for each scale. For the ground truth y_k and predictions $\hat{y}_k$ for any scale k and N number of samples, the MSE loss is defined as:

$$\mathcal{L}_k = \frac{1}{N} \sum_{n=1}^{N} (y_n^k - \hat{y}_n^k)^2 \tag{1}$$

We have employed three loss variants in our methodology. They are defined as follows:

– **Fixed**: For any scale k, the previous scales' weights are frozen and the total loss is defined as:

$$\mathcal{L}_{Fixed} = \mathcal{L}_k \tag{2}$$

– **Shallow-Supervision**: For any scale k, the total loss is defined as the sum of the MSE loss of that scale and the previous scale. Here, during the training of each scale, the loss is backpropagated until the previous scale.

$$\mathcal{L}_{Shallow-Supervision} = \begin{cases} \mathcal{L}_k + \mathcal{L}_{k-1}, & \text{if } k > 0 \\ \mathcal{L}_k, & k = 0 \end{cases} \tag{3}$$

– **Deep-Supervision**: For any scale k, the total loss is defined as the summation of the MSE loss of the initial scale to that scale.

$$\mathcal{L}_{Deep-Supervision} = \sum_{i=0}^{k} \mathcal{L}_i \tag{4}$$

In addition to the aforementioned correspondence loss, we have incorporated a segmentation loss for the Unet-Backbone models. Consequently, the cumulative loss is quantified using the subsequent formula:

$$\mathcal{L}_{total} = \alpha * \mathcal{L}_{seg} + (1 - \alpha) * \mathcal{L}_{PDM} \tag{5}$$

In this context, α represents an empirically determined hyperparameter designed to balance the weights between the segmentation and correspondence losses. $\mathcal{L}_{seg}$ corresponds to the binary cross-entropy (BCE) loss between the original and the predicted segmentation, and $\mathcal{L}_{PDM}$ refers to any one of the aforementioned loss variants (Fixed, Shallow-Supervision, Deep-Supervision).

3.5   Evaluation Metric

We use two key metrics to evaluate the effectiveness of the proposed methodology: Root Mean Square Error (RMSE) and surface-to-surface distance (in mm). RMSE is calculated as the square root of the average squared differences between

the predicted and actual observations. Specifically, we average the RMSE for the
x, y, and z coordinates, where N is the total number of 3D correspondences:

$$RMSE = \frac{1}{3}\left(RMSE_x + RMSE_y + RMSE_z\right) \tag{6}$$

where, for N sets of ground truth and predicted correspondence points at scale
k, $RMSE_x = \sqrt{\frac{|y_{nx}^k - \hat{y}_{nx}^k|_2^2}{N}}$ and similar for y and z coordinates.

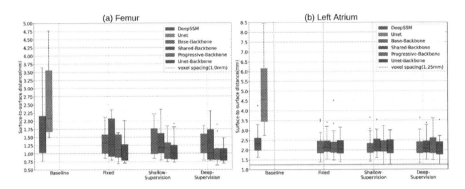

Fig. 2. Surface-to-surface distance comparison of our proposed models with DeepSSM
[5] in (a) femur and (b) left atrium dataset. The black line in each boxplot marks the
median value, and the blue horizontal line represents the voxel spacing of the images.
(Color figure online)

Surface-to-surface distance is computed by converting the ground truth and
predicted points to meshes and calculating the euclidean distance from each
vertex of the ground truth mesh to the closest face of the predicted mesh. The
reported values are the average of the vertex-wise surface-to-surface distance
from the ground truth to the predicted shapes.

3.6 Training Procedure

We have employed a multiscale, progressive training strategy to ensure better
convergence. This means we train one scale at a time, and only after that scale
reaches convergence do we move on to the next scale. We have used a Cosine
Annealing learning rate scheduler [22] to update each epoch's learning rate.
The rapid change in the learning rate of this scheduler has helped to make
sure the learning process is not stuck at a local minimum during training. The
initial learning rate is set to 0.001, and Adam optimization is used. Each scale
is trained for a maximum of 50 epochs with a batch size of six. However, to
avoid overfitting, we have employed an early stopping strategy, where we stop
the training if the validation loss is not improved after 15 consecutive epochs.

In the case of the Unet-Backbone models, the segmentation component is
trained first for five epochs to ensure a good shape prior to the correspondence

prediction. Then each scale is trained as previously explained. The value of the α parameter for the $\mathcal{L}_{total}$ is empirically set to 0.1.

The training process is implemented in PyTorch, and training is performed on a 12th Gen Intel(R) Core(TM) i9-12900K Desktop with 128 GB RAM and NVIDIA RTX A5000 GPU.

Table 1. The comparison between DeepSSM and the proposed models on the Deep Supervision loss in test data in terms of RMSE.

Dataset	RMSE on Test Data (Mean ± Standard Deviation)				
	DeepSSM	Base-Backbone	Shared-Backbone	Progressive-Backbone	Unet-Backbone
Femur	1.37 ± 0.72	1.15 ± 0.52	1.07 ± 0.41	0.93 ± 0.34	**0.78 ± 0.21**
Left Atrium	1.72 ± 0.8	1.65 ± 0.42	1.62 ± 0.45	1.55 ± 0.43	**1.48 ± 0.28**

4   Results

4.1   Femur

We have trained each model for three different loss functions as described in Sect. 3.4. The surface-to-surface distance comparison is shown in a boxplot in Fig. 2(a). The y-axis shows the two baselines (DeepSSM [5] and Unet for segmentation) and the three loss variants. For each loss variant column, different boxplots denote different model architectures. The blue and orange boxplot represents the baseline DeepSSM and Unet results, whereas the green, red, purple, and brown boxplots represent the Base-Backbone, Shared-Backbone, Progressive-Backbone, and Unet-Backbone, respectively.

We observe a consistent trend in Fig. 2(a) across all model architectures, namely that deep supervision enables the model to make more accurate predictions of correspondence points, resulting in more accurate shapes. This suggests that the progressive training strategy is benefitting from the deep supervision, as the gradients from the later scales, which capture fine-scale shape features, are used to fine-tune the earlier scales. This allows for improved conditions for the input signal for the finer scales, as the scales are not independent; each training iteration contributes to learning the fine-shape features.

The model performance remains consistently high across various architectures. Generally, the Base-Backbone models have shown slightly better results than the baseline, indicating that progressive architectures can yield improved outcomes. Furthermore, the Shared-Backbone and Progressive-Backbone models outperform the Base-Backbone. Notably, multitasking with a progressive backbone proves most effective in enhancing performance, as evidenced by the results of the Unet-Backbone. We have also sought to compare our approach with standard segmentation architecture, which calculates the surface-to-surface

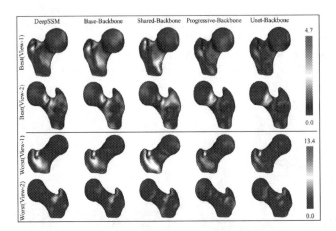

Fig. 3. Reconstruction error of the models' output is shown as a heatmap on the ground truth meshes for DeepSSM's best and worst output in the test data. The models reported in the figure are trained with Deep-Supervision losses.

distance between original and predicted segmentations. The proposed SSM models fared much better than segmentation-based models in reconstructing shapes from images.

We have compared the proposed models with the DeepSSM in terms of RMSE (Table 1) which, unlike surface-to-surface distance, captures whether or not the points are in correspondence. We can see a significant improvement in RMSE error for the Progressive-Backbone (32.12%) and Unet-Backbone models (43.06%). This improvement shows the superiority of the proposed models in the test data.

Additionally, we quantitatively evaluate the performance of our proposed models for 3D mesh reconstruction by comparing the reconstruction errors via heatmaps on the ground truth meshes. Specifically, we select the best and worst outputs of DeepSSM on the test data based on surface-to-surface distance and compare them with the proposed models' predictions. From this analysis, we generate error maps for our models' prediction on the ground truth mesh for the selected samples. The results of the comparison are shown in Fig. 3. Our findings show a significant improvement in the proposed models' prediction, particularly for the Progressive-Backbone and Unet-Backbone models for both cases.

Downstream Task - Group Differences: It is clinically significant to capture the statistical morphological difference between the CAM-FAI shape and the typical femur bone shape. In this experiment, we have employed models trained with Deep-Supervision loss. Our approach involves the construction of two groups - one for controls and one for pathology (CAM-FAI) - and computing the difference between their means (μ_{normal} and μ_{cam}). By doing so, we were able to showcase this difference on a mesh, which is known as group difference [18]. To achieve this, we utilize the ShapeWorks' PDM model, DeepSSM and the

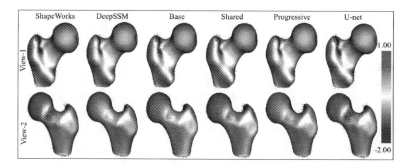

Fig. 4. Group Difference comparison of our proposed models with the original DeepSSM [5] and ShapeWorks [9].

proposed models' predicted particles, using the entire data for testing and training. Our findings demonstrate a strong similarity between the ShapeWorks and the proposed models' group differences, particularly for the Progressive-Backbone and Unet-backbone models (Fig. 4). This suggests that the proposed models can effectively obtain correspondences without the need for heavy preprocessing and segmentation steps. This ability to characterize the CAM deformity is crucial in observing the expected outcome of femur anatomy smoothly exhibiting inward motion around the CAM lesion as observed in clinical practice. Our results indicate that the proposed models have the potential to be a valuable tool in the analysis of femur anatomy and can aid in the diagnosis and treatment of CAM deformities.

4.2   Left Atrium

The left atrium MRI dataset presents significant variations in intensity and quality, influenced by the topological differences related to the arrangements of pulmonary veins. Similar to the femur dataset, we have trained our models for all loss variants explained in 3.4. The results are shown in Fig. 2(b). We can see that the Deep-Supervision loss helps in the case of this dataset as the surface-to-surface distance is much better for this loss compared to the Fixed and Shallow-Supervision.

All of the models proposed outperformed the baseline results. Interestingly, the SSM-based methods seem to be generating better 3D shapes compared to the standard segmentation baseline. The high variability of the dataset is likely contributing to the Unet model's underwhelming performance. Regarding different model architectures, the proposed models have demonstrated similar performance for a specific loss type, with Unet-Backbone slightly edging ahead. The trend in performance is consistent with the femur dataset, where the Progressive-Backbone and Unet-Backbone models surpass the Base-Progressive and Shared-Progressive models. By examining Table 1, it is apparent that the suggested models have shown significant improvement, particularly the Progressive-Backbone

(9.88%) and Unet-Backbone (13.95%) models concerning RMSE in the test data. These enhancements in both evaluation metrics highlight the advantageous impact of the proposed training techniques, especially when dealing with complex datasets.

Furthermore, we have conducted a thorough analysis of the reconstruction error comparison for the best and worst output of DeepSSM in the test data with respect to surface-to-surface distance. Our findings indicate that the output for both the best and worst case of the proposed models have significantly enhanced DeepSSM's outputs, as illustrated in Fig. 5. This outcome is consistent with the femur dataset, which further validates the effectiveness of the proposed models in improving the accuracy of surface reconstruction.

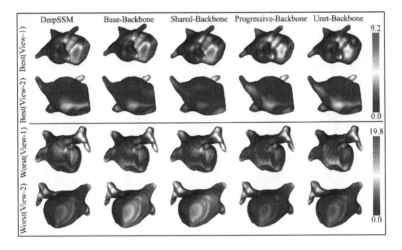

Fig. 5. Reconstruction error of the models' output is shown as a heatmap on the ground truth meshes for DeepSSM's best and worst output in the test data.

Downstream Task - Atrial Fibrillation Recurrence Prediction: Atrial Fibrillation (AF) is a medical condition characterized by an irregular heartbeat. To treat AF, doctors often use a therapeutic procedure called catheter ablation. Unfortunately, some patients may experience a recurrence of AF even after undergoing ablation. The left atrium dataset includes binary labels for each patient indicating whether they had AF recurrence following ablation.

To train a multi-layer perceptron (MLP) model for classifying AF recurrence, we utilized PCA scores from ground truth data and the latent features of the encoder-decoder network for both the DeepSSM and proposed models. Our experiments employ the same training, validation, and test sets, and we use the Deep-Supervision trained models. The results, presented in Table 2, demonstrate a significant performance improvement compared to the DeepSSM. Notably, our Unet-Backbone model even outperforms the ShapeWorks accuracy. We believe that the Unet-Backbone's encoding of image-based features, not available in PDM, contributes to its success in this downstream task.

Table 2. The comparison between baselines and the proposed models' accuracy on AF recurrence.

Metric (%)	ShapeWorks	DeepSSM	Base-Backbone	Shared-Backbone	Progressive-Backbone	Unet-Backbone
Accuracy	63.33	56.66	58.7	60.0	61.66	**73.33**

5   Ablation Studies

We have conducted an ablation study to analyze the impact of different components within our Unet-Backbone architecture. The study focuses on three key areas: the decoder, the bottleneck, and a fusion of the bottleneck and decoder features. The decoder plays a crucial role in reconstructing the spatial information lost during the encoding process. However, when used alone, it may not capture complex object details due to the lack of context. In contrast, the bottleneck condenses the input image information into a more manageable form, making it easier to extract high-level features. However, relying only on the bottleneck for feature extraction may result in a loss of information, especially for larger and more complex inputs. Our study shows that the fusion of bottleneck and decoder features produces the best results in the Unet-Backbone models (Fig. 6). This approach combines the strengths of the decoder and the bottleneck, resulting in a more robust representation.

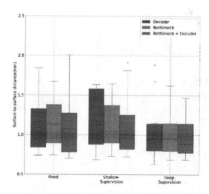

Fig. 6. Surface-to-surface distance comparison of different features in femur dataset.

6   Conclusion

Performing statistical shape modeling directly on images is a difficult task. Many image quality complications, such as artifacts, spatial resolution, signal-to-image ratio, etc., make it challenging to perform shape modeling. Hence, our work proposes a multiscale training methodology to learn the features gradually. The proposed training method utilizes multi-tasking based progressive learning and deep supervision to provide better performance. We have tested our methodology on two different datasets with different types of images (CT and MRI scans), and the proposed models provide improved results in both cases. This training method can be integrated into any deep learning-based shape models and achieve better performance. These contributions will help accelerate the adoption of automated statistical shape modeling from images in clinical use cases.

Acknowledgements. The National Institutes of Health supported this work under grant numbers NIBIB-U24EB029011, NIAMS-R01AR076120, and NIBIB-R01EB016701. The content is solely the responsibility of the authors and does not necessarily represent the official views of the National Institutes of Health.

References

1. Adams, J., Bhalodia, R., Elhabian, S.: Uncertain-deepssm: From images to probabilistic shape models. In: International Workshop on Shape in Medical Imaging, pp. 57–72. Springer (2020)
2. Adams, J., Elhabian, S.: From images to probabilistic anatomical shapes: A deep variational bottleneck approach. arXiv preprint arXiv:2205.06862 (2022)
3. Adams, J., Elhabian, S.: Fully bayesian vib-deepssm. arXiv preprint arXiv:2305.05797 (2023)
4. Beg, M.F., Miller, M.I., Trouvé, A., Younes, L.: Computing large deformation metric mappings via geodesic flows of diffeomorphisms. Int. J. Comput. Vision **61**(2), 139–157 (2005)
5. Bhalodia, R., Elhabian, S., Adams, J., Tao, W., Kavan, L., Whitaker, R.: Deepssm: a blueprint for image-to-shape deep learning models. arXiv preprint arXiv:2110.07152 (2021)
6. Bhalodia, R., Elhabian, S.Y., Kavan, L., Whitaker, R.T.: DeepSSM: a deep learning framework for statistical shape modeling from raw images. In: Reuter, M., Wachinger, C., Lombaert, H., Paniagua, B., Lüthi, M., Egger, B. (eds.) ShapeMI 2018. LNCS, vol. 11167, pp. 244–257. Springer, Cham (2018). https://doi.org/10.1007/978-3-030-04747-4_23
7. Bhalodia, R., et al.: Deep learning for end-to-end atrial fibrillation recurrence estimation. In: 2018 Computing in Cardiology Conference (CinC), vol. 45, pp. 1–4. IEEE (2018)
8. Biffi, C., Cerrolaza, J.J., Tarroni, G., Bai, W., De Marvao, A., Oktay, O., Ledig, C., Le Folgoc, L., Kamnitsas, K., Doumou, G., et al.: Explainable anatomical shape analysis through deep hierarchical generative models. IEEE Trans. Med. Imaging **39**(6), 2088–2099 (2020)
9. Cates, J., Elhabian, S., Whitaker, R.: Shapeworks: particle-based shape correspondence and visualization software. In: Statistical Shape and Deformation Analysis, pp. 257–298. Elsevier (2017)
10. Cates, J., Fletcher, P.T., Styner, M., Shenton, M., Whitaker, R.: Shape modeling and analysis with entropy-based particle systems. In: Karssemeijer, N., Lelieveldt, B. (eds.) IPMI 2007. LNCS, vol. 4584, pp. 333–345. Springer, Heidelberg (2007). https://doi.org/10.1007/978-3-540-73273-0_28

11. Davies, R.H., Twining, C.J., Cootes, T.F., Waterton, J.C., Taylor, C.J.: A minimum description length approach to statistical shape modeling. IEEE Trans. Med. Imaging **21**(5), 525–537 (2002)

12. Fayek, H.M., Cavedon, L., Wu, H.R.: Progressive learning: a deep learning framework for continual learning. Neural Netw. **128**, 345–357 (2020)

13. Fuessinger, M.A., Schwarz, S., Neubauer, J., Cornelius, C.P., Gass, M., Poxleitner, P., Zimmerer, R., Metzger, M.C., Schlager, S.: Virtual reconstruction of bilateral midfacial defects by using statistical shape modeling. J. Cranio-Maxillofacial Surg. **47**(7), 1054–1059 (2019)

14. Gao, T., Du, J., Dai, L.R., Lee, C.H.: Snr-based progressive learning of deep neural network for speech enhancement. In: Interspeech, pp. 3713–3717 (2016)

15. Gardner, G., Morris, A., Higuchi, K., MacLeod, R., Cates, J.: A point-correspondence approach to describing the distribution of image features on anatomical surfaces, with application to atrial fibrillation. In: 2013 IEEE 10th International Symposium on Biomedical Imaging, pp. 226–229. IEEE (2013)

16. Gerig, G., Styner, M., Jones, D., Weinberger, D., Lieberman, J.: Shape analysis of brain ventricles using spharm. In: Proceedings IEEE Workshop on Mathematical Methods in Biomedical Image Analysis (MMBIA 2001), pp. 171–178. IEEE (2001)

17. Goparaju, A., et al.: Benchmarking off-the-shelf statistical shape modeling tools in clinical applications. Med. Image Anal. **76**, 102271 (2022)

18. Harris, M.D., Datar, M., Whitaker, R.T., Jurrus, E.R., Peters, C.L., Anderson, A.E.: Statistical shape modeling of cam femoroacetabular impingement. J. Orthop. Res. **31**(10), 1620–1626 (2013)

19. Karras, T., Aila, T., Laine, S., Lehtinen, J.: Progressive growing of gans for improved quality, stability, and variation. arXiv preprint arXiv:1710.10196 (2017)

20. Li, C., Zia, M.Z., Tran, Q.H., Yu, X., Hager, G.D., Chandraker, M.: Deep supervision with intermediate concepts. IEEE Trans. Pattern Anal. Mach. Intell. **41**(8), 1828–1843 (2018)

21. Liu, Y., Lew, M.S.: Learning relaxed deep supervision for better edge detection. In: Proceedings of the IEEE Conference on Computer Vision and Pattern Recognition, pp. 231–240 (2016)

22. Loshchilov, I., Hutter, F.: Sgdr: Stochastic gradient descent with warm restarts. arXiv preprint arXiv:1608.03983 (2016)

23. Park, J.J., Florence, P., Straub, J., Newcombe, R., Lovegrove, S.: Deepsdf: learning continuous signed distance functions for shape representation. In: Proceedings of the IEEE/CVF Conference on Computer Vision and Pattern Recognition, pp. 165–174 (2019)

24. Ronneberger, O., Fischer, P., Brox, T.: U-net: Convolutional networks for biomedical image segmentation. In: Medical Image Computing and Computer-Assisted Intervention-MICCAI 2015: 18th International Conference, Munich, Germany, October 5–9, 2015, Proceedings, Part III 18, pp. 234–241. Springer (2015)

25. Styner, M., et al.: Framework for the statistical shape analysis of brain structures using spharm-pdm. Insight J. (1071), 242 (2006)

26. Wang, L., Lee, C.Y., Tu, Z., Lazebnik, S.: Training deeper convolutional networks with deep supervision. arXiv preprint arXiv:1505.02496 (2015)

27. Wu, Y., Lin, Y., Dong, X., Yan, Y., Bian, W., Yang, Y.: Progressive learning for person re-identification with one example. IEEE Trans. Image Process. **28**(6), 2872–2881 (2019)

28. Zhang, Y., Chung, A.C.S.: Deep supervision with additional labels for retinal vessel segmentation task. In: Frangi, A.F., Schnabel, J.A., Davatzikos, C., Alberola-

López, C., Fichtinger, G. (eds.) MICCAI 2018. LNCS, vol. 11071, pp. 83–91. Springer, Cham (2018). https://doi.org/10.1007/978-3-030-00934-2_10

29. Zhao, Z., Taylor, W.D., Styner, M., Steffens, D.C., Krishnan, K.R.R., MacFall, J.R.: Hippocampus shape analysis and late-life depression. PLoS ONE **3**(3), e1837 (2008)

30. Zhou, Y., et al.: Multi-task learning for segmentation and classification of tumors in 3d automated breast ultrasound images. Med. Image Anal. **70**, 101918 (2021)

Muscle Volume Quantification: Guiding Transformers with Anatomical Priors

Louise Piecuch[1,2(✉)] , Vanessa Gonzales Duque[1,3] , Aurélie Sarcher[2] ,
Enzo Hollville[5,6] , Antoine Nordez[2,4] , Giuseppe Rabita[5] ,
Gaël Guilhem[5] , and Diana Mateus[1]

[1] Nantes Université, École Centrale Nantes LS2N, UMR CNRS 6004, Nantes , France
[2] Nantes Université, Movement - Interactions - Performance, MIP, IP UR 4334 UFR
STAPS, Nantes, France
Louise.Piecuch@ls2n.fr
[3] Technical University of Munich (TUM), Munich, Germany
[4] Institut Universitaire de France (IUF), Paris, France
[5] Laboratory Sport Expertise and Performance, INSEP, Paris, France
[6] Fédération Française de Badminton, Saint-Ouen, France

Abstract. Muscle volume is a useful quantitative biomarker in sports,
but also for the follow-up of degenerative musculo-skelettal diseases. In
addition to volume, other shape biomarkers can be extracted by segmenting the muscles of interest from medical images. Manual segmentation is
still today the gold standard for such measurements despite being very
time-consuming. We propose a method for automatic segmentation of 18
muscles of the lower limb on 3D Magnetic Resonance Images to assist
such morphometric analysis. By their nature, the tissue of different muscles is undistinguishable when observed in MR Images. Thus, muscle
segmentation algorithms cannot rely on appearance but only on contour cues. However, such contours are hard to detect and their thickness
varies across subjects. To cope with the above challenges, we propose a
segmentation approach based on a hybrid architecture, combining convolutional and visual transformer blocks. We investigate for the first
time the behaviour of such hybrid architectures in the context of muscle
segmentation for shape analysis. Considering the consistent anatomical
muscle configuration, we rely on transformer blocks to capture the long-
range relations between the muscles. To further exploit the anatomical
priors, a second contribution of this work consists in adding a regulari-

---

This work has been supported by the European Regional Development Fund (FEDER),
the Pays de la Loire region on the Connect Talent scheme (MILCOM Project), Nantes
Métropole (Convention 2017-10470) and FULGUR (Team: 30 researchers, Grant:1.9
M€) which benefits from a French Research Agency aid (reference ANR-19-STPH-
0003). FULGUR is part of the perspective of the Paris 2024 Olympic and Para-
lympic Games in collaboration with the French Federations of Athletics, Rugby and Ice
Sports, Universities of Nantes, Côte d'Azur, Savoie Mont Blanc, Jean Monnet Saint-
Etienne, Saclay, the Mines Saint Etienne school, the french Alternative Energies and
Atomic Energy Commission (CEA), the French National Centre for Scientific Research
(CNRS), Natural Grass and Super Sonic Imagine.

C. Wachinger et al. (Eds.): ShapeMI 2023, LNCS 14350, pp. 173–187, 2023.
https://doi.org/10.1007/978-3-031-46914-5_14

sation loss based on an adjacency matrix of plausible muscle neighbour-hoods estimated from the training data. Our experimental results on a unique database of elite athletes show it is possible to train complex hybrid models from a relatively small database of large volumes, while the anatomical prior regularisation favours better predictions.

Keywords: Vision transformers · Muscle segmentation · MRI · Anatomical prior

1   Introduction

Skeletal muscles are composed of muscle fibers, usually arranged in bundles sur-rounded by connective tissue. Different to other organs in the body, their shape can change relatively fast under physical training, injuries or under the effect of certain diseases. Therefore, the evolution, shape and volume of muscles have been studied in the sports and medical literature as biomarkers [8,15,20,21]. Such measurements can be extracted in a non-intrusive way through medical imaging. Magnetic Resonance Imaging (MRI) is well suited for the task for its ability to image soft tissues with high contrast [13,15]. An important intermedi-ate step to go from images to biomarkers is the segmentation. Once segmented it is possible to make comparisons between athletes, find trends according to discipline, sex, height, weight, or even within an individual, e.g. by detecting muscular asymmetries between the legs [8,20]. In this study, our main focus is the sports domain. However, muscle segmentation is also useful in the context of skeleto-muscular diseases like Duchenne's dystrophy, where monitoring mus-cle development is crucial. Therein, muscle segmentation helps track the disease progression, assess its impact on muscle tissue, or adapting treatment strategies.

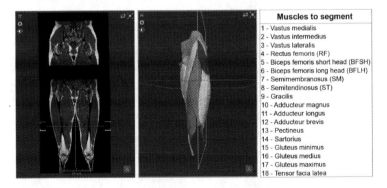

Fig. 1. Example of input MRI (**left**), and output labelmap (**middle**) The table (**right**) lists all the muscles to segment and their associated label.

We aim to segment muscles from 3D MR images for subsequent shape anal-ysis in elite sports (see Fig. 1). Previous studies relied on manual annotations, which are very time-consuming and laborious [15,20]. Indeed, segmenting MR images from elite athletes poses several challenges. In a broader context, the high number of muscles to be segmented is a major constraint. Also, these muscles are interconnected and influence each other. Unlike multi-organ segmentation where diverse labels are present, muscles have similar tissue types, so texture information is not discriminating. Instead, muscle segmentation primarily relies on often thin or imperceptible boundaries. Elite athletes, with more developed muscles and less in-between fat, present an additional challenge. Finally, working with the full 3D data preserves important contextual information but necessitates memory management for large volumes. These specific considerations highlight the need for tailored segmentation approaches that account for the distinct characteristics and requirements of sports-related imaging analysis. In order to automate the segmentation task while tackling these challenges, we rely on the UNETR architecture [11] to leverage the strengths of both Convolutional Neural Networks (CNN)s and transformers. Furthermore, medical and sports professionals have extensive knowledge of human anatomy, including muscle structure and their spatial collocation. Leveraging this expertise, we incorporate prior knowledge into the learning process through a regularization loss inspired by [6] that enforces feasible muscle adjacencies. This loss leverages our knowledge of muscle anatomy to improve the accuracy and reliability of the segmentation model.

2   Related Work

Image segmentation is relevant in various sports-related applications. Miller et al. [15], examine and compare the variations in muscle volume between male elite sprinters and sub-elite sprinters. Furthermore, the study investigates the relationship between muscle volumes, strength, and sprint performances, all based on manual segmentations. The delineation of muscle boundaries and regions of interest are also performed by human experts in [20] to characterize the hamstring muscles with a statistical shape model. Alternative methods have emerged, including semi-automatic approaches. For instance, Hansdfield et al. [8], manually revised the output of an automatic algorithm to investigate the distribution of muscle volumes in the lower limb among elite sprinters. Gilles et al. [7], focus on the registration and segmentation of hip and thigh muscles using deformable discrete models. More recently, automatic methods such as Yokota et al. [21] utilize multi-atlas techniques to automate the segmentation of hip and thigh muscles from CT scans. Cheng et al. [5] rely on a U-Net for segmenting the quadriceps and patella from MRI scans, but primarily focusing on pediatric medical applications. Ni et al. [17] proposed an automatic method for segmenting lower limb muscles of collegiate athletes (basketball, football, and soccer) using a cascaded 3-D CNN. The approach comprises two independently trained networks to address the muscle localization task on low-resolution images and the subsequent segmentation task on cropped high-resolution images. Agosti et al. [1] also propose a two-phase method for segmenting leg muscles on a database comprising both healthy subjects and patients. The initial phase relies on a classification network to identify the specific leg area being segmented, followed by a second phase involving the segmentation of the muscles of interest through a deep learning network. To the best of our knowledge, apart from the methods mentioned earlier, there are only a limited number of automated approaches specifically designed for muscle segmentation. This is particularly true in the sports domain, where the task presents its own challenges. However, given the amount of time required for a manual segmentation (∼40 h/subject in our case) there is a clear need to automate this task. Therefore, we propose an automatic method on a unique database of elite athletes, which is difficult to acquire and collect given the athletes' profiles, but also to annotate due to the significant muscle development and little adipose tissue.

The U-Net [19] architecture is considered the reference for automating segmentation tasks in medical imaging [13]. However, the emergence of transformers in recent years has opened up new possibilities. While CNNs excel at capturing local structures, they have limitations when it comes to capturing long-range relationships among different regions in an image. As CNNs go deeper, their receptive field gradually expands, leading to distinct features extracted at different stages. In contrast, transformer blocks leverage the power of Multi-head Self-attention (MSA) to establish a global receptive field, even at the lowest layer of models like the Vision Transformer (ViT). In this sense, transformer-based models, are well-suited for medical image segmentation since long-range dependencies are common within the human body. Another asset is the flexibility of their network architecture. Indeed, several architectures that combine transformers and CNNs have been proposed [13], by offering various ways to integrate transformers into U-Net like networks. Petit et al. [18] introduced transformers in the decoder of U-Net. Transformers can also be incorporated into the bottleneck section of the U-Net architecture, as demonstrated in TransUnet [4]. Another approach involves independently processing the image through transformer blocks and convolutional layers, and subsequently merging the information obtained at each encoding step, as in [12]. UNETR [11] combines the strengths of CNNs and transformers by replacing the encoder of the U-Net architecture with a series of transformer blocks. A transformer block at the input reformulates the segmentation problem as a 1D sequence-to-sequence inference task, similar to transformers in natural language processing [13]. While more recent architectures have been deemed powerful, e.g. Swin-Unetr [10], they are usually associated to higher training complexities, requiring larger computational resources and longer training times compared to traditional Transformers. Based on our data limitations and Hasany et al. [9] findings showing that UNETR captures global information fast, i.e., even at the third layer of transformers, we opt for a UNETR model.

As mentioned above, one of the challenges we face is that the tissue of different muscles appears identical when observed in MR Images, making texture information irrelevant. On the contrary, contextual information can be a major asset. Multiple approaches can be employed to incorporate such relevant context, including modeling it with a loss function. Such functions can be constructed based on the morphology of the objects being segmented e.g. star-shaped [16] or vertebrae like [2]. However, such shape priors are difficult to apply to our segmentation problem, since simple priors do not adequately capture the muscle variations and given there is no known atlas available. From a more topological perspective, BenTaieb et al. [3] addressed the issue of region exclusion and inclusion by penalizing incorrect label hierarchy. They noticed constraints between certain regions in their specific application and developed solutions to enforce inter-region connectivity. Since the muscles are separate entities, BenTaieb's method does not apply either. Finally, Ganaye et al. [6], proposed a method based on multiple-organ and brain region adjacencies. Given that the positions of muscles remain consistent within the legs, the adjacency relationships among athletes' muscles are also expected to be preserved. Therefore, we adapt the idea of an adjacency constraint from [6] to regularize the training of our automatic segmentation method.

3   Methodology

The purpose of this work is to design an automatic tool to segment muscles from MR images of the lower-limbs. More specifically, the approach receives as input 2 MR scans of the same subject (hip and thighs) and provides as output the semantic segmentation labelmaps (a probability of each voxel to belong to one of the 18 considered muscles), as shown in Fig. 1. To address the above problem we rely on a hybrid (ViT + CNN) UNETR architecture [11], and in this way capture long range dependencies within a muscle and between muscles. To further reinforce the anatomical priors we first built an adjacency matrix from our training data, by estimating the probability of two muscles being next to each other. Then, we rely on this adjacency matrix to define a penalizing loss that forces the model to make predictions that respect the prior connectivities. Next we describe the details of the architecture and loss.

3.1   Model

Lets define the input to the model to be an image $\mathbf{x} \in \mathbb{R}^{H \times W \times D}$ (with $H \times W \times D$ the image size) and the associated ground truth labelmap as the function $\mathbf{lab} : i \in \mathbf{R}^{H \times W \times D} \longmapsto [0, ..., C]$, with C the number of labels. We also denote as $\widehat{lab}(i)$ the predicted labelmap obtained as output of the model. The chosen UNETR [11] architecture is based on a U-Net, whose encoder has been replaced by a succession of T transformer blocks. These T blocks retain global information (e.g. on fairly long muscles), thanks to self-attention modules, while the architecture keeps access to more local information through the convolutional layers. Next, we follow [11] to describe details of each block. Since transformer blocks work on 1D sequences, we convert our 3D input data $\mathbf{x}$ into a sequence of flattened non-overlapping patches x_v^N of equal resolution ($P \times P \times P$), as shown in Fig. 2; thus, the sequence has length $N = (H \times W \times D)/P^3$. A linear layer, $E \in \mathbb{R}^{(P^3 \cdot C) \times K}$, is then used to project each patch into a K dimensional embedding space, which is the same throughout the transformers layers. A 1D learnable positional embedding $E_{pos} \in \mathbb{R}^{N \times K}$ is added to the sequence of the

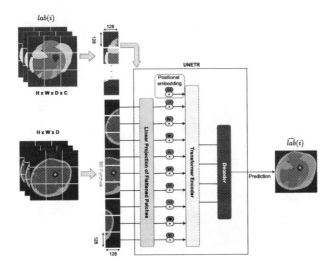

Fig. 2. Overview of our method during training phase for muscle segmentation from MR volumes.

projected patch embeddings in order to keep the spatial information and help reconstruct back the image. We denote the result of patch projections and positional embedding as:

$$z_0 = [x_v^1 E; x_v^2 E; ...; x_v^N E] + E_{pos} \tag{1}$$

After the embedding, the sequence z_0 passes through a stack of transformer blocks as shown in Fig. 3. A typical transformer block is composed of a multi-head self-attention (MSA) and a multi-layer perceptron (MLP) (c.f. Eq (6) in [11]). The data is then passed through a normalisation layer, $Norm()$.

$$z_t' = MSA(Norm(z_{t-1})) + z_{t-1}, \qquad t = 1...T, \tag{2}$$

$$z_t = MLP(Norm(z_t')) + z_t', \qquad t = 1...T, \tag{3}$$

The UNETR architecture incorporates a direct connection between the transformer encoder and the CNN decoder block through skip-connections at different resolutions, enabling the calculation of the final semantic segmentation output. In the architecture bottleneck, a deconvolutional layer is employed to increase the resolution of the transformed feature map by a factor of 2. This upscaled feature map is then concatenated with the feature map from the previous transformer output (e.g. z_{t_9} in Fig. 3). Next, consecutive $3 \times 3 \times 3$ convolutional layers are applied, followed by an upsampling using a deconvolutional layer, until the output reaches the original input resolution. Finally, a $1 \times 1 \times 1$ convolutional layer with a softmax activation generates the voxel-wise semantic prediction.

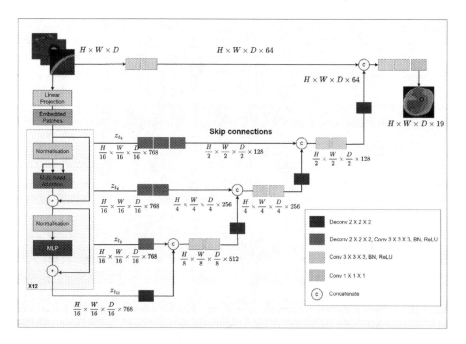

Fig. 3. Architecture of UNETR, modified figure from [11]. The transformer blocks containing MSA and MLP blocks is repeated 12 times. The output of the transformer blocks are z_t. Only 4 of these outputs are illustrated here in a manner of clarity.

3.2   Prior Anatomical Knowledge

The relative location of a muscle with respect to others is overall consistent across participants, especially for healthy subjects. Therefore, we propose to exploit such anatomical knowledge to further guide the training. In practice, we represent the relative positioning of the muscles with a probabilistic adjacency matrix. Similarly to [6], we employ this matrix within a regularizing loss term that penalizes predictions that do not respect the known adjacencies.

To create a probabilistic matrix, we extracted binary adjacency matrices for each subject in the database. To do this, three 4-neighbour filters are applied to the subject's manual segmentation labelmap **lab**. These derivative filters perform the difference $d_{i,j}$ between the value **lab**(i) of a given voxel i and its neighbour $j \in \mathcal{N}(i)$ (where $\mathcal{N}(i)$ corresponds to the neighborhood voxels of i), such that $d_{i,j} = \textbf{lab}(i) - \textbf{lab}(j)$. These filters are applied separately in the 3 directions of the labelmap. Any non-zero difference ($d_{i,j} \neq 0$), indicates the presence of a boundary between these two neighboring voxels. Once $d_{i,j}$ has been calculated, and the boundaries found, we associate each boundary with the respective pair of labels (muscles). We then fill 1 in the corresponding location of the $N_{muscles} \times N_{muscles}$ adjacency matrix if a boundary between a pair of labels was detected in any direction. After extracting the binary adjacency matrices for all subjects, we sum them up and normalise the result by the number of muscles. The resultant

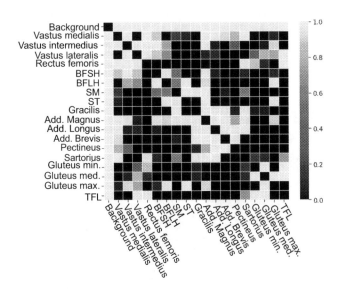

Fig. 4. Probabilistic adjacency matrix providing prior knowledge on muscle anatomy. The rows and columns of the matrix correspond to each label (muscles and background). Each element of the matrix has a value between 0 and 1. The higher the value is, the more likely the adjacency between 2 muscles.

probabilistic adjacency matrix $\mathbf{A}$ is shown in Fig. 4. The process is summed up in the following equation:

$$a_{bc}(\mathbf{lab}) = \sum_{i} \sum_{j \in \mathcal{N}(i)} \delta_{b,\mathbf{lab}(i)} \delta_{c,\mathbf{lab}(j)}, \tag{4}$$

where $b, c \in [0, ..., C]$ are 2 different labels and $\delta_{b,\mathbf{lab}(i)}$ is the Kronecker delta function equal to 1 when voxel i has label b (or 0 otherwise). Here, a_{bc} is the adjacency function calculated on the ground-truth labelmap $\mathbf{lab}$ and $\tilde{a}_{bc} = (a_{bc} > 0)$ is its binarized version, which is summed-up and normalised to obtain $\mathbf{A}$. The same function is computed during training but instead on the predicted probabilistic labelmap $\widehat{\mathbf{lab}}$, as discussed in Sect. 3.3.

Unlike [6], we have chosen to keep the ground truth matrix probabilistic to consider the variability across subjects and the likelihood of the muscles connections. Two muscles are considered adjacent if at least one of their voxels is in contact in the labelmap.

3.3 Loss Function

The loss function combines a common segmentation loss function, the softdice cross-entropy loss with a regularization loss to consider the prior anatomical information of muscle adjacency.

$$L_{final} = L_{seg}(\mathbf{lab}, \widehat{\mathbf{lab}}) + \lambda L_{NonAdjLoss}(\mathbf{lab}, \widehat{\mathbf{lab}}) \tag{5}$$

A weighting lambda is applied to the regularization loss so as not to penalise the model too much while incorporating the anatomical constraint.

Soft Dice Cross Entropy Loss is a combination of soft dice loss and cross-entropy:

$$L_{seg}(\mathbf{lab}, \widehat{\mathbf{lab}}) = (1 - \frac{2}{C}\sum_{c=1}^{C} \frac{\sum_{i=1}^{I} \mathbf{lab}, \widehat{\mathbf{lab}}}{\sum_{i=1}^{I} \mathbf{lab}^2 + \sum_{i=1}^{I} \widehat{\mathbf{lab}}^2})(-\frac{1}{I}\sum_{i=1}^{I}\sum_{c=1}^{C} \mathbf{lab} \log \widehat{\mathbf{lab}}),$$

(6)

where I is the number of voxels ($I = H \times W \times D$); C is the number of classes; $\mathbf{lab}$ and $\widehat{\mathbf{lab}}$ denote respectively, the probabilistic prediction and ground-truth encoded in one-hot.

NonAdjLoss is the proposed regularisation loss enforcing the segmentation predictions to satisfy the anatomical constraints.

$$L_{NonAdjLoss}(\mathbf{lab}, \widehat{\mathbf{lab}}) = \sum_{\forall(b,c)\in[0,...,C]} (1 - a_{bc}(\mathbf{lab}))a_{bc}(\widehat{\mathbf{lab}}),$$

(7)

where $a_{bc}(\widehat{\mathbf{lab}})$ is the adjacency function calculated during training from applying Eq. 4 to $\widehat{\mathbf{lab}}$. If the model predicts a wrong adjacency, we penalize it with the inverse of the probability of this link existing. Thus, using the network with $a_{bc}(\mathbf{lab})$ as a differentiable adjacency matrix, allows us to penalize the forbidden connectivities of any prediction.

4   Experimental Validation

4.1   Experimental Settings

Dataset. The dataset, composed of 18 3D registered MRI (pelvis and thighs) of low-limb muscles from elite-athletes, was acquired at the medical imaging centre of the INSEP. We split the dataset into 15:1:2 for training, validation and test. The MR images were manually annotated to obtain the labelmaps of the 18 muscles in Fig. 1, which took between 30 to 40 h per subject. The MRI are cropped to show only one leg for the training and inference. The average volume is $467.2 \times 450.2 \times 1556.2$ pixels for a spacing of $0.55 \times 0.55 \times 0.55$. The spacing for training is resized to $1 \times 1 \times 1$ for memory reasons. Intensities are normalised between 0 and 1 and data augmentation is performed before the training on the patches (flips, rotations, intensity). During inference, we post-process the output of all compared methods to identify the largest predicted connected component and fill any holes.

Evaluation Metrics. The first objective of this project is to recover the volume of each muscle. To this end, we mesure the volumetric error of each muscle, in cm^3 and percentage as:

$$\text{Vol}_{err_{cm^3}} = |V_{GT} - V_{pred}| \qquad \text{Vol}_{err_{\%}} = 100 \times \frac{|V_{GT} - V_{pred}|}{V_{GT}},$$

(8)

where V_{GT} and V_{pred} correspond to the ground-truth and predicted volumes of a given muscle respectively. Note that $\text{Vol}_{err_{\%}}$ takes into account the size of the muscle from which we are extracting the volume, which $\text{Vol}_{err_{cm^3}}$ does not. We also rely on the Dice Score (DSC) and the 95% Hausdorff Distance (HD95) to evaluate the performance of the model.

Implementation Details. The implementation relies on MONAI, a PyTorch-based open-source framework[1]. Training was done on an NVIDIA GeForce RTX 3090 Ti (24 GB) graphic card. Training included two phases. During the first phase, the model was trained without NonAdjLoss for 6667 epochs. Then, the model was fine-tuned with the regularisation loss for 5000 epochs. The regularization weight was set to $\lambda = 0.3$. Each model (pretrained and fine-tuned) was trained with a batch size of 1, using the AdamW optimizer and an initial learning rate of 0.0001. Full training took 48 hours (11667 epochs). The UNETR architecture was configured with 12 transformer blocks ($T = 12$) and has an embedding size of $K = 768$ [11]. To match the size of the data, we set the patch size to $128 \times 128 \times 128$.

4.2   Quantitative Results

Regarding the volumetric error (%), we report the results in Fig. 5. Most of the muscles have an error under 5% for the training set. As expected, the values for the test set are higher but in average bellow 10%. Higher errors for the Pectineus can be explained by its small size and for the Gluteus minimus by its more challenging boundaries. In addition, we compared our method against a U-Net architecture. We also investigated the impact of the regularization cost

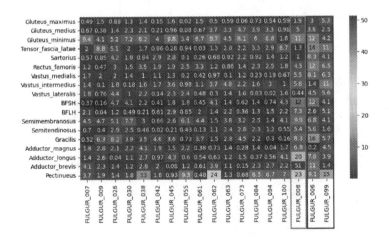

Fig. 5. Volumetric error (%) for all subjects and muscles. Validation and Test subjects are indicated in orange and red. (Color figure online)

---

[1] https://monai.io/.

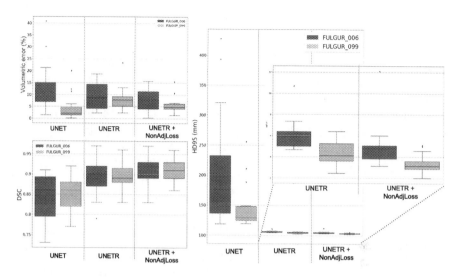

Fig. 6. Boxplot of the test results for UNET, UNETR and UNETR + NonAdjLoss. The volumetric error (%) (**upper left**), the Dice Score Coefficient (**bottom left**) and the Hausdorff Distance (mm) (**right**) of the two test subjects (FULGUR_006 and FULGUR_099) are shown here. The boxplots are computed across the 18 muscles of interest for each of the test subjects.

function on the learning process. To visualize the results, we present boxplots of the volumetric error, Dice coefficient, and Hausdorff distance 95 in Fig. 6, on the test dataset.

The methods based on UNETR show an overall decrease in the number of outliers across metrics, indicating improved performance in terms of reducing extreme errors. Moreover, both UNETR and the fine-tuned method with Non-AdjLoss reduce either the mean volumetric error or its variance. The transition to UNETR also results in a reduction of the 95HD error. For example, in the case of F006, the average Hausdorff distance decreased from approximately 175 mm to around 6mm with UNETR. Similarly, for F099, the average Hausdorff distance was reduced from 125 mm to an average of 4 mm with UNETR. Moreover, the NonAdjLoss regularization further reduces the average Hausdorff distance, resulting in an average of 5 mm for F006 and 3 mm for F099. These results confirm that the inclusion of anatomical constraints enabled the learning process to generate predictions closer to the ground truth. The NonAdjLoss regularization has effectively guided the model to capture the anatomical characteristics and spatial relationships of the muscles, leading to improved segmentations. Finally, there is an increase in Dice (DSC) with the methods incorporating transformers. We observe an average DSC of around 0.86 for U-Net, while U-Net with transformers (UNETR) achieves an average DSC of 0.9. Furthermore, when combining UNETR with NonAdjLoss, the average DSC further improves to 0.92. These results highlight the enhanced segmentation performance (Fig. 6 and 7).

4.3   Qualitative Results

Regarding the qualitative results, one notable observation is that when providing the MRI scans of both legs, the model is capable of segmenting both legs successfully, even if it was only trained on one of them as shown in Fig. 7. This can be attributed to the inherent symmetry found in human anatomy, the utilization of data augmentations during training and the sequential inference process. The model has learned to generalize well to the other leg, leveraging the common features and structures These results demonstrate the adaptability of the model to handle variations such as symmetries and multiple instances.

When we examine the predictions, we observe plausible and overall good quality labelmaps. The remaining errors, resemble human mistakes made during the manual segmentation, which unfortunately can still be found in the ground truth of this dataset. Such errors include, voxels belonging to other labels be present within a muscle, and mixing the boundaries of muscles belonging to a group (such as adductor groups). Figure 8-right shows an example of the TFL that influences the gluteus maximus in the prediction, which further explains the quantitative results for that muscle. Additionally, we observe less anatomically accurate ground-truth label shapes in small ambiguous regions. Therefore, the presence of some errors in both the manual annotations and the model predictions is expected.

We can also observe that most errors occur at the boundaries of the segmentations as shown in Fig. 8-left. This is particularly noticeable in the case of athletes since they have a significant muscle development, and the presence of adipose tissue between their muscles is reduced. However, suboptimal predictions can have a direct impact on adjacent predictions. For example, if a muscle

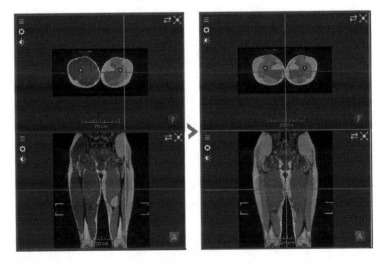

Fig. 7. Comparison of the Ground Truth (**left**) given as input to the model and the prediction of our trained model when we give an MRI with both legs (**right**).

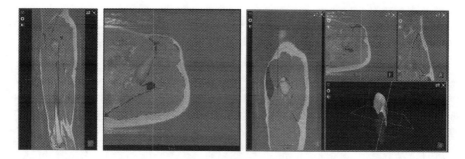

Fig. 8. Comparison of the predicted labelmap on a test subject and the GT labelmap, where blue color is the missing volume from the prediction and red color is the volume that is added comparatively to the GT (**left**). Prediction of the TFL (electric blue) and gluteus medius (turquoise green) that influence each other, supperposed with the comparison from the left part (**right**). (Color figure online)

is segmented slightly outside its boundaries, its neighboring muscle will have a reduced segmentation, resulting in a predicted decrease in muscle volume. This phenomenon occurs particularly in muscle groups such as the adductors (longus, magnus and brevis), where the boundaries are difficult to discern even to the naked eye and remain a challenge even to human experts.

5   Conclusion

We have proposed a method that leverages long-range shape dependencies and prior anatomical information to segment muscles of elite athletes. Our experimental validation demonstrates that by incorporating anatomical priors as constraints into the segmentation process, our method achieves improved accuracy and captures the nuances of muscle boundaries more effectively. By accurately segmenting muscles, our method provides a valuable tool for quantitative analysis, allowing for a more comprehensive assessments of muscle morphology. This information can be valuable in identifying potential asymmetries or variations, guiding personalized training programs, injury prevention strategies, and performance optimization in sports and athletic settings. Moreover, a prediction is significantly faster that the manual segmentation method initially applied, which took approximately 30 to 40 h per subject. Since, the significant amount of time required for manual segmentation and the challenging nature of the task have limited the size of the database, we plan to evaluate the revision time when starting from our method's predictions to confirm the acceleration of the labeling process for new subjects.

To further advance and explore potential improvements, several directions can be considered. The exploration of methods based on unlabelled data could be pursued. Another possibility is to investigate the application of newer architecture designs that combine CNNs and transformers, such as Swin Transformers [14]. A third avenue for improvement is to explore the reduction of the number of transformer blocks in the network [9], to gain insights on the optimal balance between model complexity and segmentation accuracy. Finally, we plan to study the learned positional encodings and attention maps to better understand where the model focuses during the segmentation process. Analyzing attention maps can provide valuable insights into the features and regions that contribute most significantly to accurate muscle segmentation and to identify correlations between muscle groups. This understanding can guide future refinement of the model architecture and its performance optimization. Finally, with some adaptions our method could be used for the morphological study of muscles from patient with muskuloskeletal diseases.

Acknowledgments. We would like to acknowledge the manual segmentation operators, Iwen Dirouon and Eva Filleur, for their invaluable contribution. We would also like to thank Caroline Giroux for her support with the database. In conclusion, our gratitude extends to Guillaume Pelluet for his collaboration in developing the adjacency matrix code.

References

1. Agosti, A., et al.: Deep learning for automatic segmentation of thigh and leg muscles, Magnetic Resonance Materials in Physics, Biology and Medicine, pp. 1–17 (2022)
2. Al Arif, S.M.R., Knapp, K., Slabaugh, G.: Fully automatic cervical vertebrae segmentation framework for x-ray images. Comput. Methods Programs Biomed. **157**, 95–111 (2018)
3. BenTaieb, A., Hamarneh, G.: Topology aware fully convolutional networks for histology gland segmentation. In: International Conference on Medical Image Computing and Computer Assisted Interventions (MICCAI), pp. 460–468 (2016)
4. Chen, J., et al.: Transunet: transformers make strong encoders for medical image segmentation. arXiv preprint arXiv:2102.04306 (2021)
5. Cheng, R., Crouzier, M., Hug, F., Tucker, K., Juneau, P., McCreedy, E., Gandler, W., McAuliffe, M.J., Sheehan, F.T.: Automatic quadriceps and patellae segmentation of mri with cascaded u2-net and sassnet deep learning model. Med. Phys. **49**(1), 443–460 (2022)
6. Ganaye, P.A., Sdika, M., Triggs, B., Benoit-Cattin, H.: Removing segmentation inconsistencies with semi-supervised non-adjacency constraint. Med. Image Anal. **58**, 101551 (2019)
7. Gilles, B., Moccozet, L., Magnenat-Thalmann, N.: Anatomical modelling of the musculoskeletal system from MRI. In: Larsen, R., Nielsen, M., Sporring, J. (eds.) MICCAI 2006. LNCS, vol. 4190, pp. 289–296. Springer, Heidelberg (2006). https://doi.org/10.1007/11866565_36
8. Handsfield, G., Knaus, K., Fiorentino, N., Meyer, C., Hart, J., Blemker, S.: Adding muscle where you need it: non-uniform hypertrophy patterns in elite sprinters. Scandinavian J. Med. Sci. Sports **27**(10), 1050–1060 (2017)
9. Hasany, S.N., Petitjean, C., Meriaudeau, F.: A study of attention information from transformer layers in hybrid medical image segmentation networks. In: SPIE Medical Imaging: Image Processing, vol. 12464 (2023)

10. Hatamizadeh, A., Nath, V., Tang, Y., Yang, D., Roth, H.R., Xu, D.: Swin unetr: Swin transformers for semantic segmentation of brain tumors in mri images. In: Int. MICCAI Brainlesion Workshop, pp. 272–284. Springer (2021). https://doi.org/10.1007/978-3-031-08999-2_22

11. Hatamizadeh, A., Tang, Y., Nath, V., Yang, D., Myronenko, A., Landman, B., Roth, H.R., Xu, D.: Unetr: Transformers for 3d medical image segmentation. In: IEEE/CVF Workshop on Applications of Computer Vision (WACV) (2022)

12. Li, H., Hu, D., Liu, H., Wang, J., Oguz, I.: CATS: complementary cnn and transformer encoders for segmentation. In: Proceedings of IEEE International Symposium on Biomedical Imaging (ISBI) (2022)

13. Li, J., Chen, J., Tang, Y., Wang, C., Landman, B.A., Zhou, S.K.: Transforming medical imaging with transformers? a comparative review of key properties, current progresses, and future perspectives. Medical image analysis, p. 102762 (2023)

14. Liu, Z., et al.: Swin transformer: hierarchical vision transformer using shifted windows. In: IEEE/CVF International Conference on Computer Vision (ICCV) (2021)

15. Miller, R., et al.: The muscle morphology of elite sprint running (2020)

16. Mirikharaji, Z., Hamarneh, G.: Star shape prior in fully convolutional networks for skin lesion segmentation. In: Frangi, A.F., Schnabel, J.A., Davatzikos, C., Alberola-López, C., Fichtinger, G. (eds.) MICCAI 2018. LNCS, vol. 11073, pp. 737–745. Springer, Cham (2018). https://doi.org/10.1007/978-3-030-00937-3_84

17. Ni, R., Meyer, C.H., Blemker, S.S., Hart, J.M., Feng, X.: Automatic segmentation of all lower limb muscles from high-resolution magnetic resonance imaging using a cascaded three-dimensional deep convolutional neural network. J. Med. Imaging **6**(4), 044009 (2019)

18. Petit, O., Thome, N., Rambour, C., Themyr, L., Collins, T., Soler, L.: U-net transformer: self and cross attention for medical image segmentation. In: International Conference on Medical Image Computing and Computer Assisted Interventions (MICCAI) (2021)

19. Ronneberger, O., Fischer, P., Brox, T.: U-Net: convolutional networks for biomedical image segmentation. In: Navab, N., Hornegger, J., Wells, W.M., Frangi, A.F. (eds.) MICCAI 2015. LNCS, vol. 9351, pp. 234–241. Springer, Cham (2015). https://doi.org/10.1007/978-3-319-24574-4_28

20. Sutherland, A.M., et al.: Statistical shape modelling reveals differences in hamstring morphology between professional rugby players and sprinters. J. Sports Sci. **41**(2), 164–171 (2023)

21. Yokota, F., Otake, Y., Takao, M., Ogawa, T., Okada, T., Sugano, N., Sato, Y.: Automated muscle segmentation from ct images of the hip and thigh using a hierarchical multi-atlas method. Int. J. Comput. Assisted Radiol. Surgery (IJCARS) **13**, 977–986 (2018)

Geodesic Logistic Analysis of Lumbar Spine Intervertebral Disc Shapes in Supine and Standing Positions

Ye Han[1]([⊠]) [iD], James Fishbaugh[1] [iD], Christian E. Gonzalez[2] [iD],
Donald A. Aboyotes[2], Jared Vicory[1] [iD], Simon Y. Tang[2] [iD],
and Beatriz Paniagua[1] [iD]

[1] Kitware, Inc., Clifton Park, NY 12065, USA
ye.han@kitware.com
[2] Washington University in St. Louis, St. Louis, MO 63110, USA

Abstract. Non-specific lower back pain (LBP) is a world-wide public health problem that affects people of all ages. Despite the high prevalence of non-specific LBP and the associated economic burdens, the pathoanatomical mechanisms for the development and course of the condition remain unclear. While intervertebral disc degeneration (IDD) is associated with LBP, there is overlapping occurrence of IDD in symptomatic and asymptomatic individuals, suggesting that degeneration alone cannot identify LBP populations. Previous work has been done trying to relate linear measurements of compression obtained from Magnetic Resonance Imaging (MRI) to pain unsuccessfully. To bridge this gap, we propose to use advanced non-Euclidean statistical shape analysis methods to develop biomarkers that can help identify symptomatic and asymptomatic adults who might be susceptible to standing-induced LBP. We scanned 4 male and 7 female participants who exhibited lower back pain after prolonged standing using an Open Upright MRI. Supine and standing MRIs were obtained for each participant. Patients reported their pain intensity every fifteen minutes within a period of 2 h. Using our proposed geodesic logistic regression, we related the structure of their lower spine to pain and computed a regression model that can delineate lower spine structures using reported pain intensities. These results indicate the feasibility of identifying individuals who may suffer from lower back pain solely based on their spinal anatomy. Our proposed spinal shape analysis methodology have the potential to provide powerful information to the clinicians so they can make better treatment decisions.

Keywords: geodesic logistic regression · intervertebral disc degeneration · longitudinal shape analysis

1 Introduction

Low back pain has a large worldwide prevalence and it affects people of all genders and ages [12]. Studies estimate that at least 75–85% of Americans experience some form of back pain during their life, with the highest prevalence happening in females and in adults aged 40–65 [11]. Work disabilities in adults aged 40 to 65 years of age costs employers an estimated \$7.4 billion/year, with LBP exacerbation accounting for a disproportionate share of lost productivity [15].

Despite of the incredible burden of nonspecific LBP, no recent research effort has yet succeeded to relate clinical variables such as pain to anatomy [21]. Other investigators have looked at LBP with ultrasound, but always while the person is in supine position [20]. Some investigators have related pain with local anatomy in the case of evident pathology using MRI, but this is more challenging in the case of nonspecific LBP.

One particularly interesting group of individuals are those that are otherwise back-healthy with no prior history of LBP, but that develop clinically significant LBP symptoms after prolonged standing [5,13,16,17]. Individuals with jobs that involve standing for periods of more than 30 min of each hour is one of the strongest predictors of work-related LBP [1,18]. Moreover, individuals who develop LBP symptoms during prolonged standing (PD - Pain Developers) are three times more likely to develop chronic LBP compared to those that do not develop pain (NPD - Non-pain developers). Diagnostic imaging routinely taken of the lower back traditionally involves having the subject lying down, which will not reflect the anatomy present during LBP development after prolongued standing. Positional MRI technology allows signal acquisition in an open configuration, and it enables the acquisition of scans reflecting the spinal anatomy of an individual while standing and other positions [19].

We propose using positional MRI techniques combined with advanced shape statistics in order to characterize the changes of the lumbar spine from supine to standing in young, back-healthy individuals that develop LBP after prolonged standing. Our hypothesis is that methods that take into account the entire anatomy of the lower back as well as its progression between different positions will be capable of relating subject-specific spinal anatomy to clinical variables such as pain.

2   Materials

2.1   Demographics

Eleven human participants (4 male/7 female) were recruited for this study with informed consent under approval of Washington University Institutional Review Board approval. The participants were between 18 and 30 years old with a Body Mass Index lower than 30 kg/m^2. These participants were recruited from the community surrounding the Saint Louis, Missouri area, and were included in our study if they reported LBP after prolonged standing in a supervised clinic [7,19].

Following the standing MRI acquisition, the participants were instructed to continue standing for up to two hours, but they were allowed to end the standing regimen if they could not tolerate the pain or discomfort. Every 15 min, the participant reported the extent of their LBP symptoms on a visual analogue scale (VAS). A VAS rating was made by marking the location along a 100 mm horizontal line corresponding to the current level of pain, with 0 being *no pain* and 100 being *worst pain*, which was collected for each patient at each time point of each repeated MRI. Table 1 shows the patient data we use in this study. The VAS values of 11 subjects are recorded in supine and standing positions at 15-min intervals.

Table 1. Subject VAS values at each time point. Highly variable VAS values are reported by each subject.

Subject id	VAS Supine	T0	T1	T2	T3	T4	T5	T6	T7
03	0	5	8	4	13	16	9	13	12
05	0	0	0	2	4	5	10	11	15
15	0	0	0	0	0	0	0	7	14
16	0	0	0	5	9	10	3	10	6
18	0	0	0	0	0	0	0	8	
19	0	0	0	0	0	0	0	17	11
25	0	0	0	0	3	0	7	20	8
28	0	9	0	17	25	20	10	19	
29	0	0	5	20	17	23	28	34	44
31	0	0	0	0	0	0	0	5	6
33	0	0	0	0	3	3	3	5	4

2.2   MRI Acquisition and Segmentation

Images of the lumbar spine (L1-S1) were obtained using the 0.6T Open Upright® MRI (Fonar, New York, NY) system. A 3-plane localizer was used to acquire sagittal T2 weighted images with the following acquisition parameters: repetition time = 610 ms, echo time = 17 ms, field of view = 24 cm, acquisition matrix = 210×210, slice thickness = 3 mm, no gap, scan duration = 2 min). This sequence was optimized for reducing scan time and motion artifacts [19]. The table was adjusted to a horizontal position of 180°. The participant was positioned in supine for 10 min prior to the first scan. The MRI table then was moved to a vertical position with a 84° table tilt where participants were then standing. The table tilt of 84° stablizes the participant and prevent motion artifacts during the imaging in standing. Participants were told to stand normally without leaning on the sides of the magnet, back of the scanner or on the VersaRest™ during the scan in standing.

The exported DICOM files were then manually segmented using 3D Slicer [9] by a single expert rater. Each sagittal slice was evaluated and the intervertebral disc (IVD) contour was defined for each lumbar segment and the combined into a set of discrete volumes. Figure 1 shows an example snapshot of the MRI image and the segmented IVDs.

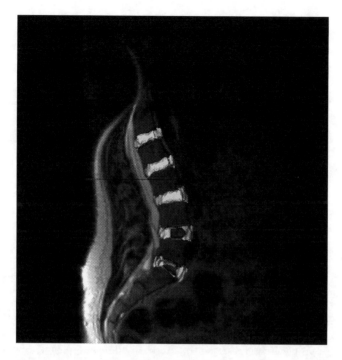

Fig. 1. An MRI acquisition of the lumbar spine region. Two of the segmented IVD shapes have holes on the side due to the limited size of scanning window.

2.3   Correspondence Estimation

The first step in establishing correspondence is the construction of a prototype shape complex, as in   [4]. The prototype shape defines the topology as well as the density of sampling across the surface of each shape, and serves as the common shared representation across the population. A mesh of the prototype shape complex for the intervertebral discs is shown in the left of Fig. 2. For each subject, the prototype shape complex is aligned to the supine observation by diffeomorphic surface registration. This brings the prototype shape complex into the subject-specific coordinate system and closely matches the geometry of the raw observations. This has the added benefit of preserving the topology defined by the prototype shape, dealing with holes or missing structures in the observations, as shown with the red arrow in Fig. 2.

Next, for each subject, the aligned supine shape complex is propagated across the time-series by diffeomorphic piecewise geodesic registration. This carries the common topology and sampling to match each raw observation in the time-series. Starting from a prototype shape configuration with a dense sampling, this establishes correspondence across the population and across time [4]. The shape alignment procedure was implemented using deformetrica [2].

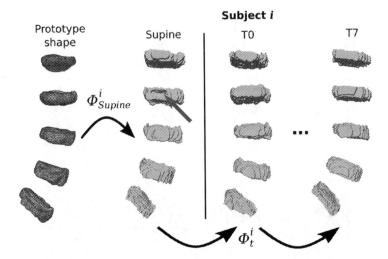

Fig. 2. Overview of correspondence estimation. The prototype shape complex is a common representation across the population which defines topology and shape sampling. First, the prototype shape complex is aligned to the supine observation of each subject by diffeomorphic registration by ϕ^i_{Supine}. The red arrow shows a hole in the raw observation which will be replaced with the consistent topology of the prototype shape. The results are propagated across the time-series by diffeomorphic piecewise geodesic registration ϕ^i_t which establish correspondence across the time-series as well as the population of subjects i. (Color figure online)

After the correspondence is established, we adopt partial procrustes alignment to obtained aligned IVDs (Fig. 3) for geodesic logistic shape analysis which is described in details in the next section.

3   Methods

3.1   Shape Space and Geodesics

The IVD shapes analyzed by geodesic logistic regression (GLR) are obtained through partial procrustes alignment and lie on shape space defined as the pre-shape space of Kendall space [10], in which translation, rotation, and similarity components are all removed. The shape space is formed as a high dimensional sphere which can be treated as a Riemannian manifold M. A geodesic on M is a zero-acceleration curve with the property that no shorter curve exists between any two points within a small neighborhood other than the geodesic. We use two manifold specific operations in this work, namely the exponential map and the log map. An exponential map $Exp(p, v) = q$ maps a shape $p \in M$ to another shape $q \in M$ along a geodesic path in the direction and magnitude of a tangent vector v. A log map $Log(p, q) = v$ is the inverse operation of the exponential map in which a unique tangent vector that maps p to q along a geodesic path

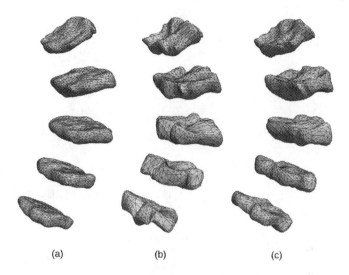

(a)   (b)   (c)

Fig. 3. Procrustes aligned IVD shapes for a single individual in (a) supine position and at time points (b) T0 and (c) T7 of the time series.

is obtained given two shapes p and q. The geodesic distance between the two shapes is then defined as the L2-norm of their log map $dist(p, q) = ||Log(p, q)||$. More rigorous and complete definitions of related concepts are available in [3].

3.2   Logistic Regression in Shape Space

A logistic regression model can be written as

$$p(x) = \frac{1}{1 + e^{-\frac{(x-u)}{s}}} \tag{1}$$

where $p(x)$ is the probability of explanatory variables x having certain label, u is the decision boundary and s is the scale parameter representing the sharpness of the decision boundary.

In shape space, we also want the decision boundary which splits the space into two subspaces. Similar to the concept of principal nested spheres [8], the splitting boundary for a $n + 1$ sphere S^{n+1} could be a n sphere S^n, the sphere that is one dimension lower than S^{n+1}. We define the decision boundary as such a sphere with two parameters $C(u, d)$, where u is a point on S^{n+1} and d is the geodesic distance between u and the decision boundary C. Thus, if we create a scalar field on Kendall space based on the geodesic distance to u, C is the isocontour with value d. In terms of value, u is equivalent to the normal vector of C.

Given the above definition of a decision boundary $C(u, d)$ on S^{n+1} and a shape x, we define the geodesic logistic regression model as

$$p(x) = \frac{1}{1 + e^{\frac{-(dist(x,u)-d)}{s}}} \tag{2}$$

where $dist(x, u)$ is the geodesic distance between x and u, and s is the scale parameter.

3.3   Parameter Estimation

Let $y \in \{0, 1\}$ be the label of corresponding x, the unified negative log-likelihood function is formulated as

$$l = -y \ln p - (1 - y) \ln(1 - p). \tag{3}$$

Given an input data set, we would like to minimize this negative likelihood function to obtain fitted parameters of the logistic regression model. To facilitate calculation, we apply the chain rule and define intermediate variable z as

$$z = \frac{dist(x, u) - d}{s}. \tag{4}$$

Then p and the partial derivatives can then be derived as:

$$p = \frac{1}{1 + e^{-z}}, \tag{5}$$

$$\frac{\partial l}{\partial p} = -\frac{y}{p} + \frac{1 - y}{1 - p}, \tag{6}$$

$$\frac{\partial p}{\partial z} = p(1 - p), \tag{7}$$

$$\frac{\partial z}{\partial u} = \frac{-\hat{Log}_u x}{s}, \tag{8}$$

$$\frac{\partial z}{\partial d} = -\frac{1}{s}, \tag{9}$$

$$\frac{\partial z}{\partial s} = -\frac{dist(x, u) - d}{s^2} = -\frac{z}{s}, \tag{10}$$

where $\hat{Log}_u x$ is the normalized tangent vector from u to x.

The shape space is not Euclidean, hence we estimate the parameters by using iterative gradient descent. The ith iteration is calculated on the local tangent hyperplane at u_i and the parameters are updated as:

$$u_{i+1} = Exp(u_i, -\alpha \frac{\partial l}{\partial u_i}), \tag{11}$$

$$d_{i+1} = d_i - \alpha \frac{\partial l}{\partial d_i},$$ (12)

$$s_{i+1} = s_i - \alpha \frac{\partial l}{\partial s_i},$$ (13)

where α is the step size.

Given a dataset with K shapes, the negative log-likelihood function L is simply the summation of the individual l's and the parameter estimation process would be the same as the above in the summation form.

4   Results and Discussions

4.1   Test on Low Dimensional Synthetic Data

To validate our GLR model, we first test our implementation on the unit 3D sphere. Points on the unit sphere represent shapes with only one 3D point and the sphere is the corresponding shape space. Figure 4 shows the progress of estimating GLR model parameters from 9 labeled input points at various iterations. The red and the green are the input points with labels 1 and 0 respectively. The blue and the black background points indicate the sub-regions around the input points labeled with 1 and 0 from the GLR model at the ith iteration. We set the optimization parameters as 0.0001 and 1×10^{-8} for the step size and the step termination tolerance. The initial d, s and u are chosen as 0.5, 1 and the Fréchet mean [14] of all input points. As shown in Fig. 4, the decision boundary moves fast in the first 5000 iterations with some overshooting due to the large gradient and the linearized iterative optimization scheme. The decision boundary then gradually conform to the input points until convergence at iteration $i = 23395$.

4.2   Analysis of IVD Data

We further applied our GLR model to the obtained IVD data set described in Sect. 2. For both tests on the cross-subject supine shapes and on the IVD shapes within individual subjects, we select the same parameters for GLR model parameter estimation: step size $= 0.01$, max iteration $= 10000$ and step termination tolerance $= 1 \times 10^{-8}$.

Cross-Subject Supine Shapes. We first test our GLR model on the supine shapes from the input data set. We adopted three types of labeling strategies to explore, under our GLR model, whether there is potential separation between individual subjects that relates their supine IVD shape characteristics to the development of LBP. First, to test separation between significant pain developers versus non-significant pain developers, we use the maximum VAS values in time series data as the labeling criteria. Thresholds of 15 and 20 are adopted $(\max(\text{VAS}) \geq 15/20)$. Second, we try separating easy pain developers from non-easy pain developers by labeling subjects using total numbers of time points that

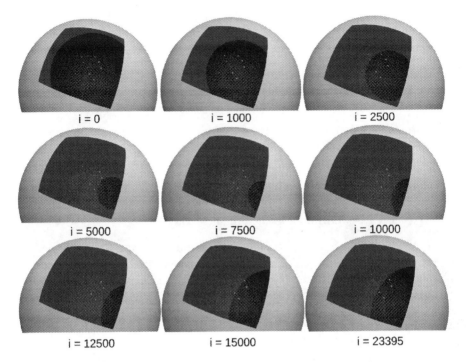

Fig. 4. The progress of estimating GLR model parameters from 9 labeled input points on a 3D sphere at various iterations. The red and the blue background points are labeled with 1, and the green and black background points are labeled with 0 in the input and the GLR model respectively. The parameter fitting iterations terminate at $i = 23395$ due to the step size being smaller than the preset step tolerance. (Color figure online)

has VAS values greater than certain threshold of low VAS value. 4 or more entries with 5 or greater VAS values within a single subject's time series data is adopted as the labeling criterion (sum(VAS≥5) ≥ 4). Last, to distinguish between early pain developers and late pain developers, number of non-zero entries is used and we choose 4 as the labeling criterion (sum(VAS = 0) ≥ 4). All the labeling criteria are chosen such that the label distribution are relatively balanced. Otherwise, a uniform label of the majority population could yield a very decent regression result.

Table 2 shows the regression results from fitted GLR model on the supine shapes using different labeling criteria. All fitted GLR models yield labeling result with 72.7% or higher accuracy, indicating the likely separation between individual subject's supine shapes under the proposed GLR model and the selected labeling criteria i.e. self reported pain intensity after standing. We deem results to be very promising considering the heterogeneity present in the input data set.

Table 2. Regression results from fitted GLR model.

Labeling criteria	Mislabeled subjects											Accuracy
	03	05	15	16	18	19	25	28	29	31	33	
max(VAS) $\geq$ 15		x		x								81.8%
max(VAS) $\geq$ 20				x			x	x				72.7%
sum(VAS$\geq$5) $\geq$ 4	x		x		x							72.7%
sum(VAS=0) $\geq$ 4						x			x	x		72.7%

Patient-Specific Supine and Time Series Data. We further applied GLR to each patient supine and standing time series data to test whether there is a possible separation of the IVD anatomy of a single patient based on pain (VAS = 0) and non-pain (VAS $\neq$ 0). The results from fitting GLR models on individual subjects are shown in Table 3. Note that the logistic regression model is different from other classification paradigms like support vector machine which searches for maximizing the separation margin between data points with different labels. Instead, the parameter estimation process would generate a GLR model that minimizes the overall negative log-likelihood function. If there is a cut-off on the GLR outputs that clearly separates the inputs with different labels, a unique distance d can then be inferred accordingly based on Eq. 2 and thus the boundary separating each population can be derived. As indicated from the results, we find almost perfect cut-offs for most of the subjects. Only three subjects have a single missclassified time point data. These results are very encouraging because our GLR formulation can consistently separate between pain and non-pain IVD shapes for almost all patients, especially considering the heterogeneous nature

Table 3. Outputs from fitted GLR model on individual time series data (mislabeled shapes in red using manual cut-offs).

Subject id	Fitted GLR model outputs (mislabeled in red)									Cut-off	Accuracy
	Supine	T0	T1	T2	T3	T4	T5	T6	T7		
03	0.621	0.966	0.971	0.977	0.980	0.973	0.978	0.978	0.975	0.7	100%
05	0.506	0.606	0.595	0.808	0.853	0.801	0.821	0.791	0.810	0.7	100%
15	0.171	0.219	0.187	0.146	0.155	0.145	0.226	0.259	0.257	0.24	100%
16	0.532	0.690	0.647	0.743	0.749	0.725	0.685	0.793	0.749	0.7	88.9%
18	0.026	0.014	0.014	0.016	0.018	0.022	0.040	0.461		0.2	100%
19	0.004	0.018	0.029	0.195	0.125	0.183	0.137	0.439	0.453	0.3	100%
25	0.338	0.448	0.323	0.457	0.497	0.467	0.463	0.484	0.475	0.47	88.9%
28	0.541	0.896	0.688	0.909	0.879	0.900	0.891	0.884		0.7	100%
29	0.185	0.013	0.999	0.999	1.000	1.000	0.999	0.997	1.000	0.5	100%
31	0.122	0.188	0.045	0.034	0.169	0.265	0.194	0.245	0.403	0.23	88.9%
33	0.185	0.589	0.519	0.544	0.682	0.701	0.684	0.633	0.761	0.6	100%

of the IVD shapes and the limitations in the data acquisition and preparation i.e. the incompletely scanned IVDs, the low resolution from the positional MRI and the manual segmentation.

4.3   Limitations and Future Work

Below we list several limitations of our current study that, if improved in future, could be further used to refine our understanding of the geometric characteristics of LBP.

1. Some of the field of views of the positional MRI scans employed in this study were too small to capture the entire IVD geometry. This means that the IVD shapes had non-closed boundaries on the left and/or right sides and thus were incomplete. We artificially closed them with flat planes in order to carry our GLR analysis.
2. The VAS values are subjectively self-reported with limited positional consistency. Subjective assessment of VAS may cause shift in the perceived VAS value over the duration of the study ($\sim$2 h) as a patient may become more used to the standing position. Meanwhile, even if the scan data was generated in a way that the patients remained in the same standing position as much as possible, minor positional adjustment may still be possible during the study. The non monotonic changes in the VAS value potentially reflect the above aspects of the data set.
3. A more sophisticated version of the GLR model may yield better fitting results on the data set. It is possible to parameterize the boundary distance d as a function of the tangent vector at u. Parametric generalization of the decision boundary, an S^n sphere, to spline based geodesic curves [6] is also feasible to be used to create decision boundary with closed topology.

5   Conclusions

In this paper we present a novel application of a GLR model targeted to analyze the anatomy of the spine of patients that suffer LBP after prolonged standing as captured by positional MRI both in supine and standing positions. The GLR model has demonstrated very positive results in delineating IVD shapes based on VAS values on the cross population supine data set as well as the individual, subject-specific supine and standing data sets.

LBP happens in otherwise healthy individuals, and it first appears after prolonged standing. The appearance of LBP during standing is a strong predictor to the development of chronic LBP, that has the potential to cause an incredible human and financial burden to individuals that suffer it. Thus, we believe our proposed tool will be important to prevent patient from suffering and loss of productivity. Also importantly, this system has the potential to optimize work performance on individuals that are required to stand for prolonged times at their jobs by predicting the time before pain develops.

We plan to continue this work by improving data acquisition and processing, increasing our sample size and refining our non-Euclidean shape modeling techniques.

Acknowledgements. This work is supported by the National Institute of Health R01EB021391 "Shape Analysis Toolbox: From medical images to quantitative insights of anatomy" (SlicerSALT).

References

1. Andersen, J.H., Haahr, J.P., Frost, P.: Risk factors for more severe regional musculoskeletal symptoms: a two-year prospective study of a general working population. Arthritis Rheumatism **56**, 1355–1364 (2007). https://doi.org/10.1002/ART.22513. https://pubmed.ncbi.nlm.nih.gov/17393441/

2. Bône, A., Louis, M., Martin, B., Durrleman, S.: Deformetrica 4: an open-source software for statistical shape analysis. In: Reuter, M., Wachinger, C., Lombaert, H., Paniagua, B., Lüthi, M., Egger, B. (eds.) ShapeMI 2018. LNCS, vol. 11167, pp. 3–13. Springer, Cham (2018). https://doi.org/10.1007/978-3-030-04747-4_1

3. do Carmo, M.P.: Differential Geometry of Curves and Sur4. Fletcher, T.: Geodesic Regression on Riemannian Manifolds. Prentice Hall (1976)

4. Fishbaugh, J., et al.: Estimating shape correspondence for populations of objects with complex topology. In: 2018 IEEE 15th International Symposium on Biomedical Imaging (ISBI 2018), pp. 1010–1013. IEEE (2018)

5. Gregory, D.E., Callaghan, J.P.: Prolonged standing as a precursor for the development of low back discomfort: an investigation of possible mechanisms. Gait Posture **28**, 86–92 (2008). https://doi.org/10.1016/J.GAITPOST.2007.10.005. https://pubmed.ncbi.nlm.nih.gov/18053722/

6. Hanik, M., Hege, H.C., Hennemuth, A., von Tycowicz, C.: Nonlinear regression on manifolds for shape analysis using intrinsic bézier splines. In: Martel, A.L., Abolmaesumi, P., Stoyanov, D., Mateus, D., Zuluaga, M.A., Zhou, S.K., Racoceanu, D., Joskowicz, L. (eds.) Medical Image Computing and Computer Assisted Intervention - MICCAI 2020, pp. 617–626. Springer, Cham (2020)

7. Hwang, C., LR, V.D., S, H.: Do changes in sensory processing precede low back pain development in healthy individuals? Clin. J. Pain **34**(6), 525–531 (06 2018). https://doi.org/10.1097/AJP.0000000000000563

8. Jung, S., Dryden, I.L., Marron, J.S.: Analysis of principal nested spheres. Biometrika **99**(3), 551–568 (2012). https://doi.org/10.1093/biomet/ass022

9. Kikinis, R., Pieper, S.D., Vosburgh, K.G.: 3D Slicer: a platform for subject-specific image analysis, visualization, and clinical support. In: Jolesz, F.A. (ed.) Intraoperative Imaging and Image-Guided Therapy, pp. 277–289. Springer, New York (2014). https://doi.org/10.1007/978-1-4614-7657-3_19

10. Klingenberg, C.P.: Walking on kendall's shape space: understanding shape spaces and their coordinate systems. Evolutionary Biology, pp. 1–19 (2020)

11. Lucas, J., Connor, E., Bose, J.: Back, lower limb, and upper limb pain among u.s. adults, 2019. NCHS Data Brief, no 415. (2019). https://doi.org/10.15620/CDC: 107894. https://stacks.cdc.gov/view/cdc/107894

12. Maher, C., Underwood, M., Buchbinder, R.: Non-specific low back pain. Lancet **389**(10070), 736–747 (2017). https://doi.org/10.1016/S0140-6736(16)30970-9. https://www.sciencedirect.com/science/article/pii/S0140673616309709

13. Nelson-Wong, E., Callaghan, J.P.: Transient low back pain development during standing predicts future clinical low back pain in previously asymptomatic individuals. Spine **39**, E379–83 (2014). https://doi.org/10.1097/BRS.0000000000000191. https://europepmc.org/article/MED/24384659
14. Nielsen, F., Bhatia, R.: Matrix Information Geometry. Springer, Berlin Heidelberg (2012). https://books.google.com/books?id=MAhygTspBU8C
15. Ricci, J.A., Stewart, W.F., Chee, E., Leotta, C., Foley, K., Hochberg, M.C.: Back pain exacerbations and lost productive time costs in united states workers. Spine **31**, 3052–3060 (2006). https://doi.org/10.1097/01.BRS.0000249521.61813. AA. https://europepmc.org/article/med/17173003
16. Sorensen, C.J., Johnson, M.B., Callaghan, J.P., George, S.Z., Dillen, L.R.V.: Validity of a paradigm for low back pain symptom development during prolonged standing. Clin. J. Pain **31**, 652–659 (2015). https://doi.org/10.1097/AJP.0000000000000148. https://pubmed.ncbi.nlm.nih.gov/25171636/
17. Sorensen, C.J., Norton, B.J., Callaghan, J.P., Hwang, C.T., Dillen, L.R.V.: Is lumbar lordosis related to low back pain development during prolonged standing? Manual Therapy **20**, 553–557 (2015). https://doi.org/10.1016/J.MATH.2015.01.001. https://pubmed.ncbi.nlm.nih.gov/25637464/
18. Sterud, T., Tynes, T.: Work-related psychosocial and mechanical risk factors for low back pain: a 3-year follow-up study of the general working population in norway. Occupat. Environ. Med. **70**(5), 296–302 (2013). https://doi.org/10.1136/oemed-2012-101116. https://oem.bmj.com/content/70/5/296 for low back pain: a 3-year follow-up study of the general working population in norway. Occupational and Environmental Medicine **70**(5), 296–302 (2013). https://doi.org/10.1136/oemed-2012-101116. https://oem.bmj.com/content/70/5/296
19. Weber, C.I., Hwang, C.T., van Dillen, L.R., Tang, S.Y.: Effects of standing on lumbar spine alignment and intervertebral disc geometry in young, healthy individuals determined by positional magnetic resonance imaging. Clin. Biomechanics **65**, 128–134 (2019). https://doi.org/10.1016/j.clinbiomech.2019.04.010. https://www.sciencedirect.com/science/article/pii/S0268003319301159
20. Yu, X., et al.: Binary classification of non-specific low back pain condition based on the combination of b-mode ultrasound and shear wave elastography at multiple sites. Front. Physiol. **14** (2023). https://doi.org/10.3389/FPHYS.2023.1176299. https://pubmed.ncbi.nlm.nih.gov/37187960/
21. Zhu, W.P., Huang, Y., Hu, P., Lin, W.: Lumbar extensor and flexor muscle structural changes in young female nurses with chronic bilateral non-specific low back pain: a case-control study. Discovery Med. **35**(444), pp. 10, 2023.24976/DISCOV.MED.202335176.45. https://pubmed.ncbi.nlm.nih.gov/37272111/

SlicerSALT: From Medical Images to Quantitative Insights of Anatomy

Jared Vicory[1]([📧])[iD], Ye Han[1][iD], Juan Carlos Prieto[2][iD], David Allemang[1][iD],
Mathieu Leclercq[2], Connor Bowley[1], Harald Scheirich[1],
Jean-Christophe Fillion-Robin[1], Steve Pizer[2][iD], James Fishbaugh[1][iD],
Guido Gerig[3][iD], Martin Styner[2][iD], and Beatriz Paniagua[1][iD]

[1] Kitware Inc, Carrboro, NC 27510, USA
jared.vicory@kitware.com
[2] University of North Carolina, Chapel Hill, NC 27599, USA
[3] New York University, Brooklyn, NY 11201, USA

Abstract. Three-dimensional (3D) shape lies at the core of understanding the physical objects that surround us. In the biomedical field, shape analysis has been shown to be powerful in quantifying how anatomy changes with time and disease. The Shape AnaLysis Toolbox (SALT) was created as a vehicle for disseminating advanced shape methodology as an open source, free, and comprehensive software tool. We present new developments in our shape analysis software package, including easy-to-interpret statistical methods to better leverage the quantitative information contained in SALT's shape representations. We also show Slicer-Pipelines, a module to improve the usability of SALT by facilitating the analysis of large-scale data sets, automating workflows for non-expert users, and allowing the distribution of reproducible workflows.

Keywords: Shape analysis · Statistics · Open-source software

1 Introduction

Statistical shape analysis is an essential area of research in computer science and mathematics, with application areas as diverse as biology, anatomy, agriculture, or paleontology. Shape represents critical morphometric features not encoded in simple derived measurements such as volume or linear distances. Over the past three decades, our research team and others have been developing shape analysis methodologies that are widely used within their own labs. However, these methods have not fully transitioned into clinical applications due to requiring the users to be experts in computer science or statistics.

SlicerSALT was created with the vision to integrate recent methodological advancements into a user-friendly toolkit. SlicerSALT meets the critical need of the broad scientific community to have access to technologies otherwise only available to specific computer science research labs, all while encouraging reproducibility and transparency. It benefits from a large international user community by making use of the 3D Slicer ecosystem as well as Kitware's open source

C. Wachinger et al. (Eds.): ShapeMI 2023, LNCS 14350, pp. 201–210, 2023.
https://doi.org/10.1007/978-3-031-46914-5_16

libraries such as the Insight Toolkit (ITK) [10] and the Visualization Toolkit (VTK) [12]. SALT follows strict quality standards for software development and testing. These factors position SlicerSALT as one of the leaders in shape analysis software development and dissemination and create conventions and quality standards for future work.

SlicerSALT fills a niche for researchers interested in a broad range of biomedical shape analysis tasks. Most other packages focus on tasks such as image segmentation (itkSNAP [16]) or specific neuroimaging pipelines (FreeSurfer [5], FSL [6]). There are only a few freely distributed toolkits for generic shape analysis tasks, including ShapeWorks [3], Deformetrica [4] or Statismo [1]. However, those packages each have a relatively narrow focus. Deformetrica deals with deformations of the 2D or 3D ambient space while ShapeWorks and Statismo are only designed for point distribution models (PDMs) or principal component analysis (PCA). SALT, on the other hand, provides users with a wide variety of tools for both creating different shape representations and performing statistical analysis to accomodate a wider range of applications.

This paper is intended to present the new features in SlicerSALT v4.0, as compared with previous versions [15]. Over the past six years SlicerSALT has provided biomedical researchers with access to sophisticated shape analysis methodology otherwise reserved for computer science experts. In this new release, we intend to expand this toolkit towards novel and interpretable shape statistics, in order to maximize the potential benefits of the geometric information contained in medical data and expand its use in order to support impactful biomedical research.

2   Shape Representations and Correspondence

In addition to standard PDM models computed using SPHARM-PDM, SlicerSALT 4.0 has improved support for skeletal shape models and producing correspondence via registration. We have also updated various visualization tools for examining shape populations for quality control.

2.1   Skeletal Representations

The functionality for creating, displaying, analyzing, and storing skeletal representations (s-reps) was completely re-engineered for SlicerSALT 4.0. These updates were necessary to make the tools easier to use as well as allowing reuse of the code in other modules.

We have developed a new SRep module that allows users to see basic information about s-reps as well as modify their display properties. The user can easily change spoke and skeletal grid thicknesses, colors, opacity, and visibility for different parts of the s-rep.

This module also adds a new Medical Reality Modeling Language (MRML) data node for s-reps. This allows modules in this and other extensions to easily use, manipulate, and display s-rep data as it would with other data types native

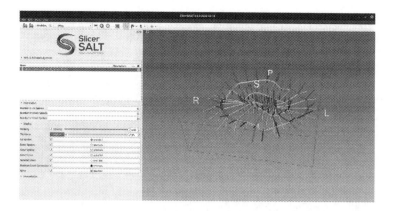

Fig. 1. Skeletal representations module in SlicerSALT with an s-rep visualization.

to 3D Slicer. Additionally, this makes it easier to save, load and share s-rep files as they can be read and written via the standard 3D Slicer Save and Load buttons as well as loaded via drag-and-drop. The new JSON-based file format is also more human readable and allows the data to be read using standard JSON parsing libraries.

S-reps are fit to objects using the SRepCreator and SRepRefiner modules. First, an input surface is given to the SRepCreator which generates an s-rep using a method based on surface curvature flow to estimate a best-fitting ellipsoid [8]. An s-rep is analytically computed from the ellipsoid and propagated back to the original surface via thin-plate splines. The SRepRefiner takes this initial s-rep and uses an optimization process to improve its fit to the original object surface as well as geometric properties such as ensuring that s-rep spokes are orthogonal to the object boundary.

Other modules in SlicerSALT that work on s-reps have been modified to support this new format. These modifications will be detailed in the following relevant sections.

2.2   Registration Based Correspondence

Common techniques for creating shape representations, such as SPHARM-PDM, attempt to create models that already have inherent correspondence across a population. This works well for objects with simple topologies. However, for objects with more complicated geometries or topologies, other approaches are required. A variety of techniques have been developed to address this problem, but many of them do not have implementations that are available for wide use and do not come with groups of associated tools for analyzing the resulting representations.

SlicerSALT 4.0 now includes a module for creating corresponding representations of objects with complex geometry or non-spherical topology [14]. It takes a mesh of a representative object as a template and deforms this template to match other objects of the population via registration to create correspondence. We use ITK's registration framework with the option of either a b-spline or diffeomorphic demons [13] registration to register a template surface mesh to each of the target objects. For this, we first create level-set representations of the template and target meshes by computing distance functions from each object. The template distance function is then registered to the target using the diffeomorphic demons registration algorithm. The resulting transform is then applied to the original template mesh, yielding a mesh that conforms to the target object's boundary but has the topology of the template for all of the population. Figure 2 shows correspondences generated using this module.

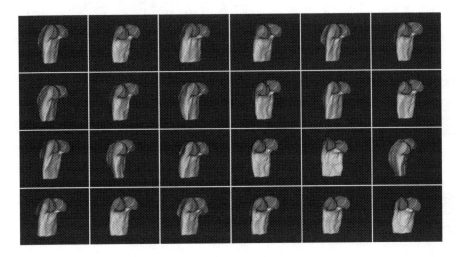

Fig. 2. Corresponding femur models computed via registration-based correspondence. Similar coloring at similar anatomical locations show the quality of the correspondence.

2.3 Shape Population Viewer

Shape Population Viewer is a module that allows the user to quickly visualize an entire population of objects simultaneously. This allows the user to quickly assess population-wide correspondences such as in Fig. 2 or verify there are no ill-formed meshes. This module has been extended to support s-reps as seen in Fig. 3.

3 Shape Analysis Methods

SlicerSALT 4.0 introduces changes to several existing shape analysis modules as well as introducing several new ones.

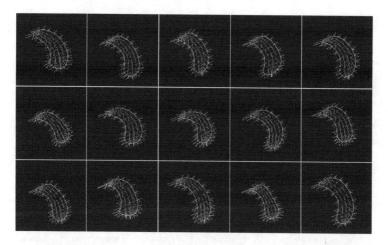

Fig. 3. ShapePopulationViewer displaying a set of 15 s-reps.

3.1 Shape Variation Analyzer

The Shape Variation Analyzer (SVA) module is designed to take in a set of corresponding PDMs or s-reps and use principal component analysis (PCA) or composite principal nested spheres (CPNS) [11] to compute a mean shape and its major modes of variation. The user can then visualize the mean shape as well as how moving along each principal component changes the shape of the object. This module allows for a quantitative comparison of the generated shape space to those created by other methods by examining the percent of variation explained by the modes of variation as well as computing generalization, specificity, and compactness measures on the distribution.

PCA. For Euclidean shape representations such as PDMs, SVA allows the computation of PCA directly on the point positions of the shape representation. SVA also allows the user to view the mean shape and explore the modes of variation of the PCA shape space and to evaluate the quality of the generated models as seen in Fig. 4.

The inputs are a list of VTK mesh files that are in correspondence. PCA results can be saved using the "Save Exploration" button and loaded using the "JSON File" entry.

Principal Nested Spheres. Because s-reps are non-Euclidean, the s-rep data must first be Euclideanized using principal nested spheres (PNS) [7] before PCA can be computed. This approach is known as composite principal nested spheres (CPNS) and has been shown to produce more compact representations with more meaningful modes of variations for s-reps. After CPNS is computed the user can interact with the mean s-rep and its modes of variation similarly to the PCA tools for PDMs. The module automatically chooses to use CPNS when

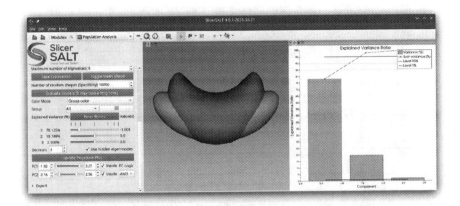

Fig. 4. A screenshot of the SVA module in SlicerSALT. The left panel shows controls, the middle shows a mean shape (gray, transparent) alongside a shape (blue) deformed along its first principal component, and the right shows a PCA scree plot. (Color figure online)

s-reps are provided so that the user can not accidentally use the wrong analysis tools.

3.2 Distance Weighted Discrimination

Distance Weighted Discrimination (DWD) is a binary classification method designed to address shortcomings with support vector machine (SVM) performance when applied to high-dimension, low-sample-size (HDLSS) data [9] . We have created a module in SlicerSALT that allows the user to perform PDM classification based on one categorical variable. Similarly to SVM, DWD performs classification by computing the distance of each sample to a separating hyperplane, with samples lying on the same side of the hyperplane being classified together. It improves over SVM by considering the effects of all of the data on the separating hyperplane rather than just a limited set of support vectors.

3.3 Deep Learning for Geometry: FlyBy CNN

Building upon recent work to start bringing powerful deep learning methodologies to 3D Slicer, SlicerSALT now includes a projection-based convolutional learner called FlyBy CNN [2]. Specifically, 2D projection snapshots are acquired at locations along a predefined path on an encompassing sphere. The collection of snapshots forms an ordered 2.5D dataset that is analyzed via existing time series CNN approaches. An implementation via LSTM (Long Short-Term Memory) networks has been developed and applied to condylar surfaces. The results showed improved disease diagnosis performance when compared to a traditional dense network. The main advantage of this approach is that it can be applied to any general surface without the need for an existing correspondence or a consistent topology across surfaces. The disadvantage is that highly folded surfaces are not well represented with a projection approach, although shape-aware snapshot pathways are a planned extension to address this.

4   Infrastructure

In addition to new methods for creating and analyzing shapes, SlicerSALT 4.0 introduces key new application and infrastructure features to further support community research efforts.

4.1   Pipelines

SlicerPipelines is a new extension that allows the creation and use of simple modules in 3D Slicer without coding. This extension was developed as part of SlicerSALT and has been added to the Slicer Extensions Index for the broader research community to use.

Pipelines are pieces of logic that take a single MRML node as input (a model, a volume, etc.) and returns a single MRML node as output. Pipelines are created by stringing together individual existing 3D Slicer modules to produce a new derived workflow. Some examples of modules that could become parts of a pipeline are converting a model to a segmentation, using the 3D Slicer Segment Editor's hollow, margin, and thresholding effects, and methods like decimation and smoothing from the 3D Slicer Surface Toolbox.

Because they currently can only take a single node as input and return a single node as output, Pipelines are best for simple, repetitive workflows (e.g. threshold an image, apply smoothing, convert to a model, repeat for the next image). For more complicated workflows that depend on multiple MRML inputs or user interaction, manually coding a new module is currently still required, though integration with the automated pipeline creation is planned for the future.

4.2   Sample Data and Tutorials

To aid new users in understanding how to use the methods deployed in Slicer-SALT, we have integrated tutorials and example data for many modules directly into the SlicerSALT interface. Tutorials are given as links to slides on Google Drive which give the user a brief introduction to the method deployed in a module as well as walking them through its typical use. For sample data, data is automatically downloaded and verified. The user is able to inspect the data to better understand the input formats required for the module. The module UI is automatically populated with paths to the downloaded data so that the user can understand how to load their own data into the module. The user can run then run the module and inspect the sample output. An example of the Tutorials panel showing sample data being downloaded and paths automatically set is shown in Fig. 5.

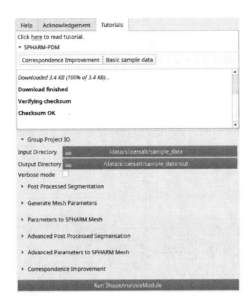

Fig. 5. The Tutorials panel showing a link to an online tutorial as well as sample data download buttons.

5   Discussion

The structures captured in 3D images coming from the biomedical fields have brought shape processing methods into the spotlight. The complex and time-varying phenomena contained in geometric structures require higher sensitivity to local variations relative to traditional markers, such as the volume of structures. In this new project period, SlicerSALT will continue providing biomedical researchers with access to sophisticated shape analysis methodology otherwise reserved for computer science experts. In addition, with the expansion towards novel and interpretable shape statistics, we will maximize the potential benefits of the geometric information contained in medical data and expand its use in order to support impactful biomedical research.

SlicerSALT development is ongoing and there are several plans for future improvements. As deep learning becomes increasingly important in the medical imaging and shape analysis communities, we plan to build upon the work described in Sect. 3.3 to introduce improved support for creating and analyzing shapes using deep learning. Our ultimate goal is to create blueprint modules for both training deep learning models and using existing models to perform inference. This will allow users to rapidly create new modules by giving them a starting point which they can customize rather than requiring module development from scratch.

To demonstrate the power of the SlicerPipelines extension for creating reusable and reproducible workflows, we plan to create example pipelines using public datasets for carrying out end-to-end shape analysis tasks. One example could be to start from a set of segmentations, create a population of PDMs using SPHARM-PDM, compute and visualize a shape space using SVA, and finally classify the inputs into two groups using DWD. These pipelines will help non-expert users quickly try various analysis methods on their own data and give more advanced researchers a way to distribute their workflows in a way that they can be easily resused.

Acknowledgements. SlicerSALT development has been funded by NIH NIBIB awards R01EB021391 and R56EB021391 as well as NHLBI award R01HL153166.

References

1. Albrecht, T., et al.: Statismo - a framework for PCA based statistical models. Insight J. pp. 1–18 (2012)
2. Boubolo, L., et al.: FlyBy CNN: a 3D surface segmentation framework. In: Medical Imaging 2021: Image Processing, vol. 11596, pp. 627–632. SPIE (2021)
3. Cates, J., Elhabian, S., Whitaker, R.: ShapeWorks: particle-based shape correspondence and visualization software. Statistical Shape and Deformation Analysis: Methods, Implementation and Applications, pp. 257–298 (2017). https://doi.org/10.1016/B978-0-12-810493-4.00012-2
4. Durrleman, S.: Deformetrica (2013). http://www.deformetrica.org
5. Fischl, B.: FreeSurfer. Neuroimage **62**(2), 774–781 (2012). https://doi.org/10.1016/j.neuroimage.2012.01.021
6. Jenkinson, M., Beckmann, C.F., Behrens, T.E.J., Woolrich, M.W., Smith, S.M.: FSL. NeuroImage **62**, 782–90 (2012). https://doi.org/10.1016/j.neuroimage.2011.09.015, http://www.ncbi.nlm.nih.gov/pubmed/21979382
7. Jung, S., Dryden, I.L., Marron, J.S.: Analysis of principal nested spheres. Biometrika **99**(3), 551–568 (2012). https://doi.org/10.1093/biomet/ass022
8. Liu, Z., Hong, J., Vicory, J., Damon, J.N., Pizer, S.M.: Fitting unbranching skeletal structures to objects. Med. Image Anal. **70**, 102020 (2021)
9. Marron, J.S., Todd, M.J., Ahn, J.: Distance-weighted discrimination. J. Am. Stat. Assoc. **102**, 1267–1271 (2007). https://doi.org/10.1198/016214507000001120, http://www.tandfonline.com/doi/abs/10.1198/016214507000001120
10. Mccormick, M., Liu, X., Jomier, J., Marion, C., Ibanez, L.: ITK: enabling reproducible research and open science. Front. Neuroinformatics **8** (2014). https://doi.org/10.3389/FNINF.2014.00013, https://pubmed.ncbi.nlm.nih.gov/24600387/
11. Pizer, S.M., et al.: Nested sphere statistics of skeletal models. In: Breuß, M., Bruckstein, A., Maragos, P. (eds.) Innovations for Shape Analysis: Models and Algorithms, pp. 93–115. Springer, Berlin, Heidelberg (2013). https://doi.org/10.1007/978-3-642-34141-0_5
12. Schroeder, W., Martin, K., Lorensen, B.: The Visualization Toolkit: An Object-oriented Approach to 3D Graphics. Kitware Inc (2006)
13. Vercauteren, T., Pennec, X., Perchant, A., Ayache, N.: Diffeomorphic demons: efficient non-parametric image registration. Neuroimage **45**(1), S61–S72 (2009)
14. Vicory, J., Allemang, D., Zukic, D., Prothero, J., McCormick, M., Paniagua, B.: An open-source solution for shape modeling and analysis of objects of challenging topologies. In: Medical Imaging 2021: Biomedical Applications in Molecular, Structural, and Functional Imaging, vol. 11600, pp. 137–142. SPIE (2021)

15. Vicory, J., et al.: SlicerSALT: shape analysis toolbox. In: Reuter, M., Wachinger, C., Lombaert, H., Paniagua, B., Lüthi, M., Egger, B. (eds.) ShapeMI 2018. LNCS, vol. 11167, pp. 65–72. Springer, Cham (2018). https://doi.org/10.1007/978-3-030-04747-4_6
16. Yushkevich, P.A., et al.: User-guided segmentation of multi-modality medical imaging datasets with ITK-SNAP. Neuroinformatics **17**(1), 83–102 (2018). https://doi.org/10.1007/s12021-018-9385-x

Predicting Shape Development: A Riemannian Method

Doğa Türkseven[1], Islem Rekik[1,2], Christoph von Tycowicz[4], and Martin Hanik[3,4(✉)]

[1] BASIRA Lab, Istanbul Technical University, Istanbul, Turkey
[2] Computing, I-X and Department of Computing, Imperial College London, London, UK
i.rekik@imperial.ac.uk
[3] Freie Universität Berlin, Berlin, Germany
[4] Zuse Institute Berlin, Berlin, Germany
{vontycowicz,hanik}@zib.de

Abstract. Predicting the future development of an anatomical shape from a single baseline observation is a challenging task. But it can be essential for clinical decision-making. Research has shown that it should be tackled in curved shape spaces, as (e.g., disease-related) shape changes frequently expose nonlinear characteristics. We thus propose a novel prediction method that encodes the whole shape in a Riemannian shape space. It then learns a simple prediction technique founded on hierarchical statistical modeling of longitudinal training data. When applied to predict the future development of the shape of the right hippocampus under Alzheimer's disease and to human body motion, it outperforms deep learning-supported variants as well as state-of-the-art.

Keywords: Shape development Prediction · Regression · Riemannian manifold

1 Introduction

Shapes of anatomical structures are of considerable medical interest, and they are encountered particularly often in the analysis of medical images. Studies have shown that they should be modeled as elements of curved manifolds—shape spaces—instead of ordinary Euclidean space [18]. Thus, it is imperative to develop methods for such spaces when working on problems involving anatomical shapes.

A particularly interesting and relevant task is the prediction of the future development of a shape—"How will an anatomical structure look like after a certain amount of time has passed?" This question is of great interest as shapes of anatomical structures are often correlated with (states of) diseases; see, e.g., [6, 16,18]. Predicting how a shape will develop in the future could thus play a significant role in diagnosis and prevention as well as aid physicians in their choice of treatment.

One important example is the relation between the shape of the hippocampus and Alzheimer's disease, as previous studies have shown that the former can be

© The Author(s), under exclusive license to Springer Nature Switzerland AG 2023
C. Wachinger et al. (Eds.): ShapeMI 2023, LNCS 14350, pp. 211–222, 2023.
https://doi.org/10.1007/978-3-031-46914-5_17

used to discriminate between Alzheimer's and normal aging [11]. Thus, if the longitudinal development of the hippocampus' shape can be predicted, a better prognosis can be achieved; and with Alzheimer's improved early diagnosis can have a serious positive impact [19].

To uncover developmental trends in populations, longitudinal studies that involve repeated observations of individuals play an essential role. The variability in such data can be distinguished as cross-sectional (i.e., between individuals) and longitudinal (i.e., within a single individual over time). The latter is highly correlated, violating the independence assumption of standard statistical tools like mean-variance analysis and regression, thus requiring inferential approaches that can disentangle cross-sectional and longitudinal effects. For shape-valued data, another challenge that warrants attention is that curved spaces lack a global system of coordinates. Assessing differences in longitudinal trends requires a notion of transport between tangent spaces to spatially align subject-specific trajectories. For manifolds, *parallel transport* has been shown to provide highly consistent transports [14,16] with improved sensitivity over other methods. Specifically, in [5] the adequateness of geodesic subject-wise models for the progression of subcortical brain structures in Alzheimer's disease and the potential for prognosis via parallel transport of individual trends has been shown.

To assess longitudinal and cross-sectional shape variation jointly, hierarchical statistical models pose an adequate and very flexible framework [12]. In recent years, various generalizations of such models to manifold-valued data have been proposed based on probabilistic [3] and least-squares theoretic [13,15,17] formulations. These approaches account for the inherent interrelations by describing each subject with its own parametric spatiotemporal model—most prominently geodesics. Additionally, the subject-specific trends are assumed to be perturbations of a population-average trend, which is referred to as the "fixed" effect and is often of primary interest.

Population-level analysis apart, a few studies have focused on shape development prediction *for individuals*. A deep learning pipeline for predicting longitudinal bone shape changes in the femur to diagnose knee osteoarthritis was introduced in [6]. This approach utilizes a spherical encoding to map a 3D point cloud of the bone into a 2D image. It thereby relies on the assumption that the femora are (approximately) star-shaped. While the proposed pipeline produces accurate predictions of future bone shape, anatomies in general (e.g., hippocampi) are not star-shaped prohibiting a spherical encoding. A further limitation of the method is that it needs 3 separate observations for its prediction—a requirement that strongly hinders early diagnosis.

A varifold-based learning approach for predicting infant cortical surface development has been proposed in [20]. The method uses regression on varifold representations [9] to learn typical shape changes and uses the latter to predict from a single baseline. Although the method partly uses the curved space of diffeomorphisms, it only employs it for pre- and post-processing. It further requires the user to set several data-dependent parameters as optimal as possible (i.e., two data-dependent kernel sizes, a weighting between terms in the loss of the regression, the concentration of time points, and the number of neighbors to be considered in a nearest neighbor search), leading to a relatively high entry-barrier for users.

In this work, we propose a novel method based on hierarchical statistical modeling to predict shape evolution. In contrast to previous geodesic approaches [5] that relied on hand-picked reference individuals, we provide a data-driven approach for learning shape progression. Representing subject-wise trends as geodesics in shape space allows us to learn from longitudinal observations while respecting within-subject correlations. After training on whole trajectories, the prediction requires only a single shape making it applicable when only the baseline observation of the longitudinal development is given. Conceptually, our prediction is very simple as it only consists of a parallel translation of the initial velocity of the mean trend and a subsequent evaluation of a geodesic. It thus provides a high degree of interpretability. Furthermore, no data-dependent parameters need to be set, which makes the method very user-friendly.

To validate our approach, we apply it to two real-world problems. We use it to predict the development of the shape of the right hippocampus under Alzheimer's. To the best of our knowledge, this is the first approach that applies longitudinal statistical modeling to the prediction of shape developments from a single baseline. We also use our method to predict human body motion as an example of an application in which large shape changes occur. In both applications, we outperform state-of-the-art by a considerable margin. Our method even slightly outperforms deep-learning-enriched variants. The source code for the geometric components used in the proposed method is available in the Morphomatics library [1]. An implementation of our proposed method can be found on https://github.com/morphomatics/ShapePrediction.

2 Method

We first recall the necessary basics from Riemannian geometry and geometric statistics; a good reference on them is [18].

A Riemannian manifold is a differentiable manifold M together with a Riemannian metric $\langle \cdot, \cdot \rangle_p$ that assigns a smoothly[1] varying scalar product to every tangent space T_pM. The metric also yields a (geodesic) distance function d on M. A further central object is the Levi-Civita connection ∇ of M. Given two vector fields X, Y on M, it is used to differentiate Y along X; the result is again a vector field, which we denote by $\nabla_X Y$. With a connection one can define a geodesic γ as a curve without acceleration, i.e., $\nabla_{\gamma'}\gamma' = 0$, where $\gamma' := \frac{\mathrm{d}}{\mathrm{d}t}\gamma$. It is a fundamental fact that each element of M has a so-called normal convex neighborhood U in which any two points $p, q \in U$ can be joined by a unique length-minimizing geodesic $[0,1] \ni t \mapsto \gamma(t; p, q)$ that does not leave U. Since it is the solution of a second-order differential equation, p, and $\gamma'(0)$ determine γ completely. Indeed, for every $p \in U$ there are the Riemannian exponential

---

[1] Whenever we say "smooth" we mean "infinitely often differentiable".

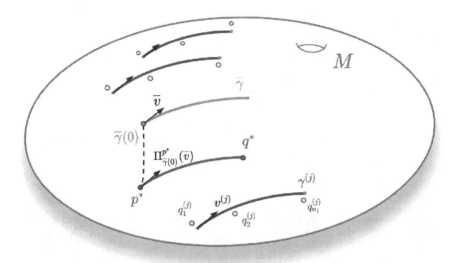

Fig. 1. Depiction of the shape prediction method in a shape space M.

$\exp_p : T_pM \to U$ and logarithm $\log_p := \exp^{-1} : U \to T_pM$ with $\exp(v) := q$ such that $\gamma'(0; p, q) = v$ and $\log_p(q) = \gamma'(0; p, q)$.

A further important fact is that in a Riemannian manifold, one usually cannot identify tangent spaces with each other. Therefore, tangent vectors must be transported explicitly along curves between points—the so-called parallel transport. This process depends on the chosen path; however, in U, we can always transport along the geodesic that connects the origin and destination. Therefore, whenever we speak of parallel-translating a vector v from some $p \in U$ to $q \in U$ transport along the geodesic from p to q is meant; the resulting vector (which is in T_qM) is denoted by $\Pi_p^q(v)$.

Whenever we work with data, means are of interest. Given $q_1, \ldots, q_n \in M$, their Frechet mean is the minimizer of the Frechet variance $F(p) := \sum_{i=1}^{n} d(p, q_i)^2$.

Interestingly, the set of all geodesics $G(U) := \{\gamma : [0, 1] \to U \mid \gamma \text{ geodesic}\}$ in U can also be given the structure of a Riemannian manifold imposing a functional-based Riemannian metric [17]. As a consequence, we can compute Frechet means of geodesics in $G(U)$ if they are sufficiently localized; we assume the latter throughout this work.

We are now ready to introduce our model for shape prediction, which we will use for hippocampi. For this let M be any shape space that is a Riemannian manifold and $U \subseteq M$ a normal convex neighborhood consisting of the shapes of interest to us. Given a shape $p^* \in U$ observed at time t_0 our goal is to predict its future form $q^* \in U$ at time point t_1. For simplicity of exposition, we assume in the following that $t_0 = 0$ and $t_1 = 1$. This can always be achieved when the times are viewed relative to an interval that contains them.

Our fundamental assumption is that p^* develops along a geodesic through U, i.e., there is $\gamma_{p^*} \in G(U)$ such that the longitudinal development of p^* at time t is given by $\gamma_{p^*}(t)$. Several works have shown that this is often an adequate choice when modeling shape developments in the medical context [16, 18]. Then, since geodesics are determined by their starting point and initial velocity, we need to find $\gamma_{p^*}'(0)$; because then

$$q^* = \exp_{p^*}(\gamma_{p^*}'(0)). \tag{1}$$

In other words, to predict the development of *any* shape $p^* \in U$ that is of interest to us we need to approximate a vector field on U that encodes the direction and speed of change that p^* undergoes until time 1.

In the following, we propose an approach that infers this vector field from data. Assume N shapes similar to p^* (i.e., close to p^* in U), which are expected to show analogous progression and are observed at a possibly varying number of time points, hence, yielding (training) data $(t_i^{(j)}, q_i^{(j)}) \in [0, 1] \times U$, for $i = 1, \ldots, n_j$ and $j = 1, \ldots, N$. Now, using geodesic regression [10] we can approximate the individual trajectories $\gamma^{(1)}, \ldots, \gamma^{(N)} \in G(U)$ of the training shapes. Utilizing the manifold structure of $G(U)$ from [13], the (Frechet) mean geodesic $\overline{\gamma}$ of $\gamma^{(1)}, \ldots, \gamma^{(N)}$ can be computed. Note that $\overline{\gamma}(0)$ and $\overline{v} := \overline{\gamma}'(0) \in T_{\overline{\gamma}(0)}M$ can be interpreted as the mean starting point and the average initial velocity of the trajectories of the training data, respectively. The geodesic $\overline{\gamma}$ is further the central fixed effect that describes the data in a geodesic hierarchical model [15, 17].

The fact that the training shapes are close to p^* suggests that the parallel transport of $\overline{v}$ to p^* is a good approximation of our target $\gamma_{p^*}'(0)$. We thus propose to use the approximation $\gamma_{p^*}'(0) \approx \Pi_{\overline{\gamma}(0)}^{p^*}(\overline{v})$ in Eq. (1). Being a comparatively simple approach, our experiments show that it can be a very good choice. The processing pipeline is shown in Fig. 1.

3 Experiments

3.1 Data and Methodology

Datasets: To evaluate our model we applied it to two data sets. The first was shape data of right hippocampi derived from 3D label fields provided by the Alzheimer's Disease Neuroimaging Initiative[2] (ADNI). The ADNI database contains, amongst others, 1632 brain MRI scans with *segmented* hippocampi. From them, we assembled three distinct groups: subjects with Alzheimer's (AD), Mild Cognitive Impairment (MCI), and cognitive normal (CN) controls. The groups contained data from 86, 201, and 116 subjects, respectively; for each subject, there were three MR images taken (approximately) six months apart. In the experiments below, we always predicted the shape after one year from the baseline. Since the experiments were performed group-wise, the fact that the groups were not balanced did not matter. Correspondence of the surfaces (2280 vertices, 4556 triangles) was established in a fully automatic manner by registering

---

[2] https://adni.loni.usc.edu.

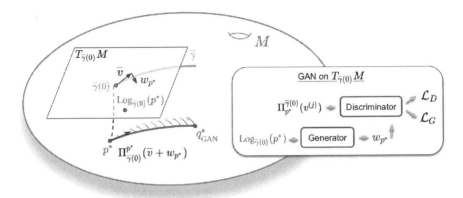

Fig. 2. GAN-enhanced prediction method.

extracted isosurfaces using the functional map–based approach of [8]. As the final preprocessing step, all meshes were aligned using generalized Procrustes analysis.

The second data set was taken from Dynamic FAUST [2], which is publicly available and contains the motion data of 10 subjects. The data is given as triangle meshes in correspondence (subject-wise and between subjects). For each subject, we used three meshes (6890 vertices, 13776 triangles) that constitute a raising of the left leg from the initial part of the "one leg loose" pattern. More precisely, we used scans 0, 20, and 40 for subjects 50002, 50004, 50020, and 50027; scans 16, 24, and 32 for subjects 50009 and 50021; scans 90, 110, and 130 for subject 50007; scans 0, 22, and 44 for subject 50022; scans 40, 48, and 56 for subject 50025; and scans 10, 30, and 50 for subject 50026.

Shape Space: We used the differential coordinate model (DCM) shape space from [21]. The DCM space works with triangular mesh representations and allows for explicit and fast computations.

Comparison Methods:

For the ADNI experiment, we used the following comparison methods. The first was the varifold-based (Varifold) method from [20]. To test whether incorporating a deep neural network improves the prediction, we also tested the following variation of our proposed method on the ADNI data: We tried[3] both a generative adversarial network (GAN) and a cyclic GAN (cGAN) to learn a correction $w_{p^*} \in T_{\overline{\gamma}(0)}M$ of $\overline{v}$. The idea was that shapes from different regions in U might show systematic differences in their development. Both the GAN and cGAN are designed to map the coordinate vector[4] of the cross-sectional difference $[\log_{\overline{\gamma}(0)}(p^*)] \in \mathbb{R}^d$ to the correction vector $[w_{p^*}] \in \mathbb{R}^d$. They were trained on the baselines $\{q_1^{(j)} \mid j = 1, \ldots, N\}$ using three-fold cross validation. We then

---

[3] Standard multi-layer perceptrons showed similar performance to GANs.

[4] We denote the *coordinate representation* in $\mathbb{R}^d$ of a tangent vector v w.r.t. a fixed but arbitrary basis by $[v]$.

used $\gamma'_{p*}(0) \approx \Pi^{p^*}_{\overline{\gamma}(0)}(\overline{v} + w_{p*})$ in Eq. (1), i.e., the prediction then became

$$q^*_{\text{GAN}} := \exp_{p*}(\Pi^{p^*}_{\overline{\gamma}(0)}(\overline{v} + w_{p*})).$$

The method using the GAN is illustrated in Fig. 2.

The generators in the GAN and cGAN networks consisted of four linear layers, with dropout layers and ReLU activation functions between them; the discriminators were composed of two linear layers without dropout. Both the GAN and cGAN used the sum of the binary cross entropy and the difference $\|\Pi^{\overline{\gamma}(0)}_{\gamma^{(j)}(0)}(v^{(j)}) - \overline{v} - w_{p*}\|_{\overline{\gamma}(0)}$ (with the norm that is induced by the Riemannian metric) as loss; for the cGAN the standard forward cycle consistency loss was also added. When referencing results obtained with the GAN and cGAN while using the DCM space as shape space, we use the notations (DCM+GAN) and (DCM+cGAN), respectively.

To evaluate how important the DCM space is to the prediction, we replaced it with the flat point distribution model (PDM) space from [7].

Finally, to further differentiate the behavior of our proposed and the varifold-based approach, we tested how well the regressed geodesics in the DCM space and the space of diffeomorphisms approximate the data. We report for all groups the averages of the mean vertex-wise error (MVE) between the mesh belonging to $\gamma^{(j)}(1)$ and the corresponding data mesh of $q_2^{(j)}$ after rigid alignment.

In the experiment with the dynamic FAUST data, we compared our method against the varifold-based method. We did not use these deep-learning refinements for the FAUST data since only ten subjects were available. (Note that an advantage of our approach is also that it can handle such small datasets.)

Software: The computations in the DCM and PDM space were performed in Morphomatics v1.1 [1]. The GAN and cGAN were implemented in PyTorch. During training, we used ADAM optimizer for both the generators and the discriminators. All computations involving varifolds were performed in Deformetrica 4.3.0rc0 [4], where we used the "landmarks" option as attachment type for a fair comparison since all meshes were in correspondence. In the ADNI experiment, we used the L-BFGS algorithm in Deformetrica, and gradient ascent or the Faust data as L-BFGS had stability problems there.

Parameter Settings: In the ADNI experiment (where measurements are in millimeters), we used Varifold with a smoothing kernel width of 2.5, a deformation kernel width of 5, $t_0 = 0.5$, a concentration of time points of 1, and the number of points used from the cloud was set to 25. The GAN and cGAN used a dropout probability of 0.2; the learning rates of both the generators and discriminators were set to 0.0001. For each fold, we trained the network with 400 epochs. In the FAUST experiment (where measurements are in meters), we used Varifold with a smoothing kernel width of 0.1, a deformation kernel width of 0.1, $t_0 = 0.5$, a concentration of time points of 1; the number of points used from the cloud was also set to 25.

All parameters were determined through exploration. All results were obtained through three-fold cross-validation whose averages are reported.

Evaluation Measures: To assess the effectiveness of our model, we employed the MVE by calculating how much (after rigid alignment) each vertex deviates from its corresponding ground truth vertex. Moreover, to compare the DCM and PDM spaces, we contrasted the MVE (which can be viewed as the intrinsic distance of PDM space) with the geodesic distance d (GD) of DCM space.

Table 1. Comparison of (variants of) our and the varifold-based prediction method w.r.t. average mean vertex-wise error (MVE) and average geodesic distance (GD).

Metric	Group	DCM	PDM	DCM+GAN	DCM+cGAN	Varifold
MVE	AD	**0.70 ± 0.07**	0.71 ± 0.02	0.73 ± 0.02	0.75 ± 0.02	1.23 ± 0.14
	MCI	**0.64 ± 0.05**	**0.64 ± 0.03**	0.67 ± 0.03	0.75 ± 0.04	1.27 ± 0.02
	CN	**0.71 ± 0.01**	0.72 ± 0.08	0.74 ± 0.07	0.80 ± 0.05	1.33 ± 0.06
GD	AD	**15.51 ± 0.88**	37.50 ± 0.79	16.20 ± 0.60	16.42 ± 0.47	17.63 ± 1.97
	MCI	**15.23 ± 0.87**	34.55 ± 1.81	15.91 ± 0.89	16.22 ± 0.50	18.36 ± 0.25
	CN	**16.44 ± 0.22**	38.39 ± 4.44	17.35 ± 1.66	18.35 ± 1.27	19.07 ± 1.13

3.2 Results

The results of our prediction comparison for the hippocampi are shown in Table 1. Our proposed method outperformed the other methods in all categories. Even though the proposed method is a relatively simple approach, it not only performed better than GAN methods but was also faster and did not involve hyperparameter tuning. Also, differently structured GANs did not improve the results by a noticeable amount. Furthermore, we can see that using the DCM space is superior to using the PDM space as its results are close w.r.t. MVE (the intrinsic PDM distance) but significantly worse w.r.t. the GD of the DCM space. Note that the magnitude of the prediction errors of our method is probably close to the resolution of the scanner. However, the longitudinal changes in shape over the course of one year are also only small.

In comparison to the varifold-based approach, the proposed method achieves superior results. The difference in prediction is visualized in Fig. 3 using a hippocampus from the AD group for which the MVE of our method was very close ($\approx$0.69) to its mean MVE. The MVE of the varifold-based method was slightly smaller ($\approx$1.03) than its MVE.

A reason why our approach works better than the varifold-based one could be the difference in approximation power of geodesic regression: Results for the

Table 2. Regression fidelity in terms of MVE.

Group	AD	MCI	CN
DCM	**0.16 ± 0.06**	**0.17 ± 0.16**	**0.18 ± 0.18**
Varifold	0.68 ± 0.58	0.62 ± 0.47	0.73 ± 0.74

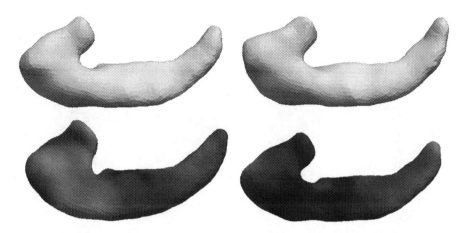

Fig. 3. Prediction comparison for a hippocampus from the AD group. Upper left: base shape at $t = 0$; upper right: the same hippocampus at $t = 1$, i.e., after one year (the ground truth for our prediction); bottom left: varifold-based prediction; bottom right: our prediction. The colors encode the vertex-wise differences to the ground truth according to the color map ($0\,\mathrm{mm}$ ▬▬▬▬ $2.25\,\mathrm{mm}$).

fitting quality of regression in DCM and the space of diffeomorphisms are shown in Table 2. Clearly, the approximation is better when the DCM space is used, thus, demonstrating an improved fidelity of DCM geodesics over diffeomorphic representations.

The results for the FAUST dataset reinforce the above findings. Our proposed method (DCM) achieves an MVE of 0.03 ± 0.003 and a GD of 0.54 ± 0.03; Varifold has an MVE of 0.07 ± 0.008 and a GD of 1.16 ± 0.04. Figure 4 shows the predictions for subject 500021. While the result of our method is close to the ground truth, the human is barely recognizable in the varifold-based prediction. A reason is the high inter- and intra-subject variability in the motion. Since point trajectories from different subjects are mixed to produce the varifold prediction, relatively unlikely leg configurations are obtained. Moreover, the formation of a male face in the varifold case highlights the advantage of a differential encoding of shape changes (i.e. tangent vectors) together with a consistent transport over combinations of absolute configurations.

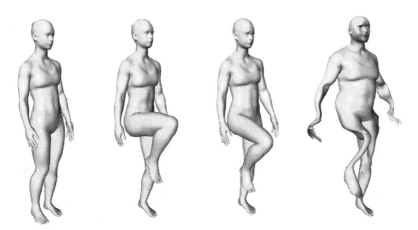

Fig. 4. Prediction of human body motion for FAUST dataset. From left to right: Baseline, ground truth, and predicted shapes using our and the Varifold method [20].

4   Conclusion

In this paper, we proposed a novel method for predicting shape development based on hierarchical statistical modeling in Riemannian shape spaces. It outperformed state-of-the-art in two experiments by a clear margin. Furthermore, it performed better than deep learning–supported variants when predicting the future development of the hippocampus shape. Our approach is thus a good fit for shapes whose progression follows geodesics (such as hippocampi) and who are well captured by population-average trends. As the latter assumption will not always be valid, it is still a promising approach to incorporate deep learning: Whenever the development depends strongly on the individual characteristics of the baseline shape, we expect that deep learning methods can be used to find adjustments to our prediction (direction) that take the dependence into account. A path for future work is thus to test this hypothesis on further anatomical structures.

Acknowledgements. This work was partially funded by grants from the European H2020 Marie Sklodowska-Curie action (grant no. 101003403) and the Scientific and Technological Research Council of Turkey under the TUBITAK 2232 Fellowship for Outstanding Researchers (no. 118C288) (https://basira-lab.com/normnets/ & https://basira-lab.com/reprime/). We are grateful for the funding by DFG (Deutsche Forschungsgemeinschaft (DFG) through Germany's Excellence Strategy – The Berlin Mathematics Research Center MATH+ (EXC-2046/1, project ID: 390685689)) and BMBF (Bundesministerium für Bildung und Forschung (BMBF) through BIFOLD - The Berlin Institute for the Foundations of Learning and Data (ref. 01IS18025A and ref 01IS18037A)).

Data collection and sharing for this project was funded by the ADNI (adni.loni.usc.edu) (National Institutes of Health Grant U01 AG024904) and DOD ADNI (Department of Defense award number W81XWH-12-2-0012). ADNI is funded by the National Institute on Aging, the National Institute of Biomedical Imaging and Bioengineering, and through generous contributions from the following: AbbVie, Alzheimer's Association; Alzheimer's Drug Discovery Foundation; Araclon Biotech; BioClinica, Inc.; Biogen; Bristol-Myers Squibb Company; CereSpir, Inc.; Cogstate; Eisai Inc.; Elan Pharmaceuticals, Inc.; Eli Lilly and Company; EuroImmun; F. Hoffmann-La Roche Ltd and its affiliated company Genentech, Inc.; Fujirebio; GE Healthcare; IXICO Ltd.; Janssen Alzheimer Immunotherapy Research & Development, LLC.; Johnson & Johnson Pharmaceutical Research & Development LLC.; Lumosity; Lundbeck; Merck & Co., Inc.; Meso Scale Diagnostics, LLC.; NeuroRx Research; Neurotrack Technologies; Novartis Pharmaceuticals Corporation; Pfizer Inc.; Piramal Imaging; Servier; Takeda Pharmaceutical Company; and Transition Therapeutics. The Canadian Institutes of Health Research is providing funds to support ADNI clinical sites in Canada. Private sector contributions are facilitated by the Foundation for the National Institutes of Health (www.fnih.org). The grantee organization is the Northern California Institute for Research and Education, and the study is coordinated by the Alzheimer's Therapeutic Research Institute at the University of Southern California. ADNI data are disseminated by the Laboratory for Neuro Imaging at the University of Southern California.

References

1. Ambellan, F., Hanik, M., von Tycowicz, C.: Morphomatics: geometric morphometrics in non-Euclidean shape spaces (2021). https://doi.org/10.12752/8544. https://morphomatics.github.io/

2. Bogo, F., Romero, J., Pons-Moll, G., Black, M.J.: Dynamic FAUST: registering human bodies in motion. In: IEEE Conference on Computer Vision and Pattern Recognition (CVPR), July 2017

3. Bône, A., Colliot, O., Durrleman, S.: Learning the spatiotemporal variability in longitudinal shape data sets. Int. J. Comput. Vis. **128**(12), 2873–2896 (2020). https://doi.org/10.1007/s11263-020-01343-w

4. Bône, A., Louis, M., Martin, B., Durrleman, S.: Deformetrica 4: an open-source software for statistical shape analysis. In: Reuter, M., Wachinger, C., Lombaert, H., Paniagua, B., Lüthi, M., Egger, B. (eds.) ShapeMI 2018. LNCS, vol. 11167, pp. 3–13. Springer, Cham (2018). https://doi.org/10.1007/978-3-030-04747-4_1

5. Bône, A., et al.: Prediction of the progression of subcortical brain structures in Alzheimer's disease from baseline. In: Cardoso, M.J., et al. (eds.) GRAIL/MFCA/MICGen -2017. LNCS, vol. 10551, pp. 101–113. Springer, Cham (2017). https://doi.org/10.1007/978-3-319-67675-3_10

6. Calivá, F., Kamat, S., et al.: Surface spherical encoding and contrastive learning for virtual bone shape aging. Med. Image Anal. **77**, 102388 (2022). https://doi.org/10.1016/j.media.2022.102388

7. Cootes, T.F., Taylor, C.J., Cooper, D.H., Graham, J.: Active shape models-their training and application. Comput. Vis. Image Underst. **61**(1), 38–59 (1995). https://doi.org/10.1006/cviu.1995.1004

8. Ezuz, D., Ben-Chen, M.: Deblurring and denoising of maps between shapes. Comput. Graph. Forum **36**(5), 165–174 (2017). https://doi.org/10.1111/cgf.13254

9. Fishbaugh, J., Durrleman, S., Prastawa, M., Gerig, G.: Geodesic shape regression with multiple geometries and sparse parameters. Med. Image Anal. **39**, 1–17 (2017). https://doi.org/10.1016/j.media.2017.03.008

10. Fletcher, P.T.: Geodesic regression and the theory of least squares on Riemannian manifolds. Int. J. Comput. Vis. **105**(2), 171–185 (2013). https://doi.org/10.1007/s11263-012-0591-y

11. Gerardin, E., Chételat, G., et al.: Multidimensional classification of hippocampal shape features discriminates Alzheimer's disease and mild cognitive impairment from normal aging. Neuroimage **47**(4), 1476–1486 (2009). https://doi.org/10.1016/j.neuroimage.2009.05.036

12. Gerig, G., Fishbaugh, J., Sadeghi, N.: Longitudinal modeling of appearance and shape and its potential for clinical use. Med. Image Anal. **33**, 114–121 (2016). https://doi.org/10.1016/j.media.2016.06.014

13. Hanik, M., Hege, H.C., von Tycowicz, C.: A nonlinear hierarchical model for longitudinal data on manifolds. In: 2022 IEEE 19th International Symposium on Biomedical Imaging (ISBI), pp. 1–5. IEEE (2022). https://doi.org/10.1109/ISBI52829.2022.9761465

14. Lorenzi, M., Pennec, X.: Geodesics, parallel transport & one-parameter subgroups for diffeomorphic image registration. Int. J. Comput. Vis. **105**(2), 111–127 (2013). https://doi.org/10.1007/s11263-012-0598-4

15. Muralidharan, P., Fletcher, P.T.: Sasaki metrics for analysis of longitudinal data on manifolds. In: Proceedings of the 2012 IEEE Conference on Computer Vision and Pattern Recognition, pp. 1027–1034. IEEE (2012). https://doi.org/10.1109/CVPR.2012.6247780

16. Nava-Yazdani, E., Hege, H.C., Sullivan, T.J., von Tycowicz, C.: Geodesic analysis in Kendall's shape space with epidemiological applications. J. Math. Imaging Vis. **60**, 549–559 (2020). https://doi.org/10.1007/s10851-020-00945-w

17. Nava-Yazdani, E., Hege, H.C., von Tycowicz, C.: A hierarchical geodesic model for longitudinal analysis on manifolds. J. Math. Imaging Vis. **64**(4), 395–407 (2022). https://doi.org/10.1007/s10851-022-01079-x

18. Pennec, X., Sommer, S., Fletcher, P.T. (eds.): Riemannian Geometric Statistics in Medical Image Analysis. Academic Press, London (2020). https://doi.org/10.1016/C2017-0-01561-6

19. Rasmussen, J., Langerman, H.: Alzheimer's disease - why we need early diagnosis. Degenerative Neurol. Neuromuscul. Dis. **9**, 123–130 (2019). https://doi.org/10.2147/DNND.S228939

20. Rekik, I., Li, G., Lin, W., Shen, D.: Predicting infant cortical surface development using a 4D varifold-based learning framework and local topography-based shape morphing. Med. Image Anal. **28**, 1–12 (2016). https://doi.org/10.1016/j.media.2015.10.007

21. von Tycowicz, C., Ambellan, F., Mukhopadhyay, A., Zachow, S.: An efficient Riemannian statistical shape model using differential coordinates. Med. Image Anal. **43**, 1–9 (2018). https://doi.org/10.1016/j.media.2017.09.004

AReg IOS: Automatic Registration on IntraOralScans

Nathan Hutin[1,2(✉)], Luc Anchling[1,2], Lucia Cevidanes[1], Felicia Miranda[1,7], Denise Curado[1], Marcela Gurgel[1], Selene Barone[1,5], Jonas Bianchi[1,4], Najla Al Turkestani[1,6], Antonio Ruellas[8], Margaret Eason[1], Kinjal Mavani[1], Juan Carlos Prieto[3], and Aron Aliaga[1]

[1] University of Michigan, Michigan, USA
[2] CPE Lyon, Lyon, France
nathan.hutin@cpe.fr
[3] University of North Carolina, Chapel Hill, USA
[4] University of the Pacific, San Francisco, USA
[5] Magna Graecia University of Catanzaro, Catanzaro, Italy
[6] King Abdulaziz University, Jeddah, Saudi Arabia
[7] Bauru Dental School, University of Sao Paulo, Bauru SP, Brazil
[8] Federal University of Rio de Janeiro, Rio de Janeiro, Brazil

Abstract. This paper presents a novel method for the automatic registration of Intra Oral Scans (IOS). Our approach uses deep learning techniques and alignment algorithms such as the Iterative Closest Point (ICP) to automatically align IOS at different time points for the same subject. For the proposed registration methods; firstly, crown segmentation is performed using the DentalModelSeg extension in 3D Slicer allowing for the identification of specific dental crowns and the determination of their centroids. The source and target IOS are initially aligned using a common set of centroids. Next, we segment a region of interest (ROI) in the palate using a deep learning algorithm based on a multi-view 3D shape analysis technique. The ROI ground truths used for training the palate segmentation algorithm were manually generated by expert clinicians and a developer. Finally, the ICP algorithm is applied to register the upper jaw scans, using the predicted ROI as the registration target. The method is designed to handle different clinical conditions. The performance of the proposed method was tested using IOS from growing subjects acquired at two-time points for each subject. The results demonstrate the effectiveness of the automated registration approach, with minimal errors observed between the automated registration and the expert clinician's registration. The average angular and linear errors ranged from 0.01 ± 1.04 to $0.32 \pm 1.03°$ and from 0.02 ± 0.13 to 0.18 ± 0.71 mm. Registration of IOS from growing subjects before and after treatment were the most challenging. The integration of the method into 3D Slicer through the SlicerAutomatedDentalTools extension enhances its accessibility and usability. The proposed method offers an automated and efficient solution for Intra Oral Scan registration. The approach showcases promising results, reducing the workload. The code is available: https://github.com/DCBIA-OrthoLab/SlicerAutomatedDentalTools.

© The Author(s), under exclusive license to Springer Nature Switzerland AG 2023
C. Wachinger et al. (Eds.): ShapeMI 2023, LNCS 14350, pp. 223–235, 2023.
https://doi.org/10.1007/978-3-031-46914-5_18

Keywords: Deep Learning · Automatic Registration · Orientation ·
Digital Dental Model

1   Introduction

In dentistry, the use of intraoral scanners has revolutionized the digitization of
gums and teeth by capturing and computing their vertex through light-based
triangulation [9]. Intraoral scanning provides an alternative to traditional dental
casts, enabling clinicians to digitize patient data in 3D, and facilitating record
keeping, treatment planning, and saving space through virtual storage. Consequently, intraoral scanning offers clinicians a new opportunity to gain deeper
insights into patient progress and make more informed decisions. The advancements in machine learning (ML) and its growing prominence in the field of
medicine further enhance this potential by facilitating faster and more efficient
task completion and decision-making processes over time.

Standardized orientation plays a crucial role in normalizing the position of
IOS scans in the virtual world and enables accurate measurement comparisons
among patients. With the advent of intraoral scanners, dentoalveolar measurements have transitioned from 2D radiographs to 3D dental model assessments [5].
Standardized orientation plays a crucial role in normalizing the position of the
IOS scans and enables accurate measurement comparisons among patients. To
further comprehend the dentoalveolar movement, registration becomes a necessary step. Manual registration, which involves matching scans before and after
treatment, has challenges as clinicians need to identify a non-moving reference
region unaffected by treatment or growth. Some structures on the palate, including the third palatal rugae, have been identified as stable regions of the maxilla [5,7,11]. However, achieving consistent manual registration poses difficulties
due to observer variability in rugae location and landmarks placement. Besides,
approaches are time-consuming and can be prone to error. While methods have
been proposed to establish standardized registration, finding a universally applicable technique remains challenging.

The aim of this study is to evaluate and validate a newly developed open-
source automated tool for the registration of IOS. The tool incorporates the
DentalModelSeg [2,4] algorithm and a multi-view 3D shape segmentation algorithm that uses a U-Net [10] to detect a stable surface mesh ROI. This allows
for the accurate alignment of scans obtained at different time points. The study
assessed the effectiveness and reproducibility of the automated tool by comparing
its measurements to those made manually by expert users employing previously
validated manual tools.

2   Related Work

Related work in registering IOS involves addressing the challenge of selecting
reliable and stable landmarks or ROI. In the context of the maxillary arch,
the palatal rugae have been widely recognized as a stable reference region and

have been utilized for registration purposes [1, 6, 12]. Registration plays a crucial role in assessing tooth movement and involves aligning two scans by applying a transformation matrix using specialized software. However, a limitation of this method is that the registration of scans depends on the subjective perspective of the clinician, lacking a standardized approach due to the considerable individual variability in the shape of palatal rugae [6].

To overcome this limitation, several papers have proposed methods to improve the reliability and standardization of registration [1, 6, 11, 12]. One approach involves using landmarks as reference points, ensuring that all clinicians can reproduce the same registration [1]. Another approach involves using patches or ROI to guide the registration process [1, 6, 12]. These methods aim to establish consistent and reproducible registration techniques across different clinicians.

In this paper, we adopted a similar approach to segment the palate ROI using a multi-view approach. By rendering the mesh from different perspectives, we capture 2D images and use a UNET [10] for the segmentation task. The segmentation output is then mapped back to individual faces in the mesh. We apply an algorithm to label faces that were not captured during the rendering process. In the next section, we explain in detail the steps of our approach.

3   Method

3.1   Data

The overall dataset comprised 164 scans. From these, Forty-eight scans were used to evaluate the performance of the Areg method. All the scans were acquired using 3Shape Trios and iTero scanners at a University Clinic. The University of Michigan Institutional Review Board (IRB) HUM00238038 waived the requirement for informed consent and granted IRB exemption for this study.

3.2   Automated Standardized Orientation (ASO)

In the process of automated IOS orientation, a template-oriented scan is required as preliminary data to guide the algorithm. In the proposed algorithm both the non-oriented and oriented scans need to have the crowns segmented, which is achieved by DentalModelSeg, an available extension in the 3D Slicer.

Then we compute the centroids of each crown and select the 4 predefined crowns and use them to roughly orient the upper jaw the algorithm computes the centroid of the segmented crowns of four teeth that were selected on both the oriented and non-oriented scans. The right and left first molars, and the right and left first premolars. Three lines of reference are then created using these centroids: the LR (Left-Right) line passing through the second molars, the AP (Anterior-Posterior) line passing through the middle between the centroid of the premolars and perpendicular to the LR line, and the SI (Superior-Inferior) line, perpendicular to both the LR and SI lines. We compute a transformation that rotates the mesh on the plane spanned by the LR lines and their angle.

A second rotation transformation is computed with the AP lines. Finally, we translate the mesh by the difference between the centroids of the scans and apply an Iterative Closest Point (ICP) [3] algorithm to further refine the alignment of the upper arches. We compute a single transformation by concatenating the two rotations, the translation, and the ICP transform. This transformation could then be applied to the lower arch as well for future analysis (Fig. 1).

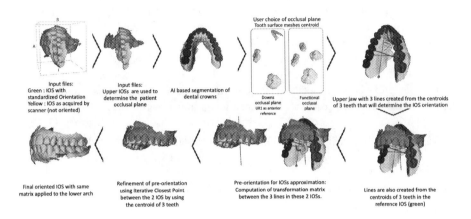

Fig. 1. Orientation workflow.

3.3   Pre-processing: Training Reference Patch

Each upper dental arch scan in the dataset was initially labeled with a surface patch in a shape similar to a Butterfly. The Butterfly patch includes a triangular mesh region on the palate. It extends over a specific area or region, encompassing the stable and reference regions, rather than isolated landmarks. The selection of the triangular mesh region considers the stability of the palate and the third rugae as a reliable references during treatment and growth. [5,7]. The Butterfly patch provides control over yaw by extending laterally, control over pitch by extending posteriorly with minimal anterior extension, as changes in anterior alveolar region are observed during treatment and growth [8]. It also offers control over roll through bilateral symmetry in the patch, aligning with the shape of the palate Fig. 2d.

First, an outline is generated (Figs. 2a, b and c), and then the patch is filled (Fig. 2d). The contour of the anterior part is formed between the left and right first premolars, while the posterior part is created between the left and right first molars, with the length of the posterior part being longer than the anterior part (Fig. 2a). The lateral contour is formed using a Bézier curve [13] connecting the endpoints of the anterior and posterior lines and the midpoint of the posterior line (Fig. 2b).

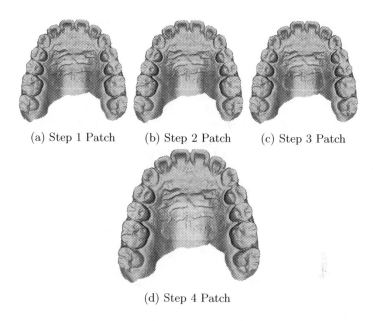

(a) Step 1 Patch   (b) Step 2 Patch   (c) Step 3 Patch

(d) Step 4 Patch

Fig. 2. Creation Butterfly's patch.

$$\text{Bézier curve} : \forall t \in [0,1], B(t) = \sum_{i=0}^{n} \binom{n}{i} (1-t)^{n-i} t^i P_i \tag{1}$$

Next, the Bézier curve is symmetrically projected with respect to the lateral lines (Fig. 2c). The lines are defined by assigning labels to the vertices, where the yellow color represents label 1 and the blue color represents label 0. The lines are given a radius, which facilitates the use of an algorithm for filling the patch. The filling algorithm starts by labeling one vertex inside of the outline, and then the label propagation algorithm is applied to propagate the label throughout the mesh until the entire patch is filled. Starting from the first vertex, its neighboring vertices with label 0 are identified and labeled as 1. Then, the algorithm continues to identify neighboring vertices with label 0 and converts them to label 1 until there are no more neighboring vertices with label 0 remaining. This process ensures that the entire patch is filled (Fig. 2d).

3.4 Training Butterfly Patch

The first step of the training involves normalizing the meshes by rescaling and translating them to fit within the unit sphere. The meshes are also oriented to ensure consistent viewing of the palate from the same viewpoint. To increase the amount of data and improve the robustness of the neural network, data augmentation techniques are applied. This includes random rotation and random translation. The random orientation is limited up to 360°C around a vector belonging to the set $\{(x, y, 1)|\forall(x, y) \in [-0.25, 0.25]^2\}$ to ensure the palate remains visible. Similarly, the vector of translation belongs to the set $\{(x, y, z)|\forall(x, y, z) \in [-0.25, 0.25]^3\}$ to maintain visibility. Additionally, the normals of the vertices are computed to provide texture to the mesh.

To simplify the problem from 3D meshes to 2D images, rendering is performed using the PyTorch3D framework. The rendering process involves using the FoVPersepectiveCameras with an OpenGL perspective and NDC coordinate system to capture images of size 320×320 pixels. The HardPhongShader is used as the shader, resulting in images with 4 channels. The first 3 channels represent the texture, while the 4th channel represents the depth map. Three cameras are positioned at the top of the palate at different positions, pointing towards the central point. The rendered images are then fed to the neural network for further processing.

The neural network used in this training process is UNet, a CNN imported from MONAI. It consists of 5 down-sampling steps and 5 up-sampling steps, with a kernel size of 3×3 and a stride of 2. The number of features increases progressively from 16 up to 256. The ground truth for the training is the segmentation of the patch, created using the method explained in the previous paragraph. The segmentation of the patch is linked to the image using the pixel-to-face mapping, which associates the faces with the corresponding pixels in the rendered image.

The loss for the UNet prediction is computed using the DiceCELoss from MONAI, which is a linear combination of the Dice coefficient and cross-entropy. The training of the UNet model is optimized using AdamW, which includes regularization and weight decay regularization with a learning rate of 1e−4. The training process was conducted on a system equipped with an NVIDIA RTX A6000 GPU with 48 GB of RAM memory and CUDA version 11.7. The system also includes an Intel® Xeon® Gold 6226R CPU with 32 cores and a frequency of 2.90 GHz. The training process took approximately 3 days to complete (Fig. 3).

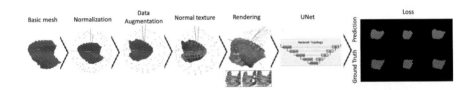

Fig. 3. Pipeline Training.

3.5   Butterfly Patch Prediction

First, the mesh undergoes a normalization process, which involves orienting, rescaling, and translating. This step aims to place the mesh within the unit sphere and ensure optimal visibility of the palate by the cameras. During the rendering process, a larger number of cameras is used compared to the training phase. This is done to capture more information from the different areas of the rugae region and palate region.

Next, the trained UNet model predicts the patch on the image, and the pixel-to-face mapping is utilized to establish the correspondence between the mesh and the images, enabling vertex labeling. After this step, there may be holes present in the predicted Butterfly patch due to the limited coverage of the mesh faces by the image pixels.

To address these holes, an "island approach" is employed, which assigns the closest-connected label to fill the gaps. Additionally, a morphological closing operation is performed to further ensure the complete filling of any remaining small holes in the patch (Fig. 4).

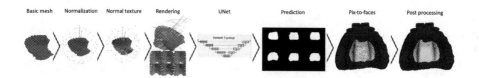

Fig. 4. Pipeline Prediction.

3.6   Automatic Registration (AReg)

Once the butterfly patch is predicted for the upper dental arches in both T1 and T2 scans, the Iterative Closest Point (ICP) algorithm is applied to register the T1 and T2 upper scans using the patches, automatically created in the previous steps, as ROI for registration. The T1 scan serves as the source (fixed), while the T2 scan serves as the target (moving). To reach this final automated step, it was necessary to initially perform steps 3.3 to 3.5 in the process (Fig. 5).

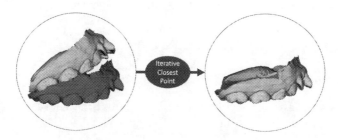

Fig. 5. Iterative Closest Point (ICP) using the predicted butterfly patch in the T1 (pink) and T2 (yellow) scans. (Color figure online)

3.7   Evaluation Metrics

Differences between measurements obtained with the automated process and the manual orientation/registration process were used to assess the error of the proposed methods. The quantifications were computed in the Automated Quantitative 3D Components (AQ3DC) extension of 3D Slicer. Landmarks placed in tooth surfaces were used to calculate the measurements.

Two types of measurements were computed for orientation and registration: linear (in millimeters) and angular (in degrees). Linear distance was calculated between selected landmarks for each component: right-left, anterior-posterior, superior-inferior and the 3D distance. For the angular measurement, two lines (one created at T1/pre-treatment and another at T2/post-treatment) were used to calculate the angular errors of yaw, pitch and roll.

Orientation. The dataset used to evaluate the orientation error consisted of 48 scans, all of which were oriented by an expert clinician. One scan was selected as the gold standard for the method, while the remaining 47 scans were used to calculate the error.

Butterfly Patch. The accuracy of the butterfly patch was determined by calculating the Dice coefficient for segmentation. The dataset used for the evaluation consisted of 48 scans that were not included in the training and validation tasks. The patch was developed together by a developer and a clinician, with the developer responsible for creating the patch and the clinician verifying the accuracy of the patch's superimposition.

Registration. The computation of the registration error calculation included a dataset of 24 growing patients (mean age of 8 years and 2 months) who were not included in the validation or training dataset during the patch learning process. The scans of these patients were registered manually by an expert clinician. The time interval between the T1 and T2 scans was 12 months.

4   Result

4.1   Orientation

Table 1 shows the average angular error $\pm$ standard deviation of the orientation method. The average angular error ranged from 0.44 ± 0.48 (Roll) to 1.19 ± 2.33 (Pitch) degrees.

Table 1. Angular error for the automatic orientation method.

Landmarks	Yaw	Pitch	Roll
Mid UR6O UL6O - UR1O	$0.75° \pm 1.15°$	NA	NA
UR6O - UR6CB	NA	$-1.19° \pm 2.33°$	$-0.44° \pm 0.48°$
UL6O UL6CB	NA	$-0.78° \pm 2.30°$	$-0.44° \pm 0.49°$
UR1O UR1CB	NA	$-1.05° \pm 2.29°$	$-0.69° \pm 0.68°$

Yaw, Pitch and Roll orientation components. Yaw rotation around the S-I axis
Pitch rotation around the R-L axis Roll rotation around the A-P axis

4.2 Butterfly Patch

The Dice coefficient for the segmented Butterfly patch was between 0.75 to
0.95 Fig. 6a. In addition, the confusion matrix shows that the butterfly patch
could be predicted with an agreement above 97.7% Fig. 6b.

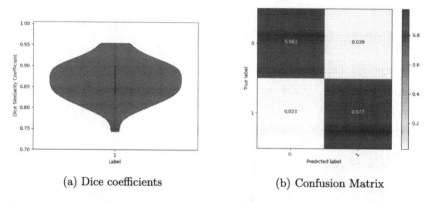

(a) Dice coefficients   (b) Confusion Matrix

Fig. 6. Dice coefficient (a) and Confusion Matrix (b) for the segmented Butterfly patch.
The label 1 represent the Butterfly patch and the label 0 the rest of the segmentation.

4.3 Registration

Tables 2 show the average error angular and linear average errors $\pm$ standard
deviations of the automated registration method respectively. The average angular error ranged from 0.44 ± 0.48 (Roll) to 1.19 ± 2.33 (Pitch) degrees.

The average angular and linear errors ranged from 0.01 ± 1.04 (Roll) to $0.32 \pm$
1.03 (Yaw) degrees and from 0.02 ± 0.13 (Right-Left) to 0.18 ± 0.71 (Superior-
Inferior) millimeters (Figs. 7 and 8 and Table 3).

Table 2. Angular error for the automatic registration method.

Landmarks	Yaw	Pitch	Roll
Mid UR6O UL6O - UR1O	$-0.32° \pm 1.03°$	NA	NA
UR6O - UR6CB	NA	$0.10° \pm 1.28°$	$0.01° \pm 1.04°$
UL6O - UL6CB	NA	$0.23° \pm 1.44°$	$-0.02° \pm 1.05°$
UR1O - UR1CB	NA	$0.18° \pm 1.29°$	$0.01° \pm 1.19°$

Yaw, Pitch and Roll orientation components. Yaw rotation around the S-I axis
Pitch rotation around the R-L axis Roll rotation around the A-P axis

Table 3. Linear error for the automatic registration method (in mm).

Landmarks	R-L	A-P	S-I	3D Distance
UR1O T1 - UR1O T2	-0.13 ± 0.56	-0.04 ± 0.35	-0.18 ± 0.71	0.52 ± 0.32
UR6MBT1 - UR6MBT2	0.02 ± 0.14	0.11 ± 0.52	-0.09 ± 0.50	0.39 ± 0.28
UL6MBT1 - UL6MBT2	0.02 ± 0.13	-0.03 ± 0.46	-0.03 ± 0.46	0.39 ± 0.30

Linear: Right (R) displacement (+), Left (L) displacement (−), Anterior (A)
displacement (+), Posterior (P) displacement (−), Superior (S) displacement (+),
Inferior (I) displacement (−)

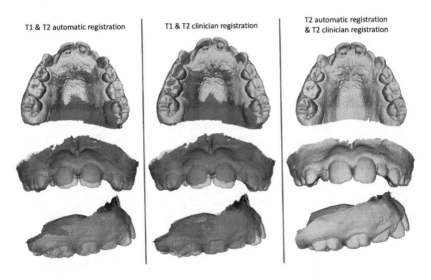

T1 & T2 automatic registration   T1 & T2 clinician registration   T2 automatic registration & T2 clinician registration

Fig. 7. Demonstration of automatic registration.

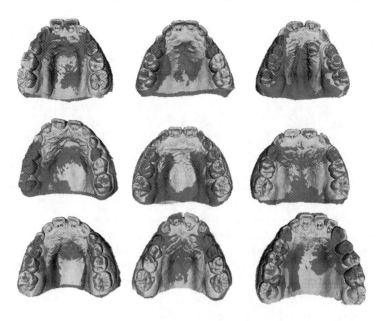

Fig. 8. Automatic Registration of 9 patients.

5   Discussion

ASO, which is already deployed on 3D Slicer as the SlicerAutomatedDentalTools extension, offers a more flexible approach compared to the proposed method. It allows users to choose the teeth for orientation and the gold standard scan. Additionally, ASO has the capability to orient the mandibular arch based on the orientation of the maxillary arch.

The crown segmentation is a fundamental step in the orientation process, as it helps identify the centroids of the teeth. The DentalModelSeg extension in 3D Slicer is used for crown segmentation, and it performs better in segmenting the crown of adult patients rather than those with growth and in the mixed dentition. It should be also noted that the performance of DentalModelSeg is better for incisors teeth compared to molars. In our method, we utilize the first premolars and first molars.

AReg introduces a new automatic registration method for Intra Oral Scan that can be applied to various types of cases. It has been observed that the registration works well for non-growing patients, but it becomes more challenging for growing patients or when there is a significant time difference between T1 and T2 scans. The automatic registration may be questionable under such circumstances and need a double check by the user in view of the standard deviation.

One limitation of our method is that if the palate is missing in the scan, registration becomes impossible as the neural network relies on the presence of the palate to predict the ROI. Several methods for registration of maxillary IOS have been reported [11]. Only a few of them tried to control the challenge of having pitch, yaw and roll rotations [1,6,12]. In addition, there is no method that reported automated registration of IOS. Further research needs to be done with other automated methods to compare our results.

AReg presents a novel approach to register Intra Oral Scans, which exhibits low error rates. However, it is important for users to review the results due to the variability observed in some challenging cases. The method is available in the SlicerAutomatedDentalTools extension of 3D Slicer. The results demonstrate promising performance, minimizing errors in scan alignment and registration. Compared to manual registration, the automated approach offers improved speed, efficiency, and reduced human errors. Integration into 3D Slicer software, specifically SlicerAutomatedDentalTools, enhances accessibility and usability for researchers and clinicians. The open-source nature of the code facilitates collaboration and potential future enhancements. Challenges remain for growing patients or significant time gaps between scans, warranting further research and refinement.

6   Conclusion

This paper presents an automated method to register Intra Oral Scans using advanced algorithms, including crown segmentation, neural network predictions, and the Iterative Closest Point (ICP) algorithm. Overall, the proposed automated registration method provides a promising and efficient solution for Intra Oral Scan registration, with potential applications in treatment planning, monitoring, and assessment in dentistry (Table 4).

Table 4. Description of landmarks.

Landmark name	Landmark description
UR1O	Center of the incisal edge of the Upper Right permanent central incisor (UR1)
UR1CB	Midpoint of the cervico-buccal part of the crown of the UR1
UR6MB	Mesial buccal cusp tip of the Upper Right permanent first molar (UR6)
UR6O	Occlusal limit of the mesiobuccal groove of the UR6
UR6CB	Cervical limit of the buccal groove of the UR6
UL6MB	Mesial buccal cusp tip of the Upper Left permanent first molar (UL6)
UL6O	Occlusal limit of the mesiobuccal groove of the UL6
UL6CB	Cervical limit of the buccal groove of the UL6

References

1. Aliaga-Del Castillo, A., et al.: Comparison and reproducibility of three methods for maxillary digital dental model registration in open bite patients. Orthod. Craniofac. Res. **25**(2), 269–279 (2022). https://doi.org/10.1111/ocr.12535. https://onlinelibrary.wiley.com/doi/abs/10.1111/ocr.12535

2. Ben-Hamadou, A., et al.: 3DTeethSeg'22: 3D teeth scan segmentation and labeling challenge. arXiv preprint arXiv:2305.18277 (2023)
3. Besl, P., McKay, N.D.: A method for registration of 3-D shapes. IEEE Trans. Pattern Anal. Mach. Intell. **14**(2), 239–256 (1992). https://doi.org/10.1109/34.121791
4. Boubolo, L., et al.: FlyBy CNN: a 3D surface segmentation framework. Proc. SPIE Int. Soc. Opt. Eng. **11596**, 79 (2021). https://doi.org/10.1117/12.2582205
5. Chen, G., et al.: Stable region for maxillary dental cast superimposition in adults, studied with the aid of stable miniscrews. Orthod. Craniofac. Res. **14**(2), 70–79 (2011). https://doi.org/10.1111/j.1601-6343.2011.01510.x. https://onlinelibrary.wiley.com/doi/abs/10.1111/j.1601-6343.2011.01510.x
6. Garib, D., et al.: Superimposition of maxillary digital models using the palatal rugae: does ageing affect the reliability? Orthod. Craniofac. Res. **22**(3), 183–193 (2019). https://doi.org/10.1111/ocr.12309. https://onlinelibrary.wiley.com/doi/abs/10.1111/ocr.12309
7. Jang, I., et al.: A novel method for the assessment of three-dimensional tooth movement during orthodontic treatment. Angle Orthod. **79**(3), 447–453 (2009)
8. Pazera, C., Gkantidis, N.: Palatal rugae positional changes during orthodontic treatment of growing patients. Orthod. Craniofac. Res. **24**(3), 351–359 (2021). https://doi.org/10.1111/ocr.12441. https://onlinelibrary.wiley.com/doi/abs/10.1111/ocr.12441
9. Richert, R., et al.: Intraoral scanner technologies: a review to make a successful impression. J. Healthc. Eng. **2017**, 1–9 (2017). https://doi.org/10.1155/2017/8427595
10. Ronneberger, O., Fischer, P., Brox, T.: U-Net: convolutional networks for biomedical image segmentation. In: Navab, N., Hornegger, J., Wells, W.M., Frangi, A.F. (eds.) MICCAI 2015. LNCS, vol. 9351, pp. 234–241. Springer, Cham (2015). https://doi.org/10.1007/978-3-319-24574-4_28
11. Stucki, S., Gkantidis, N.: Assessment of techniques used for superimposition of maxillary and mandibular 3D surface models to evaluate tooth movement: a systematic review. Eur. J. Orthod. **42**(5), 559–570 (2019). https://doi.org/10.1093/ejo/cjz075
12. Vasilakos, G., Schilling, R., Halazonetis, D., Gkantidis, N.: Assessment of different techniques for 3D superimposition of serial digital maxillary dental casts on palatal structures. Sci. Rep. **7**, 5838 (2017). https://doi.org/10.1038/s41598-017-06013-5
13. Zhang, J.: C-Bézier curves and surfaces. Graph. Model. Image Process. **61**(1), 2–15 (1999). https://doi.org/10.1006/gmip.1999.0490. https://www.sciencedirect.com/science/article/pii/S1077316999904902

Modeling Longitudinal Optical Coherence Tomography Images for Monitoring and Analysis of Glaucoma Progression

James Fishbaugh[1]([✉]) [iD], Ronald Zambrano[2] [iD], Joel S. Schuman[3] [iD],
Gadi Wollstein[2] [iD], Jared Vicory[1] [iD], and Beatriz Paniagua[1] [iD]

[1] Kitware, Inc., Clifton Park, NY 12065, USA
james.fishbaugh@kitware.com
[2] NYU Grossman School of Medicine, New York City, NY 10016, USA
[3] Wills Eye Hospital, Philadelphia, PA 19107, USA

Abstract. Glaucoma causes progressive visual field deterioration and is the leading cause of blindness worldwide. Glaucomatous damage is irreversible and greatly impacts quality of life. Therefore, it is critically important to detect glaucoma early and closely monitor progression to preserve functional vision. Glaucoma is routinely monitored in the clinical setting using optical coherence tomography (OCT) for derived measures such as the thickness of important visual structures. There is not a consensus of what measures represent the most relevant biomarkers of glaucoma progression. Further, despite the increasing availability of longitudinal OCT data, a quantitative model of 3D structural change over time associated with glaucoma does not exist. In this paper we present an algorithm that will perform hierarchical geodesic modeling at the imaging level, considering 3D OCT images as observations of structural change over time. Hierarchical modeling includes subject-wise trajectories as geodesics in the space of diffeomorphisms and population level (glaucoma vs control) trajectories are also geodesics which explain subject-wise trajectories as deviations from the mean. Our preliminary experiments demonstrate a greater magnitude of structural change associated with glaucoma compared to normal aging. Our algorithm has the potential application in patient-specific monitoring and analysis of glaucoma progression as well as a statistical model of population trends and population variability.

Keywords: longitudinal shape analysis · hierarchical modeling · diffeomorphic regression · optical coherence tomography · glaucoma progression

1 Introduction

Glaucoma causes progressive and irreversible visual field deterioration and is the leading cause of blindness worldwide [17]. Therefore, it is critically important to

© The Author(s), under exclusive license to Springer Nature Switzerland AG 2023
C. Wachinger et al. (Eds.): ShapeMI 2023, LNCS 14350, pp. 236–247, 2023.
https://doi.org/10.1007/978-3-031-46914-5_19

detect glaucoma early and closely monitor patients in order to slow progression and preserve functional vision. Optical coherence tomography (OCT) is currently used clinically in order to monitor disease progression. Regularly used imaging biomarkers derived from OCT include the thickness of the retinal nerve fiber layer (RNFL) [3,4,9,13,25], as well as optic nerve head, lamina cribrosa and macula measurements, but there is not a consensus whether one or a combination of measures increase the sensitivity in detecting damage [15,29]. Previous methodology and analysis strategies have not focused on quantitative models of 3D structural and shape change associated with disease progression using OCT.

With continuing advances in OCT technology, an increasing number of clinical studies collect longitudinal repeated scans of individual subjects to track glaucoma progression. However, to our knowledge, nobody has developed strategies to account for the inherent correlation within OCT observations of a single subject. This is crucial, as cross-sectional approaches, which assume that all observations are independent, may give misleading or wrong results such as those observed as Simpson's paradox.

In this work, we propose a longitudinal modeling scheme which treats observations as 3D OCT images themselves, rather than modeling derived measures. Such high dimensional observations limit the use of traditional statistical methods and require specialized approaches depending on the data representation. Here we consider a hierarchical modeling approach composed of a subject-wise level and a population level. At the subject level, individual trajectories are modeled as geodesics in the space of diffeomorphisms, which are transformations which deform OCT images over time. At the population level, a group-wise geodesic trend describes the average or mean trajectory taking into account the subject-wise trajectories.

We believe that models of continuous glaucoma progression based on trajectories of OCT structural images can be used for comparison, monitoring, and prediction of patient progress. The work presented here has the potential to have a positive impact in the ophthalmology research community, providing new clinically relevant information about this degenerative disease as well as support the development of better preventative strategies so glaucoma patients can preserve their vision longer.

2   Methodology

Hierarchical models are widely used in statistics when data comes with a hierarchical structure, such as longitudinal data of subjects from different groups (e.g. disease vs. control). Such data can be modeled at the subject-specific level, following individual trajectories and also at the population level, taking into account the group membership of each individual. Longitudinal study design also comes with several challenges associated with modeling and analysis. It is common for participants to reschedule imaging sessions, miss one or more scans over the duration of the study, or dropout completely. This leads to a dataset with missing data, a different number of observations per subject, and acquisitions at staggered time points.

Hierarchical and mixed-effects models are ubiquitous in Euclidean statistics and the medical imaging research community has made considerable effort in extending methods to high dimensional spaces such as Riemannian manifolds, for their application to 3D medical images and the shapes derived from them. The majority of previous work focuses on shapes with manifold representations. This includes methods based on Kendall's shape space [18], where shapes are parameterized by landmarks in anatomical correspondence and mixed-effects models are designed for this data representation [6,22]. Additionally, several methods are considered intrinsic Riemannian models which can be implemented for a variety of manifold representations [11,12,14,20,21,23]. The manifold of diffeomorphic transformations has seen considerable attention in longitudinal modeling based on relatively low dimensional shape representations [2,7], with relatively few methods focusing on 3D medical images which contain millions of voxels and are computationally expensive to work with. In contrast to methods with explicit models, learning based methods for longitudinal image analysis have been proposed [5,16,19,26,30] as data driven approaches.

Here we propose to model structural change of OCT by a hierarchical model of image change based on the large deformation diffeomorphic metric mapping (LDDMM) framework [1]. Image change is modeled as a smooth deformation of the ambient image space by diffeomorphic flow which is smooth and has a smooth inverse. The smoothness properties of diffeomorphisms are a natural choice for modeling anatomical change as they preserve topology, preventing any image folding or deletion of structures. By the analogy with linear hierarchical models, we will model all time trajectories as geodesics, which are parameterized by an initial/baseline image (intercept) and a tangent vector (slope).

The hierarchical model consists of two levels: a subject-wise level and a population level. The subject-wise level describes individual trajectories by a geodesic trend for each subject. Likewise, the population level incorporates the subject-wise geodesics in the estimation of a geodesic which represents the mean trend of the population. An overview of our longitudinal modeling framework is shown in Fig. 1. Next, we present the formulation for the subject-wise and population models first suggested in [28].

2.1   Subject-Wise Level

At this level, each of M total subjects $i = 1, ..., M$ comes with a time-series of images $\mathbf{I}_i = \{I_{i,t_0}, ..., I_{i,t_{N-1}}\}$ consisting of $N \geq 2$ images. The dynamics of subject-wise image evolution are modeled as the geodesic flow of diffeomorphisms of an initial image $\hat{I}_i(0)$ with trajectory $\hat{I}_i(t) = \hat{I}_i(0) \circ \phi_i^{-1}(t)$. Subject-specific trajectory $\phi_i^{-1}(t)$ is a geodesic defined by a tangent vector $m_i(0)$ located at $\hat{I}_i(0)$ which in practice is a voxel-wise vector field referred to as momenta [24,27]. For ease of understanding, by analogy with linear regression, we will use the shorthand *intercept* for the initial image $\hat{I}_i(0)$ and *slope* for the momenta vector field $m_i(0)$.

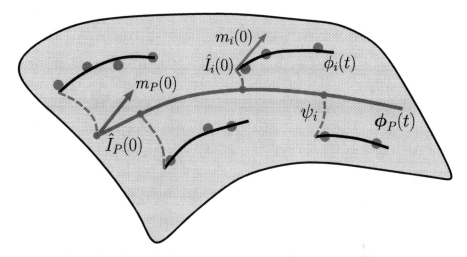

Fig. 1. Overview of the hierarchical geodesic model. Each dark gray circle represents a time-indexed image from a given subject. Subject-wise trajectories are geodesics parameterized by a initial image $\hat{I}_i(0)$ (intercept) and tangent vector $m_i(0)$ (slope) which define the image trajectory $\phi_i(t)$. The red line represents the population level model which captures the average trajectory, also parameterized by an initial image $\hat{I}_P(0)$ and tangent vector $m_P(0)$. The population level geodesic takes subject-wise intercepts and slopes into account and also seeks to minimize residual distances ψ_i. (Color figure online)

The subject-wise intercepts and slopes are estimated to minimize the least squares criterion

$$E(\hat{I}_i(0), m_i(0)) = \gamma \sum_{j=0}^{N-1} d^2(\hat{I}_i(t) - I_{i,t_j}) + Reg(m_i(0)) \qquad (1)$$

where d^2 is a squared distance between images, Reg is a regularity norm term on the initial momenta, and γ is a hyper-parameter to balance the trade-off between data-matching and regularity. Intuitively, we seek the geodesic trajectory $\hat{I}_i(t)$ which closely matches image observations I_{i,t_j} according to the image distance term d^2. Subject-wise intercepts and slopes can be obtained by M geodesic regression estimation procedures, as in [8,27].

2.2   Population Level

At the population level, we seek to estimate a mean geodesic trajectory that takes all subject-wise geodesic trends into account. As shown in Fig. 1, the population geodesic $\phi_P(t)$ is defined in the same way as subject-wise geodesics, with an initial image $\hat{I}_P(0)$ and momenta $m_P(0)$. This gives the trajectory of the population average as $\hat{I}_P(t) = \hat{I}_P(0) \circ \phi_P^{-1}(t)$, which also minimizes residual geodesic distances between ψ_i and $\phi_P(t)$. The goal is to estimate the ψ_i's, intercept $\hat{I}_P(0)$, and slope $m_P(0)$ which minimizes

$$
\begin{aligned}
E(\psi_i, \hat{I}_P(0), m_P(0)) = \gamma_{\hat{I}_P(0)} \sum_{i=1}^{M} \Big(d^2(\psi_i \cdot \hat{I}_P(t_i) - \hat{I}_i(0)) + Reg(\psi_i) \Big) \\
+ \gamma_{m_P(0)} \sum_{i=1}^{M} d^2(\psi_i \cdot \phi_P(t_i) \cdot m_P(0) - m_i(0)) \\
+ Reg(m_P(0))
\end{aligned}
\tag{2}
$$

where the first term represents the distance between the population model and the intercepts of the subject-wise models $\hat{I}_i(0)$, the second term represents the distance between the population level slope the subject-wise slopes, and the last term is the regularity norm term on the population level initial momenta. The weights $\gamma_{\hat{I}_P(0)}$ and $\gamma_{m_P(0)}$ balance the contribution of the intercept and slope terms, respectively. Intuitively, the two terms incorporate not only the difference between the population model and the subject-wise initial images, but also includes differences in subject-wise trajectories via the slope tangent vectors. The parameters of the population model can be estimated by gradient descent including constraints that ψ_i's are geodesics, as in [28].

3   Experimental Results

3.1   Data Description

To validate hierarchical diffeomorphic geodesic modeling for analysis of structural change of OCT, we utilize a clinical OCT dataset acquired at NYU Langone Health. The dataset consists of 10 controls and 10 patients with glaucoma. Each subject has at least 3 time-points covering a timespan of approximately 2 to 4 years, with most subjects having 5 or more OCT observations. OCT images are acquired from a clinical Cirrus scanner with dimension $200 \times 200 \times 1024$. All images were downsampled to $100 \times 100 \times 512$ for full 3D analysis, and additionally all scans are preprocessed by subject-wise affine alignment.

3.2   Subject-Wise Geodesic Regression

First, we validate the diffeomorphic geodesic regression model on subject-specific trajectories. Here, we test for feasibility of model fit given clinical OCT acquired overall several years, with challenges such as different fields of view during imaging and noise, which is common in OCT. This is essential to establish a baseline for the model in capturing possible structural change associated with disease progression.

We estimate the subject-wise diffeomorphic models for all controls and glaucoma patients, which involves estimating the baseline image and initial momenta parameters described in Sect. 2.1. From these intercept and slope parameters,

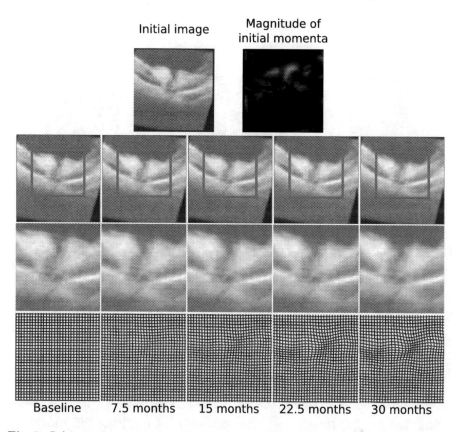

Fig. 2. Subject-wise trajectory of a glaucoma subject over a 30 month observation period. Top row) The estimated model parameters of initial image $\hat{I}_i(0)$ and the magnitude of the initial momenta vectors $m_i(0)$. 2nd row) Five frames of the continuous trajectory of image change shows an increase in the size of the optic cup. 3rd row) A zoomed in view of the trajectory which highlights tissue deformation. Bottom row) A deforming grid over time highlights the non-linear structural changes over the course of 30 months follow-up from baseline.

the full trajectory of OCT structural change can be computed as in geodesic shooting in the LDDMM framework [1]. Figure 2 shows a model estimated for a randomly selected glaucoma subject. The model parameters are shown in the top row, with the initial image along with the initial momenta, which highlight areas in the image which change most over time. Here we display 5 frames of the continuous model of OCT structural change which cover the timespan of 30 months from the baseline acquisition. A zoomed in view shows deformation around the optic cup, as the structure increases in size over time. The non-linear changes are most evident in the bottom row, which shows a deforming grid of increasing complexity over time.

Initial image   Magnitude of   Final
                 initial momenta   deformation

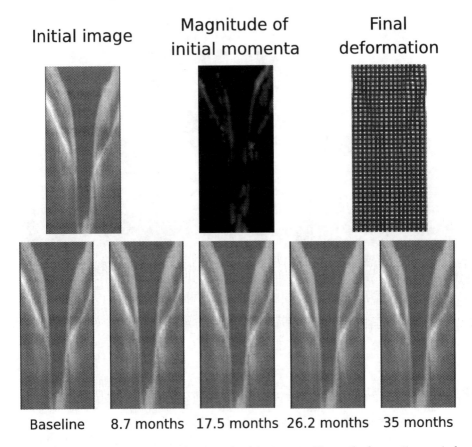

Baseline   8.7 months   17.5 months   26.2 months   35 months

Fig. 3. Subject-wise trajectory of a control subject over a 35 month observation period. Top row) The estimated model parameters of initial image $\hat{I}_i(0)$ and the magnitude of the initial momenta vectors $m_i(0)$, along with the deformation grid corresponding to the final time-point at 35 months Bottom row) Five frames of the continuous trajectory highlight consistent anatomy and very little measurable structural change.

We also illustrate the trajectory of a randomly chosen control subject over time, shown in Fig. 3. There is essentially no discernible structural change over the 35 months, even when the continuous trajectory is viewed as an animation the structure is consistent. A zoomed in view of the deformation at the final time-point of 35 months also shows negligible deformation.

3.3 Longitudinal Modeling at the Population Level

With subject-wise models consisting of initial images and initial momenta estimated, we can estimate population level parameters by minimizing Eq. 2. At this level, we estimate population-wise initial images and initial momenta which describe the trajectory of the average of the population, which takes into account the subject-wise intercepts and slopes.

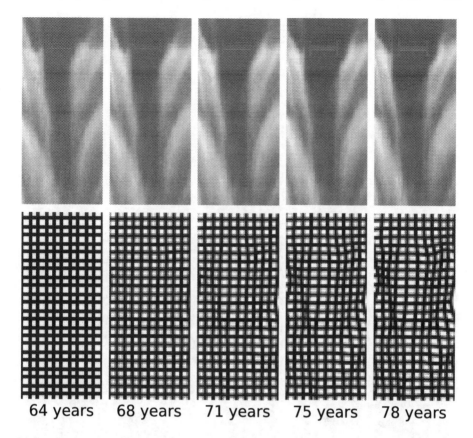

64 years   68 years   71 years   75 years   78 years

Fig. 4. Population-level model of glaucoma subjects. Top row) Several snapshots of the average trajectory of OCT over time associated with glaucoma. The red line helps to illustrate the magnitude of shape change. Bottom row) Deformation of the image grid starting from identity. Note there is a larger magnitude of structural change in the glaucoma population compared to the control population (Fig. 5), suggesting the hierarchical model is able to capture the effect of glaucoma in addition to normal aging. (Color figure online)

The average trajectory for the glaucoma population is shown in Fig. 4, which shows 5 frames of the estimated continuous image sequence from 64 to 78 years. The red line serves as a visual guide to see a widening over time associated with the glaucoma population. The grid deformation is also shown in Fig. 4 to illustrate structural change starting from the identity at the initial image. Note that this model represents an unknown mixture of age and glaucoma effects on the optic disc.

To explore the impact of age separate from the effect of glaucoma, we also estimate a population level model for the control subjects. This average trajectory can be considered as a reference model of aging. Several frames of the estimated OCT sequence are shown in Fig. 5. We observe there is age related

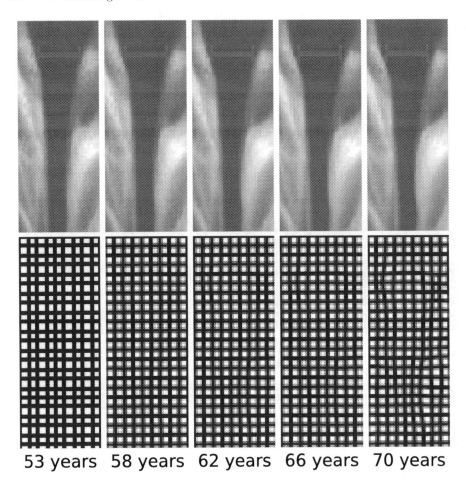

53 years 58 years 62 years 66 years 70 years

Fig. 5. Population-level model of the control group. Top row) Several snapshots of the average trajectory of OCT over time associated with normal aging. Bottom row) Deformation of the image grid starting from identity. Note there is a lower magnitude of structural change in the control population compared to the glaucoma population.

change in structure, shown in the OCT sequence as well as the deforming grid starting from identity. In comparison to the average population level model from the glaucoma group, the structural change due to normal aging is of lower magnitude. This promising result suggest hierarchical modeling of OCT has the potential to capture structural changes of key anatomical structures associated with glaucoma progression. Future work will focus on validation on a large cohort to measure the impact of between-subject variability on population level models and better distinguish between age and disease effects.

4   Conclusion

In this paper, we propose hierarchical diffeomorphic geodesic modeling for the longitudinal analysis of 3D OCT. Our driving clinical application is the study of glaucoma, with the goal to develop and test methods and tools which can be used to characterize and predict disease progression. Hierarchical modeling consists of a subject-wise level which captures the within-subject correlation of repeated OCT acquisitions. To model changes associated with group membership, the hierarchical model has a population level model which takes into account all subject-wise trends in the estimation of a mean trajectory. Experimentation on a real clinical OCT cohort of controls and glaucoma patients show a larger magnitude of structural change associated with glaucoma patients than with normal aging. Results show that longitudinal models of shape change captured by OCT imaging have the potential to monitor and quantify the progression of glaucoma.

Our experiments show that population-level models are sensitive to between-subject variability. This can be seen in the relatively blurry appearance and some ghosting artifacts in population models as compared to much sharper subject-wise models. This is due to anatomical variability, but also due to the choice of initial alignment. We chose to use an affine alignment strategy to normalize for size variability and isolate more subtle non-linear anatomical changes. However, due to natural anatomical variability and differences associated with image acquisition, such as different fields of view, this topic needs further investigation. One possible improvement would involve cropping to a consistent population wide region of interest by a clinical expert and pre-alignment driven by segmentation based landmarks. Future work will also focus on decoupling age and disease effects on structural change.

Acknowledgements. This work is supported by the National Institute of Health R01EB021391 (SlicerSALT), R01EY013178, and an unrestricted grant from Research to Prevent Blindness.

References

1. Beg, M.F., Miller, M.I., Trouvé, A., Younes, L.: Computing large deformation metric mappings via geodesic flows of diffeomorphisms. Int. J. Comput. Vis. **61**(2), 139–157 (2005)
2. Bône, A., Colliot, O., Durrleman, S.: Learning distributions of shape trajectories from longitudinal datasets: a hierarchical model on a manifold of diffeomorphisms. In: Proceedings of the IEEE Conference on Computer Vision and Pattern Recognition, pp. 9271–9280 (2018)
3. Budenz, D.L., Fredette, M.J., Feuer, W.J., Anderson, D.R.: Reproducibility of peripapillary retinal nerve fiber thickness measurements with stratus oct in glaucomatous eyes. Ophthalmology **115**, 661.e4–666.e4 (2008). https://doi.org/10.1016/j.ophtha.2007.05.035

4. Carpineto, P., Ciancaglini, M., Zuppardi, E., Falconio, G., Doronzo, E., Mastropasqua, L.: Reliability of nerve fiber layer thickness measurements using optical coherence tomography in normal and glaucomatous eyes. Ophthalmology **110**, 190–195 (2003). https://doi.org/10.1016/s0161-6420(02)01296-4

5. Couronné, R., Vernhet, P., Durrleman, S.: Longitudinal self-supervision to disentangle inter-patient variability from disease progression. In: de Bruijne, M., et al. (eds.) MICCAI 2021. LNCS, vol. 12902, pp. 231–241. Springer, Cham (2021). https://doi.org/10.1007/978-3-030-87196-3_22

6. Datar, M., Muralidharan, P., Kumar, A., Gouttard, S., Piven, J., Gerig, G., Whitaker, R., Fletcher, P.T.: Mixed-effects shape models for estimating longitudinal changes in anatomy. In: Durrleman, S., Fletcher, T., Gerig, G., Niethammer, M. (eds.) STIA 2012. LNCS, vol. 7570, pp. 76–87. Springer, Heidelberg (2012). https://doi.org/10.1007/978-3-642-33555-6_7

7. Durrleman, S., Pennec, X., Trouvé, A., Braga, J., Gerig, G., Ayache, N.: Toward a comprehensive framework for the spatiotemporal statistical analysis of longitudinal shape data. Int. J. Comput. Vis. **103**(1), 22–59 (2013)

8. Fishbaugh, J., Prastawa, M., Gerig, G., Durrleman, S.: Geodesic image regression with a sparse parameterization of diffeomorphisms. In: Nielsen, F., Barbaresco, F. (eds.) GSI 2013. LNCS, vol. 8085, pp. 95–102. Springer, Heidelberg (2013). https://doi.org/10.1007/978-3-642-40020-9_9

9. Garcia-Martin, E., Pinilla, I., Idoipe, M., Fuertes, I., Pueyo, V.: Intra and interoperator reproducibility of retinal nerve fibre and macular thickness measurements using cirrus Fourier-domain oct. Acta Ophthalmol. **89**, e23–e29 (2011). https://doi.org/10.1111/j.1755-3768.2010.02045.x

10. Gürses-Ozden, R., Teng, C., Vessani, R., Zafar, S., Liebmann, J.M., Ritch, R.: Macular and retinal nerve fiber layer thickness measurement reproducibility using optical coherence tomography (OCT-3). J. Glaucoma **13**, 238–244 (2004). https://doi.org/10.1097/00061198-200406000-00012

11. Han, Y., et al.: Hierarchical geodesic polynomial model for multilevel analysis of longitudinal shape. In: Frangi, A., de Bruijne, M., Wassermann, D., Navab, N. (eds.) Information Processing in Medical Imaging, IPMI 2023. LNCS, vol. 13939, pp. 810–821. Springer, Cham (2023). https://doi.org/10.1007/978-3-031-34048-2_62

12. Hanik, M., Hege, H.C., von Tycowicz, C.: A nonlinear hierarchical model for longitudinal data on manifolds. In: 2022 IEEE 19th International Symposium on Biomedical Imaging (ISBI), pp. 1–5. IEEE (2022)

13. Hong, S., Kim, C.Y., Lee, W.S., Seong, G.J.: Reproducibility of peripapillary retinal nerve fiber layer thickness with spectral domain cirrus high-definition optical coherence tomography in normal eyes. Jpn. J. Ophthalmol. **54**, 43–47 (2010). https://doi.org/10.1007/s10384-009-0762-8

14. Hong, S., Fishbaugh, J., Wolff, J.J., Styner, M.A., Gerig, G.: Hierarchical multigeodesic model for longitudinal analysis of temporal trajectories of anatomical shape and covariates. In: Shen, D., et al. (eds.) MICCAI 2019. LNCS, vol. 11767, pp. 57–65. Springer, Cham (2019). https://doi.org/10.1007/978-3-030-32251-9_7

15. Hood, D.C., Raza, A.S., de Moraes, C.G.V., Liebmann, J.M., Ritch, R.: Glaucomatous damage of the macula. Prog. Retin. Eye Res. **32**, 1–21 (2013). https://doi.org/10.1016/j.preteyeres.2012.08.003

16. Huang, C., et al.: DADP: dynamic abnormality detection and progression for longitudinal knee magnetic resonance images from the osteoarthritis initiative. Med. Image Anal. **77**, 102343 (2022)

17. Jonas, J.B., Aung, T., Bourne, R.R., Bron, A.M., Ritch, R., Panda-Jonas, S.: Glaucoma. Lancet (London, England) **390**, 2183–2193 (2017). https://doi.org/10. 1016/S0140-6736(17)31469-1

18. Kendall, D.G.: Shape manifolds, procrustean metrics, and complex projective spaces. Bull. Lond. Math. Soc. **16**(2), 81–121 (1984)

19. Kim, H., Sabuncu, M.R.: Learning to compare longitudinal images. arXiv preprint arXiv:2304.02531 (2023)

20. Kim, H.J., Adluru, N., Suri, H., Vemuri, B.C., Johnson, S.C., Singh, V.: Riemannian nonlinear mixed effects models: analyzing longitudinal deformations in neuroimaging. In: Proceedings of the IEEE Conference on Computer Vision and Pattern Recognition, pp. 2540–2549 (2017)

21. Muralidharan, P., Fletcher, P.T.: Sasaki metrics for analysis of longitudinal data on manifolds. In: 2012 IEEE Conference on Computer Vision and Pattern Recognition, pp. 1027–1034. IEEE (2012)

22. Nava-Yazdani, E., Hege, H.-C., von Tycowicz, C.: A geodesic mixed effects model in Kendall's shape space. In: Zhu, D., et al. (eds.) MBIA/MFCA - 2019. LNCS, vol. 11846, pp. 209–218. Springer, Cham (2019). https://doi.org/10.1007/978-3-030-33226-6_22

23. Nava-Yazdani, E., Hege, H.C., von Tycowicz, C.: A hierarchical geodesic model for longitudinal analysis on manifolds. J. Math. Imaging Vis. **64**(4), 395–407 (2022)

24. Niethammer, M., Huang, Y., Vialard, F.-X.: Geodesic regression for image time-series. In: Fichtinger, G., Martel, A., Peters, T. (eds.) MICCAI 2011. LNCS, vol. 6892, pp. 655–662. Springer, Heidelberg (2011). https://doi.org/10.1007/978-3-642-23629-7_80

25. Paunescu, L.A., et al.: Reproducibility of nerve fiber thickness, macular thickness, and optic nerve head measurements using StratusOCT. Invest. Ophthalmol. Vis. Sci. **45**, 1716–1724 (2004). https://doi.org/10.1167/iovs.03-0514

26. Ren, M., Dey, N., Styner, M., Botteron, K., Gerig, G.: Local spatiotemporal representation learning for longitudinally-consistent neuroimage analysis. Adv. Neural. Inf. Process. Syst. **35**, 13541–13556 (2022)

27. Singh, N., Hinkle, J., Joshi, S., Fletcher, P.T.: A vector momenta formulation of diffeomorphisms for improved geodesic regression and atlas construction. In: 2013 IEEE 10th International Symposium on Biomedical Imaging, pp. 1219–1222. IEEE (2013)

28. Singh, N., Hinkle, J., Joshi, S., Fletcher, P.T.: Hierarchical geodesic models in diffeomorphisms. Int. J. Comput. Vis. **117**, 70–92 (2016)

29. Tatham, A.J., Medeiros, F.A.: Detecting structural progression in glaucoma with optical coherence tomography. Ophthalmology **124**, S57–S65 (2017). https://doi.org/10.1016/j.ophtha.2017.07.015

30. Zhao, Q., Liu, Z., Adeli, E., Pohl, K.M.: Longitudinal self-supervised learning. Med. Image Anal. **71**, 102051 (2021)

IcoConv: Explainable Brain Cortical Surface Analysis for ASD Classification

Ugo Rodriguez[✉], Juan-Carlos Prieto, and Martin Styner

Department of Psychiatry UNC, University of North Carolina, Chapel Hill, NC, USA
ugo.rdgz@gmail.com, jprieto@med.unc.edu, styner@email.unc.edu

Abstract. In this study, we introduce a novel approach for the analysis and interpretation of 3D shapes, particularly applied in the context of neuroscientific research. Our method captures 2D perspectives from various vantage points of a 3D object. These perspectives are subsequently analyzed using 2D Convolutional Neural Networks (CNNs), uniquely modified with custom pooling mechanisms.

We sought to assess the efficacy of our approach through a binary classification task involving subjects at high risk for Autism Spectrum Disorder (ASD). The task entailed differentiating between high-risk positive and high-risk negative ASD cases. To do this, we employed brain attributes like cortical thickness, surface area, and extra-axial cerebral spinal measurements. We then mapped these measurements onto the surface of a sphere and subsequently analyzed them via our bespoke method.

One distinguishing feature of our method is the pooling of data from diverse views using our icosahedron convolution operator. This operator facilitates the efficient sharing of information between neighboring views. A significant contribution of our method is the generation of gradient-based explainability maps, which can be visualized on the brain surface. The insights derived from these explainability images align with prior research findings, particularly those detailing the brain regions typically impacted by ASD. Our innovative approach thereby substantiates the known understanding of this disorder while potentially unveiling novel areas of study.

1 Introduction

Autism Spectrum Disorder (ASD) is a condition related to the development of the brain that causes differences in neurological functioning. Subjects with ASD have deficits in social communication skills and behavior, the presence of repetitive behavior, restricted interests, hyper- or hypo-sensitivity to sensory stimuli, and an insistence on sameness or strict adherence to routine [21, 24].

Diagnosis of ASD is difficult as there is no medical/lab test to reliably assess the condition. Various studies have found the average age of diagnosis to be 3 years, although the vast majority of children manifest developmental problems between 12 and 24 months, with some showing abnormalities before 12 months [3]. Early diagnosis of ASD before 2.5 years of age is associated with considerable benefits for children who may "outgrow" the condition through therapy [10].

© The Author(s), under exclusive license to Springer Nature Switzerland AG 2023
C. Wachinger et al. (Eds.): ShapeMI 2023, LNCS 14350, pp. 248–258, 2023.
https://doi.org/10.1007/978-3-031-46914-5_20

Detecting ASD bio-markers from neuroimaging data is a challenging task owing to the considerable variability in cortical shape and functional organization across individuals, which hinders the ability to make accurate comparisons of brains [11, 14]. There is a clear need for precision analysis tools that are robust to these factors and the discovery of distinct features to characterize ASD.

The main contribution of this paper can be summarized into two key aspects. Firstly, it introduces a new deep-learning framework for general shape analysis that utilizes a multi-view approach. Secondly, it incorporates an explainability component that identifies crucial brain regions at the vertex level and visualizes them on the cortical surface highlighting relevant brain regions for the classification task.

The proposed method was evaluated using Precision, Recall, F1-score, and Accuracy metrics through five fivefold cross-validation experiments on a cohort of High-Risk Positive (HR+) ASD versus High-Risk Negative (HR−) ASD patients. We implement different pooling layers for our model and compare them against Spherical-U-Net [37] and Spectformer [2], methods designed for brain shape analysis and spectral analysis respectively. We also test our approach against a Random Forest classifier that uses learned shape features and demographic information combined.

2 Related Work

2.1 ASD Classification

Several studies have addressed the question of ASD classification using Machine Learning (ML) models. The majority of studies used the ABIDE I/II [13] data set which includes resting state functional (rsfMRI), structural T1/T2 (sMRI) Magnetic Resonance Images, and Diffusion (dMRI) Magnetic Resonance Images. This data set contains data from individuals with autism spectrum disorder (ASD) and typically developing individuals. We point the reader to [8, 19] for a comprehensive review of the literature on ASD classification. It has been demonstrated that different machine learning models can effectively distinguish between individuals with typical development and those with ASD. However, the data used in our study differs, as it includes HR subjects who have not yet developed the condition. This adds a layer of complexity to the analysis, as it requires the identification of biomarkers that can reliably predict ASD in the future.

2.2 3D Shape Analysis

Among the different approaches for shape analysis, learning-based methods are currently the most sophisticated ones. There are mainly 4 types of learning-based methods: multi-view, volumetric, parametric, and multi-layer-perceptrons (MLP).

Multi-view approaches adapt state-of-the-art 2D CNNs to work on 3D shapes as the arbitrary structures of 3D models, which are usually represented by point clouds or triangular meshes, are incompatible with convolutional operators that require regular grid-like structures. By rendering 3D objects from different view- points, features are extracted using 2D CNNs [18, 32]. On the other hand, volumetric approaches use 3D voxel grids to represent the shape and apply 3D convolutions to learn shape features [28,

35]. Parametric methods require shapes with spherical topology and the convolution is applied directly to the spherical representation of the shape [15, 37]. Finally, other approaches consume the point clouds directly and implement multi-layer-perceptrons and/or transformer architectures [23, 34].

Our method falls in the multi-view category. We render the object and capture 2D images from different viewpoints following an icosahedron subdivision. The multiple captures ensure coverage of the whole object. One of the primary benefits of multi-view approaches is their ability to operate on surfaces with any topology, including those with missing data or holes. Our method is tested on spheres derived from brain cortical gray/white matter surfaces.

2.3 Explainable Artificial Intelligence

ML systems are becoming increasingly ubiquitous and they outperform humans on a variety of specific tasks. There is increasing concern related to the deployment of such complex applications that have a direct impact on human lives. Such systems must be able to explain the basis for their decision to any impacted individual in terms understandable to a layperson, this is especially the case in the field of medical imaging. Explainability methods fall into 3 categories: visualization, model distillation, and intrinsic [26]. To the best of our knowledge, we found 2 methods for cortical surface analysis and explainability. First, a perturbation-based method for geometric deep learning of retinotopy through systematic manipulations of the input data and measurement of changes in the model's output [27]. Second, NeuroExplainer [36] a method that uses spherical surfaces of the brain hemispheres with cortical attributes (thickness, mean curvature, and convexity), and a spherical convolution block in an encoder/decoder architecture that propagates the vertex-wise attributes and captures fine-grained explanations for a classification task.

Our explainability model is agnostic to the input data and does not require systematic perturbations to produce explanations. Moreover, it does not require subsampling the input data through encoder/decoder architectures and does not require shapes with a spherical topology or specialized operators such as spherical convolutions. In our experiments, we use spheres of +160,000 vertex at full resolution.

3   Materials

Infants at high and low familial risk for ASD were enrolled at four clinical sites (University of North Carolina, University of Washington, Washington University, and Children's Hospital of Philadelphia) [12, 30]. HR infants had an older sibling with a clinical diagnosis of ASD, corroborated by the Autism Diagnostic Interview–Revised [17]. LR infants had a typically developing older sibling and no first- or second-degree relatives with intellectual/psychiatric disorders [9]. Infants were assessed at 6, 12, and 24 months with magnetic resonance imaging (MRI) scans and a behavioral battery that included measures of cognitive development [20] and adaptive functioning [31]. DSM-IV-TR criteria [1] and the Autism Diagnostic Observation Schedule–Generic [16] were administered to all participants at 24 months. The Autism Diagnostic Interview–Revised was administered at 24 months to all parents of high-risk infants and to all low-risk infants with clinical concerns. At 24 months, infants were classified as having ASD based on expert clinical judgment and all available clinical information.

In our experiments, we use a subset of HR infants only and compare a group of 760 HR+ v.s. 202 HR−. We include demographic data in our analysis by combining image features and demographics through a separate branch that concatenates with the output of the features computed by the NN.

The demographics include gender, visit age for MRI, volume measurements for subcortical structures (amygdala, hippocampus, lateral ventricles), intracranial volume (ICV), and cerebrum and cerebellum volume.

4   Method Description

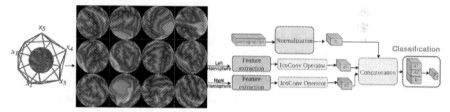

Fig. 1. Architecture for the ASD classification task. To initiate our analysis, we begin by capturing views of the unique characteristics of each cerebral hemisphere - the left and the right - as they are projected onto the spherical surface. The vantage point follow an icosahedron subdivision. We use a feature extraction network (resnet18, SpectFormer) on each individual view. We experiment with different IcoConv (IcoConv for icosahedron and convolution) operators that pool the information from all views. Finally, we concatenate the left/right outputs and normalized demographics. We perform a final linear layer for the classification

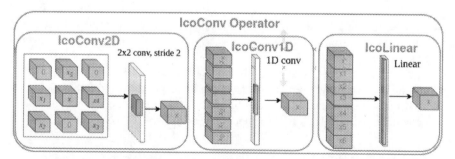

Fig. 2. Different IcoConv operators. IcoConv2D arranges the features extracted from adjacent views in 3 × 3 grid and performs an additional 2D Convolution. IcoConv1D aranges the features and performs a 1D Convolution followed by Average/Max poolo ing. IcoLinear stacks the features and performs a Linear layer

4.1 Rendering the 2D Views

The Pytorch3D[1] framework allows rendering and training in an end-to-end fashion. The rendering engine provides a map that relates pixels (pix2face) in the images to faces in the mesh and allows rapid extraction of point data as well as setting information back into the mesh after inference. We use pix2face to extract values for the 3 brain features namely: extra-axial cerebral spinal fluid (EA-CSF), surface area (SA), and cortical thickness. The EACSF features are precomputed via a probabilistic brain tissue segmentation, cortical surface reconstruction, and streamline-based local EA-CSF quantification [7]. SA and CT are precomputed via CIVET [6]. The pix2face map allows us to extract the vertex information and map them into 2D images set to 224px resolution. These images are then fed to the NN for feature extraction.

4.2 Architecture

We developed a novel NN architecture called *BrainIcoNet* and perform extensive experiments with a variety of feature extraction layers and IcoConv operators.

Figure 1 shows the general architecture of our approach which consists of a feature extraction step followed by our IcoConv operator. Figure 2 shows the different IcoConv operators.

In our experimental setup, we change the number of views, evaluating both 12 and 42 perspectives, and alter the radius of the icosahedron to adjust the proximity to the 3D object. A smaller radius sets the view closer to the 3D object thereby restricting the breadth of the captured view but acquiring finer detailed information.

The captured views are then fed to the feature extraction layer. We use two distinct branches, each dedicated to a specific hemisphere. The assumption is that the left and right hemispheres exhibit unique features that should be treated separately.

Each branch uses resnet18 or a SpectFormer block for feature extraction. The features are then arranged and passed to the IcoConv block. We experiment with 2D/1D Convolutions and a Linear layer. The IcoConv2D operator was designed to allow sharing of information across adjacent views only. As demonstrated by the results, the explainability maps are localized and corroborate previous findings about brain regions affected by ASD.

Finally, we use a linear layer for the binary classification task.

Additional experiments were conducted including demographic information which is normalized and concatenated to the left/right brain hemispheres. We train a random forest classifier and perform a feature importance analysis.

4.3 Training the Models

We perform a fivefold cross-validation training for every model. We use a series of augmentation techniques including random rotations of the input sphere, a dropout layer with $p = 20\%$ just before our linear layer for classification, and Gaussian noise applied

---

[1] https://pytorch3d.org/

on each image as well as the coordinates of the sphere points, *i.e.*, a small perturbation which is then normalized back on the spherical surface.

Training is done on an NVIDIA RTX6000 GPU with a batch size of 10, learning rate $1e-4$, the AdamW optimizer, and use the early stopping criteria to stop training automatically (patience 100) and keep the best performing model. To account for the highly imbalanced nature of our dataset classes during training, we utilize a sampling approach and ensure that each batch is balanced during training.

4.4 Explainability Maps

To find out what are the relevant areas for the classification task we use Grad-Cam [29]. This technique utilizes the gradients of the classification score with respect to the final feature map, thus, identifying which regions of the image contribute to the final classification score. We project each explainability map back to the 3D-object/sphere and apply a median filter with neighboring vertices to remove noise. The projection of these maps onto the spherical surface enables the visualization of explainability maps directly on the inflated cortical surfaces. This approach simplifies the task of identifying the regions impacted by the condition under study. By utilizing this 3D spatial representation, we are effectively able to correlate the intricate details from our explainability maps with specific locations on the brain's surface, providing a clear illustration of affected areas.

5 Results

The results in this section are computed using the test set from each fold and are reported for the whole population.

Table 1 shows the mean and standard deviation for precision, recall, f1-score, and accuracy for the 5 folds.

We perform extensive experiments with S-Unet, each IcoConv operator, different feature extraction layers namely resnet18 and SpectFormer, and increasing the number of views and reducing the radius to capture finer details.

The task of classifying HR+ v.s. HR− subjects presents a challenge. This is largely due to the fact that at this early stage, the brains of the subjects often do not exhibit explicit or easily distinguishable characteristics associated with the condition. Consequently, subtle nuances and variations may be critical in this classification task, underscoring the need for advanced and sensitive analytical methods.

We underscore that this dataset is highly imbalanced and achieving a high recall ensures that the model does not merely predict the majority class and miss the minority class instances.

The best performing model is the IcoConv2D with 42 views and the explainability maps are generated with it.

The explainability maps are shown in Fig. 3. Interestingly, the model favors features from the right hemisphere over the left ones. Furthermore, our findings support previous research [12] that highlights the significance of similar brain regions sensitive to this classification task.

Table 1. Classification report for the 5 folds. We report the mean and standard deviation for each metric. 42V = 42 views (icoshedron subdivision level 2). MAR = Macro Average Recall

Approach	Class	Precision	Recall	F1 Score	MAR	Accuracy
S-Unet	No	0.83 ± 0.02	**0.88 ± 0.05**	0.85 ± 0.02	0.585 ± 0.04	0.76 ± 0.03
	ASD	0.37 ± 0.13	0.29 ± 0.07	0.33 ± 0.08		
	ASD					
IcoCo2D	No	0.83 ± 0.01	**0.88 ± 0.05**	**0.86 ± 0.03**	0.58 ± 0.05	0.76 ± 0.04
	ASD	0.38 ± 0.19	0.28 ± 0.07	0.32 ± 0.11		
	ASD					
IcoCo1D	No	0.84 ± 0.02	0.87 ± 0.05	0.85 ± 0.03	0.595 ± 0.05	0.76 ± 0.05
	ASD	0.37 ± 0.15	0.32 ± 0.11	0.34 ± 0.11		
	ASD					
IcoCoLinear	No	0.84 ± 0.03	0.84 ± 0.04	0.84 ± 0.01	0.605 ± 0.07	0.75 ± 0.02
	ASD	0.37 ± 0.12	0.37 ± 0.17	0.37 ± 0.13		
	ASD					
IcoCoLinear$_{inf}$	No	**0.85 ± 0.01**	0.85 ± 0.04	0.85 ± 0.02	0.62 ± 0.04	0.75 ± 0.03
	ASD	0.39 ± 0.11	0.39 ± 0.07	0.39 ± 0.09		
	ASD					
Spect 42V	No	0.84 ± 0.03	0.86 ± 0.05	0.85 ± 0.02	0.605 ± 0.02	0.76 ± 0.03
	ASD	0.38 ± 0.12	0.35 ± 0.07	0.36 ± 0.05		
	ASD					
SpectICo 42V	No	**0.85 ± 0.02**	0.83 ± 0.05	0.84 ± 0.02	0.62 ± 0.02	0.74 ± 0.03
	ASD	0.37 ± 0.1	**0.41 ± 0.05**	0.39 ± 0.05		
	ASD					
IcoCo2D 42V	No	**0.85 ± 0.03**	0.86 ± 0.05	0.85 ± 0.02	**0.635 ± 0.04**	**0.77 ± 0.02**
	ASD	**0.42 ± 0.08**	**0.41 ± 0.1**	**0.42 ± 0.06**		
	ASD					

We use the Desikan parcellation [25] to identify the affected brain regions that appear in our explainability maps. Similar activation maps appear on both hemispheres centered on the entorhinal spreading to the parahippocampal, temporal pole, and fusiform. The right hemisphere present higher activity in the lingual and occipital lobe, with some activation in the right and left for the inferior parietal and superior frontal regions. These specific areas have been reported in previous studies, the entorhinal cortex, lingual and fusiform have been reported respectively by [5, 22, 33] to be areas impacted by ASD.

Finally, we test the feature importance with a random forest classifier trained using scikit-learn version 1.1.1. Figure 4 shows that gender is the most important demographic and amygdala is the most important if the gender is removed from the analysis.

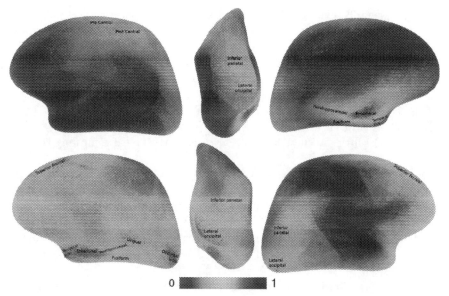

Fig. 3. Left, posterior, right views for the left hemisphere above and right hemisphere below. The gradcam maps are generated using only the correctly classified HR + subjects and using HR + as the target class. It indicates that features from the right hemisphere are preferred over the left ones. The name of the area are based on this labeling map [25]

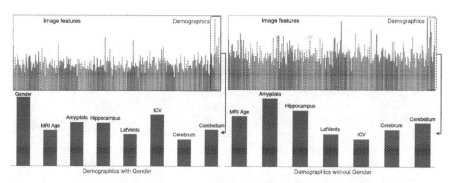

Fig. 4. The top left figure shows a plot of importance for features concatenated with normalized demographic values. The bottom left is only demographics to highlight that gender is the most important feature for the Random Forest classifier. The right plot shows an experiment with the gender removed from the analysis and shows that the amygdala is the most important feature in the demographics for the classification task

6   Conclusion

In conclusion, we have created a framework for shape analysis and explainability that is agnostic to the neural network model and the shape topology of the input meshes.

Our first contribution is a novel approach for shape analysis that does not require shapes with specific spherical topology or any form of subsampling of the mesh. Our shape analysis framework offers a significant advantage as it can handle meshes that are not in spherical topology or have holes, which is a requirement for S-Unet. We demonstrate this crucial feature by performing an experiment using subject specific inflated cortical surfaces.

Our second contribution is the visualization of explainability maps on complex shapes such as cortical surfaces. We tested our approach on a challenging classification task using subjects with high risk of developing autism and comparing HR+ v.s. HR−.

Our study utilized distinct neural networks for each hemisphere of the brain. Our results reveal that certain characteristics in the right hemisphere of the brain play a significant role in the classification of ASD. Our approach has identified brain regions and corroborate previous findings [5, 22, 33] of ASD-affected brain regions.

Finally, our approach allows including demographic information and highlights the amygdala volume as an important predictor for ASD. This finding also been corroborated in a previous study [4].

In future work, we will extend our analysis to other neuro-psychiatric disorders such as schizophrenia, attention deficit hyper-activity disorder, and bipolar.

References

1. Association, A.P., et al.: Diagnostic and statistical manual of mental disorders, text revision (dsm-iv-tr®) (2010)
2. Badri Narayana Patro, V.P.N., Agneeswaran, V.S.: Spectformer: frequency and attention is what you need in a vision transformer (2023)
3. Barbaro, J., Dissanayake, C.: Autism spectrum disorders in infancy and toddlerhood: a review of the evidence on early signs, early identification tools, and early diagnosis. J. Dev. Behav. Pediatr. **30**(5), 447–459 (2009)
4. Bellani, M., Calderoni, S., Muratori, F., Brambilla, P.: Brain anatomy of autism spectrum disorders II. Focus on amygdala. Epidemiol. Psychiatr. Sci. **22**(4), 309–312 (2013)
5. Blatt, G.J.: The neuropathology of autism. Scientifica (Cairo) **2012**, 703675 (2012)
6. Boucher, M., Whitesides, S., Evans, A.: Depth potential function for folding pattern representation, registration and analysis. Med. Image Anal. **13**(2), 203–214 (2009)
7. Deddah, T., Styner, M., Prieto, J.: Local extraction of extra-axial CSF from structural MRI. In: Medical Imaging 2022: Biomedical Applications in Molecular, Structural, and Functional Imaging, vol. 12036, pp. 29–34. SPIE (2022)
8. Eslami, T., Almuqhim, F., Raiker, J.S., Saeed, F.: Machine learning methods for diagnosing autism spectrum disorder and attention-deficit/hyperactivity disorder using functional and structural MRI: a survey. Front. Neuroinform. **14**, 575999 (2021)
9. Estes, A., et al.: Behavioral, cognitive, and adaptive development in infants with autism spectrum disorder in the first 2 years of life. J. Neurodev. Disord. **7**(1), 1–10 (2015)
10. Gabbay-Dizdar, N., et al.: Early diagnosis of autism in the community is associated with marked improvement in social symptoms within 1–2 years. Autism **26**(6), 1353–1363 (2022)
11. Glasser, M.F., et al.: A multi-modal parcellation of human cerebral cortex. Nature **536**(7615), 171–178 (2016)
12. Hazlett, H.C., et al.: Early brain development in infants at high risk for autism spectrum disorder. Nature **542**(7641), 348–351 (2017)

13. Heinsfeld, A.S., Franco, A.R., Craddock, R.C., Buchweitz, A., Meneguzzi, F.: Identification of autism spectrum disorder using deep learning and the abide dataset. NeuroImage Clin. **17**, 16–23 (2018)

14. Kong, Y., Gao, J., Xu, Y., Pan, Y., Wang, J., Liu, J.: Classification of autism spectrum disorder by combining brain connectivity and deep neural network classifier. Neurocomputing **324**, 63–68 (2019)

15. Liu, M., Yao, F., Choi, C., Sinha, A., Ramani, K.: Deep learning 3D shapes using ALT-AZ anisotropic 2-sphere convolution. In: International Conference on Learning Representations (2018)

16. Lord, C., et al.: The autism diagnostic observation schedule—generic: a standard measure of social and communication deficits associated with the spectrum of autism. J. Autism Dev. Disord. **30**(3), 205–223 (2000)

17. Lord, C., Rutter, M., Le Couteur, A.: Autism diagnostic interview-revised: a revised version of a diagnostic interview for caregivers of individuals with possible pervasive developmental disorders. J. Autism Dev. Disord. **24**(5), 659–685 (1994)

18. Ma, C., Guo, Y., Yang, J., An, W.: Learning multi-view representation with LSTM for 3-D shape recognition and retrieval. IEEE Trans. Multimedia **21**(5), 1169–1182 (2018)

19. Moridian, P., et al.: Automatic autism spectrum disorder detection using artificial intelligence methods with MRI neuroimaging: a review. arXiv preprint arXiv:2206.11233 (2022)

20. Mullen, E.M., et al.: Mullen scales of early learning. AGS Circle Pines, MN (1995)

21. Nietzel, M., Wakefield, J.: American psychiatric association diagnostic and statistical manual of mental disorders. Contemp. Psychol. **41**, 642–651 (1996)

22. Pierce, K., Redcay, E.: Fusiform function in children with an ASD is a matter of "who." Biol. Psychiatry **64**(7), 552–560 (2008)

23. Qi, C.R., Su, H., Mo, K., Guibas, L.J.: PointNet: deep learning on point sets for 3D classification and segmentation. In: Proceedings of the IEEE Conference on Computer Vision and Pattern Recognition, pp. 652–660 (2017)

24. Rahman, M.M., Usman, O.L., Muniyandi, R.C., Sahran, S., Mohamed, S., Razak, R.A.: A review of machine learning methods of feature selection and classification for autism spectrum disorder. Brain Sci. **10**(12), 949 (2020)

25. Desikan, R.S., et al.: An automated labeling system for subdividing the human cerebral cortex on MRI scans into gyral based regions of interest. Neuroimage **31**, 968–980 (2006)

26. Ras, G., Xie, N., van Gerven, M., Doran, D.: Explainable deep learning: a field guide for the uninitiated. J. Artif. Intell. Res. **73**, 329–397 (2022)

27. Ribeiro, F.L., Bollmann, S., Cunnington, R., Puckett, A.M.: An explainability framework for cortical surface-based deep learning. arXiv preprint arXiv:2203.08312 (2022)

28. Riegler, G., Osman Ulusoy, A., Geiger, A.: OctNet: learning deep 3D representations at high resolutions. In: Proceedings of the IEEE Conference on Computer Vision and Pattern Recognition, pp. 3577–3586 (2017)

29. Selvaraju, R.R., Das, A., Vedantam, R., Cogswell, M., Parikh, D., Batra, D.: Grad-CAM: why did you say that? arXiv preprint arXiv:1611.07450 (2016)

30. Shen, M.D., et al.: Increased extra-axial cerebrospinal fluid in high-risk infants who later develop autism. Biol. Psychiatry **82**(3), 186–193 (2017)

31. Sparrow, S., Balla, D., Cicchetti, D.: Vineland Scales of Adaptive Behavior, Survey form Manual. American Guidance Service, Circle Pines, MN (1984)

32. Su, H., Maji, S., Kalogerakis, E., Learned-Miller, E.: Multi-view convolutional neural networks for 3d shape recognition. In: Proceedings of the IEEE International Conference on Computer Vision, pp. 945–953 (2015)

33. Chandran, V.A., et al.: Brain structural correlates of autistic traits across the diagnostic divide: a grey matter and white matter microstructure study. Neuroimage Clin. **32**, 102897 (2021)

34. Wu, W., Qi, Z., Fuxin, L.: PointConv: deep convolutional networks on 3D point clouds. In: Proceedings of the IEEE/CVF Conference on Computer Vision and Pattern Recognition, pp. 9621–9630 (2019)

35. Wu, Z., et al.: 3D ShapeNets: a deep representation for volumetric shapes. In: Proceedings of the IEEE Conference on Computer Vision and Pattern Recognition, pp. 1912–1920 (2015)

36. Xue, C., et al.: NeuroExplainer: fine-grained attention decoding to uncover cortical development patterns of preterm infants. arXiv preprint arXiv:2301.00815 (2023)

37. Zhao, F., et al.: Spherical U-Net on cortical surfaces: methods and applications. In: Chung, A., Gee, J., Yushkevich, P., Bao, S. (eds.) 26th International Conference on Proceedings of the Information Processing in Medical Imaging, IPMI 2019, Hong Kong, China, 2–7 June 2019, vol. 26. pp. 855–866. Springer, Cham (2019). https://doi.org/10.1007/978-3-030-20351-1_67

DeCA: A Dense Correspondence Analysis Toolkit for Shape Analysis

S. M. Rolfe[1(✉)] and A. M. Maga[1,2]

[1] Center for Developmental Biology and Regenerative Medicine,
Seattle Children's Research Institute, Seattle, WA, USA
`sara.rolfe@seattlechidrens.org`
[2] Department of Pediatrics, Division of Craniofacial Medicine, University of Washington,
Seattle, WA, USA

Abstract. DeCA (Dense Correspondence Analysis) is an open-source toolkit for biologists that integrates biological insights in the form of homologous landmark points with dense surface registration to provide highly detailed shape analysis of smooth and complex structures that are typically challenging to analyze with sparse manual landmarks alone. In this work we demonstrate the use of DeCA by analyzing morphological differences of the skull in a dataset of 60 laboratory mice from different background strains.

Keywords: morphometrics · registration · phenotyping

1 Introduction

Concepts of "homology" and "developmental origin" are integral to many fields of biological research to establish equivalency of anatomical structures, regardless how different they may look (and possibly function) in different taxa. The former establishes similarity due to shared ancestry (e.g., bones in the forelimbs of bats and whales are homologous because they are both mammals). The latter focuses on the cellular origin of tissue types (e.g., in mammalian development basioccipital bone of the cranium originate from the mesoderm, whereas frontal bone derives from neural crest cells).

In the context of 3D shape analysis, one way to codify this biological insight is to segment individual structures based on both their evolutionary homologies and developmental origins and conduct the analysis of these units. In addition to being time consuming, these relationships may be hard to delineate in 3D scans of adult specimens. For example, while a single rigid structure, an adult mammalian skull is composed of over 20 different bones with different developmental origins [1]. While anatomical landmarks that delineate tissue boundaries remain an easy way to infuse expert knowledge of anatomy into analysis, they tend to be too sparse to capture the anatomical arrangement of a complex structure like skull. On the other hand, most dense correspondence analyses assume the equivalency of structures being analyzed (e.g., "mid face"), while this makes sense in context of a single species studies, it may or may not be correct to

C. Wachinger et al. (Eds.): ShapeMI 2023, LNCS 14350, pp. 259–270, 2023.
https://doi.org/10.1007/978-3-031-46914-5_21

assume in context of multiple species, or studies that involve organisms whose normal development is perturbed by genetic modifications (either through selective breeding or more directly editing genomes).

Here we introduce DeCA, a toolkit that allows to easily combine both approaches, in which the expert evolutionary and developmental knowledge can be incorporated in the process by annotating landmarks, from which the dense correspondence is automatically interpolated. DeCA is based on a dense surface correspondence method introduced in Hutton, et al. [2]. This and similar methods have been generally accepted for analysis of shape and asymmetry of the human face, which has a combination of smooth shapes and a small number of identifiable features that can be reliably landmarked [3–6].

DeCA is implemented as an extension on the 3D Slicer platform to provide the maximum ease of use for the biologists working with genetic mutants of model organisms or with multiple species. The DeCA module contains workflows to guide rigid alignment, generating a mean model from a group, creation of registered, mirrored models, and execution of dense surface registration for shape or symmetry analysis. It also supports error checking, extraction of dense semi-landmark sets from the points correspondences and visualization of heatmaps showing average and individual differences in shape. In this paper, we demonstrate the shape analysis workflow for a dataset of digitized skulls from 60 laboratory mice of differing genetic backgrounds and compare the results produced by DeCA results to typical sparse landmark analysis.

2 Methods

Each step in the shape analysis workflow in this work is implemented in the DeCA module and is freely available (https://github.com/smrolfe/DeCA). All analysis and evaluations can be run entirely within the 3D Slicer application via user-friendly module interfaces [7]. The DeCA module is shown in Fig. 1. The DeCA module is implemented as a tabbed workflow, where each tab is modular and can be used separately or combined into customized workflows.

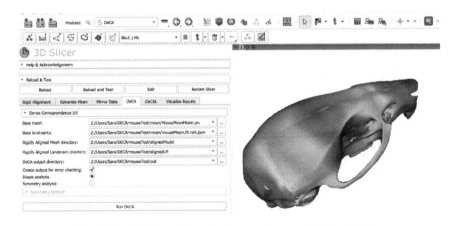

Fig. 1: The DeCA workflow implemented as a module in 3D Slicer.

Placing Manual Landmarks: The first step in establishing the dense surface mapping is to place a small number of manual landmark points on each of the specimens. The selection of landmark points is an integral step in the experiment design as the landmarks define the reference frame for the shape differences measured and provides a way to capture expert knowledge of points of biological homology in a dataset. The first consideration is to select points that can be reliably placed at true locations of homology. The distribution of the points can also impact the analysis, as points disproportionately selected from one area of the model will result in the alignment of this region to dominate the surface mapping. Regions containing high curvature or variability that is not well modeled by a smooth curve should be resampled to improve the surface alignment. The number of manual landmarks needed is dependent on the geometry of the sample. The user may need to experiment to find the correct number of landmark points to optimally align the shapes for their application. The method outlined in the "Error checking" section can be used to identify and troubleshoot alignment issues due to insufficient landmark sets.

To produce landmark position files in a format compatible with the DeCA module, placing the landmarks in 3D Slicer is recommended. 3D Slicer supports the creation of landmark templates with a fixed set of pre-named points to standardize this process. For large datasets, initializing the landmark positions using the automated landmarking module ALPACA, provided as part of the SlicerMorph extension to 3D Slicer, and adjusting the placement by hand can help significantly reduce manual landmarking time and avoid mistakes [8, 9].

Rigid Alignment of Models: The first tab in the DeCA workflow will align all specimens to a selected reference subject using a rigid body transform (rotation and translation only). The output of this step is a directory of transformed models and landmark files in aligned space.

DeCA Point Correspondence Assignment for Shape Analysis: The DeCA tab produces the set of point correspondences for a directory of rigidly aligned models. A reference model can be selected, or alternatively, an average template model can be generated from the dataset (see mean template building section). The models are aligned using Procrustes alignment to their mutual mean, removing positional and scale differences. The aligned landmarks are then transformed to the mean shape using a thin plate spline (TPS) transformation [10]. After this step, the landmarks are in absolute alignment and a transform is interpolated between the points by minimizing the bending energy of a theoretical spline placed between points. This transform preserves the homology at the manual landmark locations and while optimizing for smoothness and minimization of variation in the neighborhood around the landmark points. The final step to assign point correspondences for each point on the reference model to a point on each subject model using the iterative closest point (ICP) algorithm. The point correspondences are transformed back to the rigidly aligned subject space, generating a vector for each point on the template surface between the spatial location of the corresponding points on the template and model. The magnitude of each vector is extracted and saved in a point array of the output result model.

DeCA Point Correspondence Assignment for Asymmetry Analysis: In the DeCA tab, the user can select "Symmetry analysis" instead of the default "Shape analysis" from the Dense Correspondence options. When the "Symmetry analysis" is selected and directories are required containing for the mirrored landmarks and models for each subject in the analysis (see creating mirrored data section). The DeCA point correspondences will be run for both the original and mirrored models. For each point on the reference model, the asymmetry vector is defined between corresponding points on the subject mesh and its mirrored copy and represents the transformation due to asymmetry. The magnitude of these vectors is displayed as a heatmap on the template image to visualize the spatial distribution of asymmetry.

Error Checking: The DeCA tab provides an option to create an output directory where the TPS-warped meshes in Procrustes aligned space are saved for each subject and the reference template. Comparing these deformably warped meshes before the point correspondences are assigned in the ICP step allows the users to confirm that the surfaces of the model are in close alignment. If the surfaces are misaligned at this step, more manual landmarks should be added to improve the accuracy of the ICP step.

Mean Template Building: In the "Generate Mean" tab, a group mean template is generated from the rigidly aligned models and landmarks. A reference model is chosen from the dataset. We recommend that this reference subject is close to the group mean and that the mesh has good connectivity since the mesh vertices will serve as the index of point correspondences. The DeCA point correspondence is run to generate the point correspondences between the reference model and every subject in the dataset. The mean template model is constructed by moving each point in the reference model to the mean position of the corresponding points. The mean template output from this step can be directly used as the reference model in the DeCA tab. Optionally, the mean template model can first be smoothed to remove unwanted local shape variation or decimated to reduce the number of point correspondences calculated to simplify the mesh and improve computational time.

Creating Mirrored Data: The DeCA symmetry analysis pipeline requires aligned, mirrored models for each specimen in the dataset, with corresponding landmarks. The Mirror Data tab guides the user through creating these outputs. The user selects the axis of symmetry to mirror the model across and specifies the point ordering of the new mirrored point lists, in which the order of bilaterally paired points is reversed while centerline points are left in place.

DeCAL: DeCAL is an extension of DeCA that can be used to generate dense semi-landmarks from the point correspondences. The DeCA point correspondences are assigned across a dataset, and then spatially filtered at a threshold set by the user to select the number of points needed, and a landmark position file is output for each subject and the base mesh. Although the DeCAL point correspondences are guided by the position of the manual landmarks, these manual landmark positions are not included in the output semi-landmark set. The DeCAL points can be used in traditional analysis of landmarks including GPA, PCA and other statistical methods. GPA and PCA analysis can be run directly in 3D Slicer using the GPA module from the SlicerMorph extension [9]. Other statistical methods can be accessed by exporting the landmarks into R and using functions implemented in toolboxes such as geomorph and Morpho [11, 12]. DeCAL is also a useful tool for evaluating the shape information provided by DeCA, by comparing the relationship between landmark analysis of the DeCAL and manual landmark sets.

Visualization: In the Visualize Results tab, the user can select a result model from a previous run of DeCA for shape or symmetry analysis. The results model is stored as the template mesh geometry, with an array of the shape difference or asymmetry magnitude for each subject stored at each point on the template mesh. The user can browse through the subject magnitude arrays and display the selected subject array as a heat map on the surface of the results model. Further visualization controls, such as color map selection and color legend display are available in the 3D Slicer "Models" module.

3 Experimental Results

Data: To demonstrate the use of the DeCA toolbox, we have used the shape analysis workflow to visualize the shape differences between models extracted from microCT scans of skulls from 60 inbred mouse strains. Mouse models are key to understanding the genotype-phenotype interactions. Such studies are routinely carried out on any of a large number of inbred background strains. While the difference in anatomy between background strains can have an impact on the morphological traits measured, the differences in morphology are not well understood. This dataset was collected to quantify and characterize shape differences between commonly used background strains of laboratory mice with the goal of better understanding of how this piece of the experimental design influences the definition of "normal morphology" and to better inform comparisons of studies done on different background strains.

The microCT scans were manually annotated by an expert with a set of 51 anatomical landmarks. This number reflects both the complexity of the skull geometry and the number of available points that can be reliably placed at known points of homology (i.e. skull sutures). A higher density of points was placed on the cranial due to the larger number of identifiable features. This also incorporates the expert knowledge that this region should dominate the image registration, as it is expected to contain the least variation.

DeCA for Assessment of Shape Differences
The DeCA toolkit was used to create a rigidly aligned set of images. A reference strain with good connectivity (C57BL/6J) was selected as the base model for generating the mean model and landmark points. Heatmaps showing the mean magnitude and standard deviation across the entire group are shown in Fig. 2. As expected, the lowest variation is seen at the cranial base. The largest differences in shape are seen at the posterior of the skull, premaxilla, and nasal region, on the order of 0.5mm. There is also a localized area of high levels of shape difference near the squamosal suture, capturing height differences of the zygomatic arch.

The magnitude arrays were extracted for each subject and ranked by the highest mean values. The top 5 specimens with the highest average magnitude of shape difference

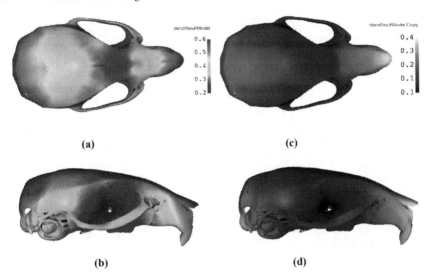

Fig. 2: Mean (a and b) and standard deviation (c and d) of the shape difference magnitude over all mouse strains. Color legend units are in mm.

are reported in Table 1. The average magnitude for an anatomical feature can also be reported for each subject by segmenting the region of interest on the DeCA output model. In Fig. 3, the nasal bone was segmented using the Dynamic Modeler module in 3D. The top 5 specimens with the highest magnitude of shape difference in the nasal bone region are reported in Table 2. Comparing Tables 1 and 2, the specimens with the highest levels of shape difference overall also have the highest levels of shape difference in the nasal region. However, the ordering of the specimen ranked by nasal shape difference has changed, with the highest ranked specimen NZBWF1/J in Fig. 3(b) showing the larger difference in the nasal bone compared to SF/CamEiJ, the specimen with the highest overall difference. Note that in this example, where only the magnitude of shape difference is used, we do not distinguish between the direction of these differences.

Table 1. Top 5 strains ranked by overall magnitude of shape difference.

Strain name	Average magnitude (mm)
SF/CamEiJ	1.468
PERC/EiJ	1.326
NZBWF1/J	1.324
CAST/EIJ	1.151
MOLG/DnJ	1.007

Table 2. Top 5 strains ranked by nasal bone magnitude of shape difference.

Strain name	Average magnitude (mm)
NZBWF1/J	1.468
PERC/EiJ	1.326
CAST/EIJ	1.324
MOLG/DnJ	1.151
SF/CamEiJ	1.007

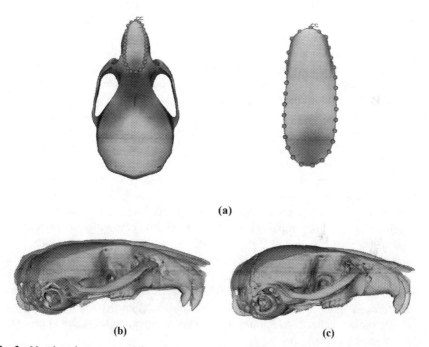

(a)

(b)   (c)

Fig. 3: Nasal region extracted from the DeCA result model used to generate shape difference rankings based only on this region (a). The top ranked specimen is the NZBWF1/J strain (purple), shown overlaid with the mean model (yellow) in (b). This specimen shows greater difference in the nasal bone region compared to the top ranked specimen for overall shape difference, (c) SF/CamEiJ which is shown (blue) overlaying with the mean template model (yellow).

DeCAL Comparison to Manual Landmark Analysis

To validate the results of the shape differences found, DeCAL was used to extract 459 semi-landmark points from the DeCA assigned point correspondences. GPA, followed by PCA were applied using the GPA module of the SlicerMorph extension within 3D Slicer. The DeCAL landmarks differ from semi-landmark points which are sampled regularly on a surface, as they represent corresponding points across the dataset. Due to this, we did not choose to use sliding semi-landmark algorithms applied prior to analysis.

Results from the DeCAL analysis are compared to GPA and PCA analysis of the 51 manual landmarks. Given that the placement of the DeCAL points is guided by the manual landmark points but includes shape information from the regions between these points, we expect that the DeCAL results would be similar, but provide more detailed information. In the case that the point correspondences have low accuracy, including such a large number of semi-landmark points would introduce noise that would be expected to obscure the shape information provided by the manual landmarks alone.

A projection of the specimens onto the first two projections of the PCA shape space is shown in Fig. 4 for both the manual and DeCAL landmark sets. In the manual landmark analysis, PC 1 accounts for 19.0 percent of the variation in the dataset and PC 2 accounts for 9.4 percent. This is comparable to the DeCAL analysis, where PC 1 accounts for 24.0 percent of the variation and PC 2 accounts for 11.8. In each projection, the outliers along each PC are consistent, with C57BL/6J at the negative extrema of PC 1 and PC 2, C57BLKS/J at the negative extrema of PC 1, SF/CamEij at the positive extrema of PC 1, and SPRET/EiJ at the positive extrema of PC 2. The positioning of these outliers differed in their separation from the primary cluster of points. With manual landmarks alone, SF/CamEij is the furthest from the remainder of the group along the PC 2 axis. When the DeCAL landmarks are used, C57BL/6J is furthest from the group along the PC2 axis. Greater separation of an outlier from the mean of the other specimen suggests that the outlier has more shape variability in regions outside of the manual landmark points.

To explore the accuracy of the use of the additional shape information provided by the DeCAL analysis, we compare at how well the deformation of the template model along the principal components matches the actual specimens at the extrema of those PC's. Figure 5 shows this comparison for SF/CamEiJ and deformation of the reference model in the positive direction along the first PC. In Fig. 5 (b) and (c), widening of the skull and shortening of the snout can be observed for both the manual landmark and DeCAL warping along PC 1 when compared to the template model in Fig. 5 (a). This is consistent with the skull morphology of SF/CamEiJ. However, the DeCAL warped model produces a more realistic warping of the nasal bone, due to the additional points that provide more localized shape information than the interpolated warping generated from the manual landmarks alone.

A similar effect can be observed when comparing the deformations along the second principal component where the basicranium becomes more flexed in the negative PC 2 direction. This can be seen in Fig. 6, in both the (c) manual landmark and (d) DeCAL warping along PC 2, but compared to the (b) C57BL/6J specimen that lies on the negative extreme of PC 2, the DeCAL warping is capable of producing a cranial base angle that is strikingly more similar to the C57BL/6J specimen.

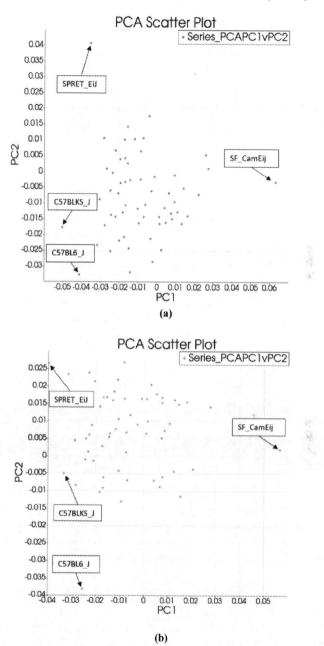

Fig. 4: Projection of each specimen onto the first two principal components for the (a) manual landmark points (b) dense semi-landmarks extracted by DeCAL.

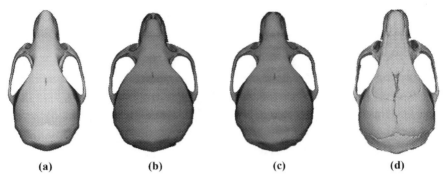

(a)   **(b)**   **(c)**   **(d)**

Fig. 5: Comparison between the (a) template model, (b) manual landmark model warped in the positive PC1 direction, (c) DeCAL model warped in the positive PC 1 direction and (d) SF/CamiEiJ specimen that is an outlier on the positive PC 1 axis. The warped models in (b) and (c) show similar changes in morphology to the specimen in (d), including wider skull and shortened snout, but the DeCAL warped model shows a more realistic warping of the nasal region.

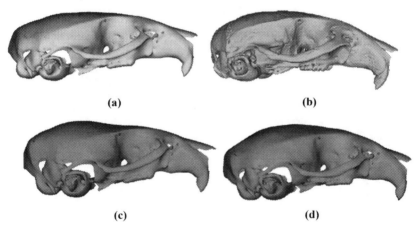

(a)   **(b)**

(c)   **(d)**

Fig. 6: Comparison between the (a) template model, (b) C57BL/6J specimen that is an outlier on the negative PC 2 axis, (c) manual landmark model warped in the negative PC 2 direction, and (d) DeCAL model warped in the negative PC 2 direction. The DeCAL warped model represents basicranial flexion seen in the C57BL/6J specimen more accurately compared to the PC2 warped model.

Discussion

In this work we have introduced a fully open-source toolkit for shape analysis via dense surface registration. While methods for dense surface registration are well established, they have typically required significant time and programming expertise to use. The DeCA module, packaged as a module on the 3D Slicer image analysis platform removes these barriers and provides a modular tool that can support a variety of workflows. We applied this to an example application to look at the shape differences in a group of mice from a large number of inbred strains. The analysis provided by DeCA was consistent with the information provided by the manual landmarks alone but offered more detailed spatial information about the distribution of the shape changes. GPA and PCA analysis of the semi-landmarks extracted by DeCAL were reliably able to estimate more realistic approximations of the variation seen in this dataset.

These experiments within a single species with some known differences help to validate DeCA for complex shape analysis applications. We plan to move towards applying DeCA for more challenging tasks such as comparisons across species which are not typically feasible with registration methods that rely on a high level of geometric similarity.

Future Development: In the current implementation of the DeCA module, the primary output is the magnitude of the shape difference vectors. These values can be viewed as a heatmap, output as an array, or averaged over a region defined on the output results model. The toolkit is still under construction and features we plan to add soon are angular measurements as described in our previous work [6] and the ability to quantify differences in pointwise geometric surface properties such as surface curvature and local mesh features such as color and texture.

Acknowledgement. Parts of this research were funded by the National Science Foundation Award [OAC 2118240] (Imageomics Institute) and National Institute of Dental and Craniofacial Research (DE027110) to AMM.

References

1. McBratney-Owen, B., Iseki, S., Bamforth, S.D., Olsen, B.R., Morriss-Kay, G.M.: Development and tissue origins of the mammalian cranial base. Dev. Biol. **322**(1), 121–132 (2008)
2. Hutton, T.J., Buxton, B.R., Hammond, P.: Dense surface point distribution models of the human face. In: Proceedings IEEE Workshop on Mathematical Methods in Biomedical Image Analysis (MMBIA 2001), pp. 153–160. IEEE (2001)
3. Claes, P., Walters, M., Vandermeulen, D., Clement, J.G.:Spatially-dense 3D facial asymmetry assessment in both typical and disordered growth. J. Anat. **219**(4), 444–455 (2011)
4. Kornreich, D., Mitchell, A.A., Webb, B.D., Cristian, I., Jabs, E.W.: Quantitative assessment of facial asymmetry using three-dimensional surface imaging in adults: validating the precision and repeatability of a global approach, pp. 126–131 (2016)
5. White, J.D., et al.: MeshMonk: open-source large-scale intensive 3D phenotyping. Sci. Rep. **9**(1), 6085 (2019)
6. Rolfe, S., Lee, S.-I., Shapiro, L.: Associations between genetic data and quantitative assessment of normal facial asymmetry. Front. Genet. **9**, 659 (2018)
7. Fedorov, A., et al.: 3D Slicer as an image computing platform for the quantitative imaging network. Magn. Reson. Imaging **30**(9), 1323–1341 (2012)
8. Porto, A., Rolfe, S., Maga, A.M.: ALPACA: A fast and accurate computer vision approach for automated landmarking of three-dimensional biological structures. Methods Ecol. Evol. **12**(11), 2129–2144 (2021)
9. Rolfe, S., et al.: SlicerMorph: an open and extensible platform to retrieve, visualize and analyze 3D morphology. Methods Ecol. Evol. **12**(10), 1816–1825 (2021)
10. Bookstein, F.L.: Principal warps: thin-plate splines and the decomposition of deformations. IEEE Trans. Pattern Anal. Mach. Intell. **11**(6), 567–585 (1989)

11. Adams, D.C., Otárola-Castillo, E.: Geomorph: an R package for the collection and analysis of geometric morphometric shape data. Methods Ecol. Evol. **4**(4), 393–399 (2013)
12. Schlager, S.: Morpho and Rvcg–shape analysis in R: R-Packages for geometric morphometrics, shape analysis and surface manipulations. In: Statistical shape and deformation analysis, pp. 217–256. Academic Press (2017)

3D Shape Analysis of Scoliosis

Emmanuelle Bourigault[1]([✉]), Amir Jamaludin[1], Emma Clark[2],
Jeremy Fairbank[3], Timor Kadir[4], and Andrew Zisserman[1]

[1] Visual Geometry Group, Department of Engineering Science, University of Oxford,
Oxford, England
{emmanuelle,amirj,az}@robots.ox.ac.uk
[2] Musculoskeletal Research Unit, University of Bristol, Bristol, England
Emma.Clark@bristol.ac.uk
[3] NDORMS, University of Oxford, Oxford, England
jeremy.fairbank@ndorms.ox.ac.uk
[4] Plexalis, Oxford, England
timor.kadir@plexalis.com

Abstract. Scoliosis is typically measured in 2D in the coronal plane, although it is a three-dimensional (3D) condition. Our objective in this work is to analyse the 3D geometry of the spine and its relationship to the vertebral canal. To this end, we make three contributions: first, we extract the 3D space curve of the spine automatically from low-resolution whole-body Dixon MRIs and obtain coronal, sagittal and axial projections for various degrees of scoliosis; second, we also extract the vertebral canal as a 3D curve from the MRIs, and examine the relationship between the two 3D curves; and third, we measure the angle of rotation of the spine and examine the correlation between this 3D measurement and the 2D curvature of the coronal projection. For this study, we use 48,384 MRIs from the UK Biobank.

Keywords: MRI · Spine Geometry · 3D/2D Correspondences

1 Introduction

Scoliosis is defined as a lateral deformation of the spine in the coronal plane, usually manually diagnosed on anteroposterior (AP) X-rays, by measuring the Cobb angle, where an angle over 10°C is considered scoliotic [4]. More recently, it has been shown that scoliosis can also be diagnosed from DXA (Dual-energy X-ray Absorptiometry) scans, which are less costly and involve a 10 times lower radiation dose than conventional X-rays [23]. However, both X-rays and DXAs do not capture the complex 3D deformation of the spine [8]. The convenience of using coronal radiographs to measure scoliosis has meant that the axial and sagittal planes have been widely disregarded.

In this work, we explore scoliosis in 3D by analysing the 3D shape of the space curve of the spine, and its relationship to the 3D space curve of the vertebral canal. For this study, we use the Dixon MRIs available in the UK Biobank.

C. Wachinger et al. (Eds.): ShapeMI 2023, LNCS 14350, pp. 271–286, 2023.
https://doi.org/10.1007/978-3-031-46914-5_22

We segment both the spine and the vertebral canal in axial slices. These segmentations allow us to extract 3D curves for the spine and canal, as illustrated in Fig. 1.

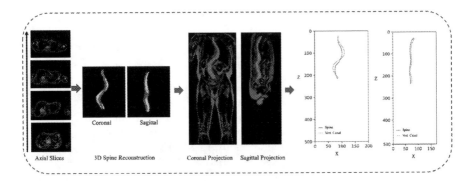

Fig. 1. Overview of the geometry pipeline. The spine (yellow) and the vertebral canal (red) are segmented in each axial slice. The centroids of the spine segments over all axial slices form a 3D space curve (similarly for the canal). The space curve is projected onto the coronal and sagittal planes, and a 2D spline curve fitted to the projected points. Curvature and angles are computed from the spline curve. (Color figure online)

Our objective is to study how the 3D spine curve deforms for a scoliotic spine, and also how the vertebral canal adapts to scoliosis. We analyse the 3D spine curve by projecting it onto coronal, sagittal and axial planes, and determine the severity of scoliosis on the coronal plane. It is worth noting that the MRIs from the UK Biobank are uniquely suitable for scoliosis measurement in 3D as there exists an established scoliosis measurement on the paired MRI to 2D DXA for the coronal projection which serves as our point of reference [2,9].

We then investigate the relationship between the spine and vertebral canal curves on the three planes, and also measure the deviation between the two curves. In addition, we measure the curvature of the coronal projection and the angle of axial rotation of the spine; and investigate their relation.

Sect. 2 outlines our method for extracting the geometry of the spine and vertebral canal from MRIs, and describes the measures we use for the analysis of the geometry. Then, Sect. 3 describes the dataset, and presents the results of the analysis, together with several visualizations of the geometry. Finally, Sect. 4 summarises the findings and the implications of this research.

1.1   Related Work

Research on the relationship between deformations on the sagittal, axial and coronal planes is still in its early phase [11,14].

The UK Biobank dataset used in this paper is of adults. However, most work on scoliosis focuses on adolescent idiopathic scoliosis (AIS), while scoliosis in adults has been relatively unexplored in past literature. Grown adults can develop degenerative scoliosis as a result of wear and tear on the discs of the spine. It has been shown that the right thoracic curves are predominant in AIS [13] but this kind of shape analysis of the spine in adult scoliosis is rare.

To date, the vast majority of scoliosis research has focused on 2D shape analysis of the spine, but not in 3D at a large scale. Limitations of 2D spine analysis arise particularly in classifying curve shape. Indeed, deviations are not limited to the coronal plane. They include twisting of the spine in multiple directions [18]. The closest work to ours is by Pasha *et al.* [15,16] in which they look at 3D curves in scoliosis; the main differences between their work and ours are: they used EOS which is quite a niche imaging modality compared to MRIs, they focused on AIS as opposed to degenerative scoliosis, and the number of samples is small (n=103).

Though we use MRIs in our work, it is worth noting that most works on spinal MRIs focus on non-scoliosis spinal disorders and as such put more emphasis on segmenting the vertebral bodies and discs individually rather than the spine as a whole [10–12, 14, 24].

2   The 3D Geometry of the Spine and Vertebral Canal

It is essential for the 3D geometrical analysis of scoliosis that we capture the whole shape of the spine. To this end, we segment the two main structures that can be seen in the axial Dixon MRIs: these are (i) the "spine" itself, which is comprised of the vertebral bodies and the intervertebral discs, and (ii) the "vertebral canal", which is the space occupied by the spinal cord and filled with cerebrospinal fluid. We do this segmentation on a per-slice basis for each axial image in a given scan volume. The centroid (a 2D point) of each segmentation can now be extracted from each slice and stacked vertically according to axial slice numbering, and spaced appropriately with the axial slice thickness, for a given volume.

A spline curve can then be fitted in 3D (or to the 2D projections), to smooth out the noise in the measurements. The full implementation details are given in the appendix, and the process is summarised in Fig. 1. The spine segmentation gives us the 3D spine curve and the vertebral canal segmentation gives us the 3D vertebral canal curve. These 3D space curves can be projected to the coronal (X,Z), the sagittal (Y,Z), as well as the axial (X,Y) plane. Figure 2 shows an example of two curves in our dataset rotating in space.

2.1   Measuring the Deviation of the Curves

Now that the 3D curves of the spine and the vertebral canal have been extracted, we can then proceed to the analysis of the curves. For a given normal spine, it can be observed that two curves should overlap when projected onto the coronal plane and should be parallel in the sagittal plane. As such, the simplest measurement that is indicative of how far away from the norm a given pair of curves are is to measure the deviation between these two curves. Simply put, to

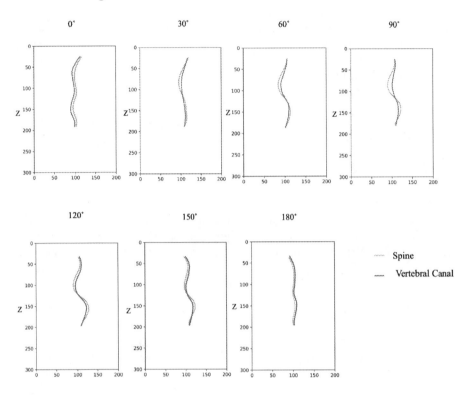

Fig. 2. Spine and Canal 2D Projections for every $30°C$ of rotation for a severe 'S shape' spine. This example is in Fig. 1. The $0°C$ projection corresponds to the sagittal projection, and the 90 projection to the coronal projection.

measure the distance, 'd', between the two curves we can simply compute the vector joining each point of the spine (x_{spine}, y_{spine}) to the vertebral canal curve (x_{canal}, y_{canal}) (see Fig. 3). If we project these vectors to the coronal plane, then the sum of their magnitudes measures the 'deviation' between the two curves. For a normal ideal spine, the 'deviation' will be zero in the coronal plane (since the spine and vertebral canal will project on top of one another), and in the sagittal plane, the point-wise difference between the two curves will have a set but constant 'deviation'.

For a given pair of spine and vertebral canal curves, we compute the point-to-point distance in the axial plane:

$$\delta_{spine-canal} = \sum_{i=1}^{N} \sqrt{(x_{i_{spine}} - x_{i_{canal}})^2 + (y_{i_{spine}} - y_{i_{canal}})^2} \qquad (1)$$

where i is the slice index and N is the total number of axial slices containing the spine and canal for a given scan.

Then, we can obtain the maximum deviation by taking the maximum of the point-wise distances of the spine-canal deviation (1). For a normal spine, the maximum deviation will be zero on the coronal plane. For the sagittal plane, a normal spine has inward curvature (lordosis) for the cervical and lumbar sections, and outward spinal curvature (kyphosis) for the thoracic section. Sagittal malalignment is as an exaggeration or deficiency of the normal lordosis or kyphosis curves. Therefore, we measure how parallel the spine and canal curves are on the sagittal plane by taking the standard deviation of the spine-canal deviations.

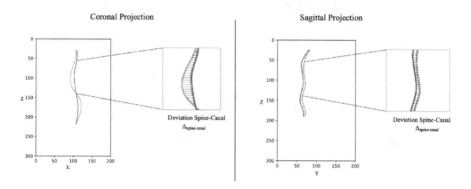

Fig. 3. Measurement of deviation between spine (yellow) and canal (red) curves. Shown is a coronal projection (left) and sagittal projection (right) and a zoom in on the maximum coronal curvature point. (Color figure online)

2.2   Curvature of the Spine Curve

The spline is continuous everywhere, as are its first and second derivatives. This is sufficient to determine the curvature κ with the standard mathematical formula:

$$\kappa = \frac{\left(y'' x' - x'' y'\right)^{\frac{3}{2}}}{\left(x'^2 + y'^2\right)} \tag{2}$$

For the results in Sect. 3 the maximum absolute curvature in the coronal plane is used to define three classes of scoliosis severity (normal, mild, severe) according to thresholds obtained on a set of 2K DXA scans annotated for Cobb angles. The threshold for scoliosis is $|\kappa| = 0.083$, mild scoliosis is: $0.083 < |\kappa| \leq 0.118$; and $|\kappa| > 0.208$ is severe scoliosis.

2.3   Angle of Spinal Axial Rotation

Aside from measuring the deviation of the two curves, we can also evaluate the lateral shift of the spine relative to the vertebral canal by measuring the angle of rotation. This is done by using two landmarks: the centroid of the spine and the centroid of the vertebral canal (see Fig. 4). The angle between the line through these centroids and the vertical is the axial rotation (under the assumption that the patient is lying on their back). Note, there are several definitions of the angle of axial rotation. They all rely on measuring the relative positions of anatomical landmarks such as the pedicles, vertebral body, and spinous processes. We use a similar approach to that of [1] and [6], but choose to detect the vertebral canal as a landmark on our axial slices as it is continuous throughout the spine.

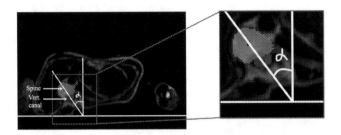

Fig. 4. The angle of axial rotation, α, is the angle between the line through the centroids of the spine (yellow) and vertebral canal (red), and the vertical direction. (Color figure online)

3   Results and Discussion

In this section, we compare the 2D projected curvature in relation to the 3D spine. We investigate how the canal curve varies with respect to the spine in Subsect. 3.2. And in Sect. 3.3, we analyse the coronal and sagittal curvatures and their relation to the angle of maximum axial rotation.

3.1   Dataset

Our dataset is comprised of 48,384 whole-body MRIs from the UK Biobank, a large open-access medical dataset with scans from more than 500,000 volunteers [21]. MRIs in the UKBiobank are of much lower resolution than standard clinical scans. Scans are resampled to be isotropic and cropped to a consistent resolution ($501 \times 156 \times 224$). The dataset is split into 80:10:10 for training (38,707), validation (4,838), and testing (4,839) for the segmentation task. 250–200-200 MRI scans are annotated for train-validation-testing for spine and vertebral canal for the baseline segmentation model and checked by an expert clinician. A part of the testing set (1,929) has been annotated by experts for Cobb angles using a modified Ferguson method in whole-body DXA scans as described in [23]. We use this annotated set to define the threshold for scoliosis in our experiment; otherwise this test set is unused in the training of our pipeline. Appendix A.1 gives details of the segmentation.

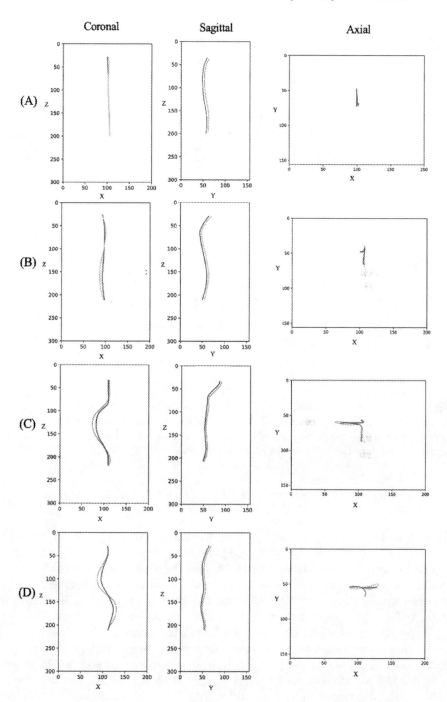

Fig. 5. Comparison of coronal, sagittal and axial 2D projections from 3D curve for normal (A), mild (B) and severe C shape (C) and severe S shape (D) scoliosis cases. Spine curve is in yellow and vertebral canal curve in red. (Color figure online)

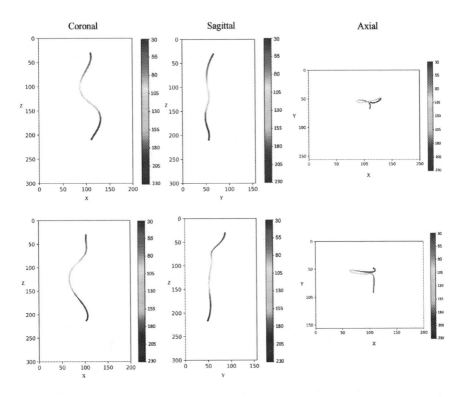

Fig. 6. Comparison of coronal, sagittal and axial 2D projections from 3D curve for a severe S shape (top), and a severe C shape (bottom) scoliosis case. The axial curves (3rd column) are more challenging to interpret. Spines are colour-coded on the z axis to visually indicate the order of the curve in the other projections.

3.2   Geometry of the Spine: Deviation of the Spine and Vertebral Canal

For a normal case, the spine and canal overlap in coronal, and are at a constant separation in sagittal (see Fig. 5). By comparing the curves of the spine and canal for normal versus scoliosis cases, we observe that the curves on coronal for scoliosis cases no longer overlap. We also observe that the vertebral canal is less curved than the spine suggesting that it deforms less than the spine. On the sagittal plane, the curves straighten from normal to scoliosis cases (see Figs. 5 and 6).

We study how the 3D deviation measurements relate to 2D. The results confirm a strong correlation in deviations between the spine and vertebral canal in 2D coronal and 3D curves (see Fig. 7A). The threshold for scoliosis is $|\kappa| = 0.083$. We define mild scoliosis as: $0.083 < |\kappa| \leq 0.118$; and $|\kappa| > 0.208$ for severe scoliosis. Distribution of spine-canal deviations can be discretised according to scoliosis severity (see Fig. 7B). This suggests that spine-canal deviations (mm) can potentially be used as another quantitative measurement of scoliosis. We then investigate how the vertebral canal is varying for different scoliosis severities ranging from normal, mild to severe C and S shape curves (Fig. 8).

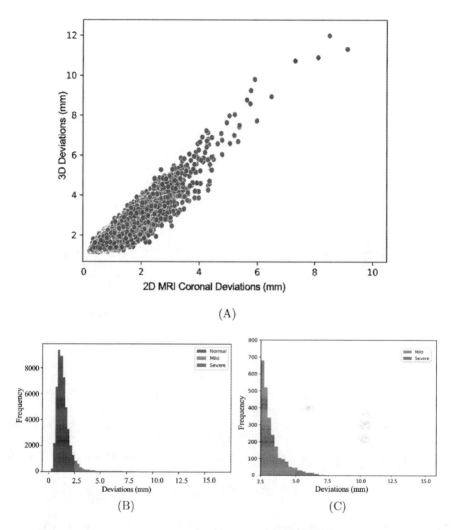

(A)

(B)   (C)

Fig. 7. (A) Scatter plot of spine-canal point-wise deviations (mm) from 2D coronal projection versus 3D (mm) (Pearson's $\rho = 0.86$, p-value < 0.05, n $= 48,384$). (B) Histogram with density function displaying the distribution of 2D spine-canal deviation values (n $= 48,384$) for normal, mild and severe scoliosis cases. (C) Zoom in on mild and severe scoliosis cases from plot in (B). The threshold for scoliosis based on human angles ($> 6°$ in whole body DXA as defined by [23]) in terms of curvature is 0.083. This threshold corresponds to 2.5 mm of spine-canal deviation.

We can now investigate the properties of the spine that are obtained from the projections of the 3D space curve (see Fig. 2 for a severe S shape curve). **Coronal vs Sagittal.** We measured the deviation of the spine and canal at the point of maximum coronal curvature. Comparing the MRI coronal and sagittal

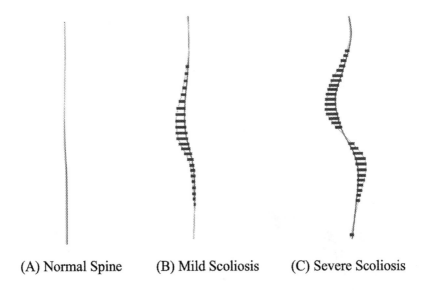

(A) Normal Spine   (B) Mild Scoliosis   (C) Severe Scoliosis

Fig. 8. Visualisation of Spine and Canal Deviations for a normal (A), mild (B) and severe (C) scoliosis cases on the coronal plane. Spine in yellow, vertebral canal in red, and deviations between the spine and vertebral canal in blue. (Color figure online)

spine-canal deviations at point of maximum coronal curvature, we observe an inverse correlation (Pearson's ρ = -0.64, n = 48,384). Curves on the sagittal plane are challenging to accurately assess due to the natural variations of the spine. We notice that severe scoliosis cases tend to have straighter spines in the sagittal plane (see Fig. 5). This inverse correlation between coronal and sagittal plane deviations is in accordance with past studies on biplanar radiographs curvature measurements [5]. Moreover, we observe a correspondence between the coronal plane and axial plane. The spine and vertebral canal deviation is greater on the axial projection for severe cases (see Fig. 5).

3.3 Curvature Measurement in MRI and Relation to Axial Plane

The correlation between maximum coronal curvature and angle of maximum axial rotation is moderately strong (ρ = 0.77, n = 48,384) which may suggest a critical role of the axial plane in relation to curvature on the coronal plane. This is in line with recent research on reconstructed 3D images [7,11]. Previous work suggested a causal link between axial deformations and onset of coronal deformations due to compensatory mechanical factors [17].

We show the scatter plot between the MRI axial angle of rotation at the point of maximum coronal curvature and the maximum of the MRI spine coronal projection in Fig. 9A, for all 48,384 scans in the UK Biobank. The correlation is relatively good (Pearson's ρ = 0.79) between coronal curvature and axial rotation at point of maximal curvature. This confirms the findings of Sect. 3.2, at a large scale.

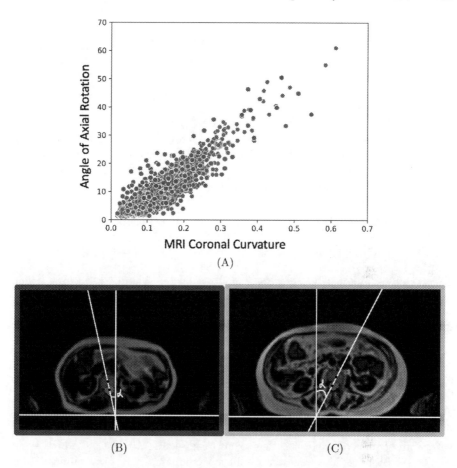

Fig. 9. (A) Scatter plot of angle of axial rotation vs MRI coronal maximum absolute curvature (Pearson's $\rho = 0.79$, n = 48,384). Angle is given in degrees. (B) and (C) Axial slices corresponding to point of max MRI coronal curvature (yellow and brown circles in (A)). Spine (red) lateral deviation is more prominent on (C) for severe scoliosis case than (B) for mild scoliosis. (Color figure online)

As a qualitative example, we compare the spine and vertebral canal masks for a mild scoliosis case (max. abs. curvature = 0.18, brown circle in Fig. 9A), and a more severe scoliosis case (max. abs. curvature = 0.29, yellow circle in Fig. 9A). Axial slices corresponding to these two cases are shown in Fig. 9B and 9C.

4   Conclusion

In this work, we investigated the geometry of scoliosis in 3D, while most prior work has focused on 2D deformations. We measured the curvature of the spine on one of the largest datasets of MRIs. One of the most remarkable outcomes of the visualizations is to see how the vertebral canal arranges itself to have less severe curvature than the spine itself. We also show that the axial plane is quite relevant for the assessment of scoliosis as suggested by the relatively high correlation between the angle of axial rotation and coronal curvature. By considering the spine as a 3D curve, we compared the projected 2D curves of the spine and canal on the coronal and sagittal plane. This efficient method could be used to measure the severity of the spine's deformation.

Ultimately, the goal of this research is to provide an accurate and consistent interpretation of spinal deformations in order to support clinicians in their decision-making process. Prior to the work in this paper, the link between coronal and sagittal curves was not well defined. Also, the role of the axial plane in relation to the coronal and sagittal planes was not yet known. However, one possible future analysis could be to use the relationship between the coronal, sagittal and axial curves as a 3D classification method, without the need to explicitly model the spine in 3D, thus facilitating its adoption in clinics.

Acknowledgements. This work was supported by the Centre for Doctoral Training in Sustainable Approaches to Biomedical Science: Responsible and Reproducible Research (SABS: R^3), University of Oxford (EP/S024093/1), and by the EPSRC Programme Grant Visual AI (EP/T025872/1). We are also grateful for the support from the Novartis-BDI Collaboration for AI in Medicine.

A   Segmentation

There are four separate aligned sequences in the MRI Dixon scans used here. These are in-phase, opposed-phase, fat-only and water-only. The fat-only and water-only sequences are best suited to our task, see Fig. 10. Note, the MRI scans in the UK Biobank have a lower resolution compared to typical clinical spine scans. We segment the spine using axial slices as they have higher resolution, and also support larger receptive fields for training the deep network.

A.1   Segmentation Network

A U-Net based network architecture is used for the segmentation task [19,26]. We use a U-Net++ [27] network with a ResNet-34 encoder. The input is 224 × 160 × 6, where we stack three adjacent MRI image slices of the spine region for the two MRI sequences (fat-only and water-only). To avoid partial volume effects, and also to benefit from more context, we ingest three adjacent slices, with the middle slice as output. The output has size 224 × 160 ×2, where 2 refers to the segmentation maps for the spine and vertebral canal.

For training, the loss function is a weighted sum of categorical cross-entropy loss [25] and dice loss [22] computed over a foreground/background/uncertain tri-map to mitigate potentially noisy boundaries in our labels which we define as ± 2px from the foreground boundary. Networks are trained for a maximum of 500 epochs with early stopping when the validation Dice does not increase

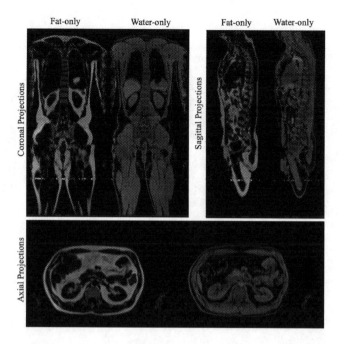

Fig. 10. Coronal, sagittal and axial projections for fat-only and water-only Dixon MRI sequences.

by e^{-4}. We use self-training to leverage the whole training set i.e. n = 38,707. Inspired from the recent work on confirmation bias reduction in self-training [3], we use an independent head for pseudo-label generation to prevent potentially inaccurate pseudo-label backpropagation (Fig. 11).

B   Spline Fitting

B.1   2D Spline Fitting

The 2D projected points (in the coronal or sagittal planes) are approximated by a piecewise cubic spline to smooth out any noise due to sampling. For this fitting, we use the method described in [2].

Using a parametrised curve, we construct polynomial piecewise cubic curves. A single cubic curve has only one inflection point, but scoliosis curves may have one or more. A solution could be to add extra control points and using higher order polynomials. However, higher order polynomials are known to be very sensitive to the locations of the control points. A common alternative in computer vision is to construct cubic curves pieced together with a greater number of inflection points. Each pair of control points form one segment of the curve, where each curve segment is a cubic with its own coefficients.

$$f_i(x) = a_i + b_i x + c_i x^2 + d_i x^3 \tag{3}$$

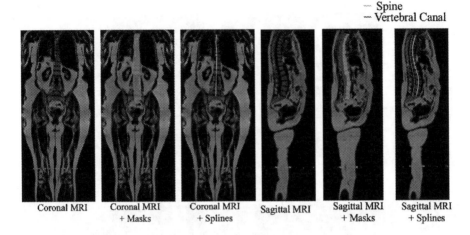

Fig. 11. Visualisation of spine and vertebral canal segmentation masks and midpoint curves on the coronal and sagittal plane.

where f is the function representing the curve between control points i and $i + 1$.
We ensure C^0, C^1, C^2 continuity conditions.

- C^0: Each segment is required to pass through its control points. That is, $f_i(x_i) = y_i$, and $f_i(x_{i+1}) = y_{i+1}$
- C^1: Each curve segment has the same slope at each junction, $f'_i(x_{i+1}) = f'_{i+1}(x_{i+1})$
- Each curve segment has the same curvature at each junction, $f''_i(x_{i+1}) = f''_{i+1}(x_{i+1})$

We improve the method in [2] by changing the uniform placement of a fixed number of knots by automatic knot selection using penalised regression splines [20]. The spline curve is composed of $n - 1$ piecewise cubic polynomials where n is the total number of knots. The number of knots is selected in the range from 2 to 10.

n is optimised using a penalty to balance goodness-of-fit and smoothness. The selection of knots is such that the model chooses from a bigger selection of functions. As the number of knots increases, the model overfits the data. Too few knots on the other hand gives a more restrictive function.

B.2   3D Spline Fitting

We now extend the 2D spline fitting to three-dimensional space. We have two systems of linear equations for x and y: $M_x \mathbf{b}_x = \mathbf{x}$ and $M_y \mathbf{b}_y = \mathbf{y}$, where $\mathbf{b}$ is the vector of curve coefficients, $\mathbf{y}$ is the vector of constants, and M is a matrix of continuity conditions i.e. C^0, C^1, and C^2. Each system is solved similarly as in 2D section above, except that we are solving two linear systems instead of one.

References

1. Aaro, S., Dahlborn, M., Svensson, L.: Estimation of vertebral rotation in structural scoliosis by computer tomography. Acta Radiol. **19**, 990–992 (1978)
2. Bourigault, E., Jamaludin, A., Kadir, T., Zisserman, A.: Scoliosis measurement on DXA scans using a combined deep learning and spinal geometry approach. In: Medical Imaging with Deep Learning (2022)
3. Chen, B., Jiang, J., Wang, X., Wan, P., Wang, J., Long, M.: Debiased self-training for semi-supervised learning (2022). 10.48550/ARXIV.2202.07136, https://arxiv.org/abs/2202.07136
4. Cobb, J.: Outline for the study of scoliosis. Instr. Course Lect. AAOS **5**, 261–275 (1948)
5. Galbusera, F., Bassani, T., Panico, M., Sconfienza, L.M., Cina, A.: A fresh look at spinal alignment and deformities: automated analysis of a large database of 9832 biplanar radiographs. Front. Bioeng. Biotech. **10**, 863054 (2022)
6. Ho, E.K., Upadhyay, S.S., Chan, F.L., Hsu, L.C.S., Leong, J.C.Y.: New methods of measuring vertebral rotation from computed tomographic scans. an intraobserver and interobserver study on girls with scoliosis. Spine **18**(9), 1173–1777 (1993)
7. Illés, T.S., Lavaste, F., Dubousset, J.: The third dimension of scoliosis: the forgotten axial plane. Orthop. Traumatol. Surg. Res. OTSR **105**(2), 351–359 (2019)
8. Illés, T.S., Tunyogi-Csapó, M., Somoskeöy, S.: Breakthrough in three-dimensional scoliosis diagnosis: significance of horizontal plane view and vertebra vectors. Eur. Spine J. **20**, 135–143 (2010)
9. Jamaludin, A., Kadir, T., Clark, E., Zisserman, A.: Predicting scoliosis in DXA scans using intermediate representations. In: MICCAI Workshop: MSKI (2018)
10. Jamaludin, A., Lootus, M., Kadir, T., Zisserman, A.: Automatic intervertebral discs localization and segmentation: a vertebral approach. In: Vrtovec, T., et al. (eds.) CSI 2015. LNCS, vol. 9402, pp. 97–103. Springer, Cham (2016). https://doi.org/10.1007/978-3-319-41827-8_9
11. Karam, M., et al.: Global malalignment in adolescent idiopathic scoliosis: the axial deformity is the main driver. Eur. Spine J. 1–13 (2022). https://doi.org/10.1007/s00586-021-07101-x
12. Khalil, Y.A., et al.: Multi-scanner and multi-modal lumbar vertebral body and intervertebral disc segmentation database. Sci. Data **9**, 97 (2022)
13. Konieczny, M.R., Senyurt, H., Krauspe, R.: Epidemiology of adolescent idiopathic scoliosis. J. Child. Orthop. **7**, 3–9 (2013)
14. Ma, Q., et al.: Coronal balance vs. sagittal profile in adolescent idiopathic scoliosis, are they correlated? Front. Pediatr. **7** (2020)
15. Pasha, S.: Data-driven classification of the 3d spinal curve in adolescent idiopathic scoliosis with an applications in surgical outcome prediction. Sci. Rep. (2018). https://doi.org/10.1038/s41598-018-34261-6
16. Pasha, S., Ecker, M., Ho, V., Hassanzadeh, P.: A hierarchical classification of adolescent idiopathic scoliosis: Identifying the distinguishing features in 3d spinal deformities. PLoS ONE (2019). https://doi.org/10.1371/journal.pone.0213406
17. Roaf, R.: Rotation movements of the spine with special reference to scoliosis. J. Bone Joint Surgery. Br. **40-B**(2), 312–332 (1958)
18. Rockenfeller, R., Müller, A.: Augmenting the cobb angle: three-dimensional analysis of whole spine shapes using bézier curves. Comput. Methods Programs Biomed. **225**, 107075 (2022). https://doi.org/10.1016/j.cmpb.2022.107075

19. Ronneberger, O., Fischer, P., Brox, T.: U-net: convolutional networks for biomedical image segmentation. In: International Conference on Medical Image Computing and Computer-Assisted Intervention, pp. 234–241 (2015)
20. Ruppert, D., Wand, M.P., Carroll, R.J.: Semiparametric Regression. No. 12, Cambridge University Press, Cambridge (2003)
21. Sudlow, C.L.M., et al.: UK biobank: an open access resource for identifying the causes of a wide range of complex diseases of middle and old age. PLoS Med. **12**, e1001779 (2015)
22. Sudre, C.H., Li, W., Vercauteren, T., Ourselin, S., Jorge Cardoso, M.: Generalised dice overlap as a deep learning loss function for highly unbalanced segmentations. In: Cardoso, M.J., et al. (eds.) DLMIA/ML-CDS -2017. LNCS, vol. 10553, pp. 240–248. Springer, Cham (2017). https://doi.org/10.1007/978-3-319-67558-9_28
23. Taylor, H., et al.: Identifying scoliosis in population-based cohorts: development and validation of a novel method based on total-body dual-energy x-ray absorptiometric scans. Calcif. Tissue Int. **92**, 539–547 (2013)
24. Windsor, R., Jamaludin, A., Kadir, T., Zisserman, A.: A convolutional approach to vertebrae detection and labelling in whole spine MRI. In: MICCAI (2020)
25. Yi-de, M., Qing, L., Zhi-bai, Q.: Automated image segmentation using improved PCNN model based on cross-entropy. In: Proceedings of 2004 International Symposium on Intelligent Multimedia, Video and Speech Processing, vol. 2004, pp. 743–746 (2004)
26. Zhang, Z., Liu, Q., Wang, Y.: Road extraction by deep residual u-net. IEEE Geosci. Remote Sens. Lett. **15**, 749–753 (2017)
27. Zhou, Z., Rahman Siddiquee, M.M., Tajbakhsh, N., Liang, J.: UNet++: a nested u-net architecture for medical image segmentation. In: Stoyanov, D., et al. (eds.) DLMIA/ML-CDS -2018. LNCS, vol. 11045, pp. 3–11. Springer, Cham (2018). https://doi.org/10.1007/978-3-030-00889-5_1

SADIR: Shape-Aware Diffusion Models for 3D Image Reconstruction

Nivetha Jayakumar[1(✉)], Tonmoy Hossain[2], and Miaomiao Zhang[1,2]

[1] Department of Electrical and Computer Engineering, University of Virginia,
Charlottesville, VA, USA
`vfb8zb@virginia.edu`

[2] Department of Computer Science, School of Engineering and Applied Science,
University of Virginia, Charlottesville, VA, USA

Abstract. 3D image reconstruction from a limited number of 2D images has been a long-standing challenge in computer vision and image analysis. While deep learning-based approaches have achieved impressive performance in this area, existing deep networks often fail to effectively utilize the shape structures of objects presented in images. As a result, the topology of reconstructed objects may not be well preserved, leading to the presence of artifacts such as discontinuities, holes, or mismatched connections between different parts. In this paper, we propose a shape-aware network based on diffusion models for 3D image reconstruction, named SADIR, to address these issues. In contrast to previous methods that primarily rely on spatial correlations of image intensities for 3D reconstruction, our model leverages shape priors learned from the training data to guide the reconstruction process. To achieve this, we develop a joint learning network that simultaneously learns a mean shape under deformation models. Each reconstructed image is then considered as a deformed variant of the mean shape. We validate our model, SADIR, on both brain and cardiac magnetic resonance images (MRIs). Experimental results show that our method outperforms the baselines with lower reconstruction error and better preservation of the shape structure of objects within the images.

1 Introduction

The reconstruction of 3D images from a limited number of 2D images is fundamental to various applications, including object recognition and tracking [12], robot navigation [44], and statistical shape analysis for disease detection [4,36]. However, inferring the complete 3D geometry and structure of objects from one or multiple 2D images has been a long-standing ill-posed problem [25]. A bountiful literature has been investigated to recover the data from a missing dimension [9,32,34,37]. Initial approaches to address this challenge focused on solving an inverse problem of projecting 3D information onto 2D images from geometric aspects [8]. These solutions typically require images captured from different viewing angles using precisely calibrated cameras or medical imaging

© The Author(s), under exclusive license to Springer Nature Switzerland AG 2023
C. Wachinger et al. (Eds.): ShapeMI 2023, LNCS 14350, pp. 287–300, 2023.
https://doi.org/10.1007/978-3-031-46914-5_23

machines [7,28]. In spite of producing a good quality of 3D reconstructions, such methods are often impractical or infeasible in many real-world scenarios.

Recent advancements have leveraged deep learning (DL) techniques to overcome the limitations posed in previous methods [5,15,27]. Extensive research has explored various network architectures for 3D image reconstruction, including UNets [30], transformers [14,22], and state-of-the-art generative diffusion models [37]. These works have significantly improved the reconstruction efficiency by learning intricate mappings between stacks of 2D images and their corresponding 3D volumes. While the DL-based approaches have achieved impressive results in reconstructing detailed 3D images, they often lack explicit consideration of shape information during the learning process. Consequently, important geometric structures of objects depicted in the images may not be well preserved. This may lead to the occurrence of artifacts, such as discontinuities, holes, or mismatched connections between different parts, that break the topology of the reconstructed objects.

Motivated by recent studies highlighting the significance of shape in enhancing image analysis tasks using deep networks [6,20,26,39,43], we introduce a novel shape-aware 3D image reconstruction network called SADIR. Our methodology builds upon the foundation of diffusion models while incorporating shape learning as a key component. In contrast to previous methods that mainly rely on spatial correlations of image intensities for 3D reconstruction, our SADIR explicitly incorporates the geometric shape information aiming to preserve the topology of reconstructed images. To achieve this goal, we develop a joint deep network that simultaneously learns a shape prior (also known as a mean shape) from a given set of full 3D volumes. In particular, an atlas building network based on deformation models [39] is employed to learn a mean shape representing the average information of training images. With the assumption that each reconstructed object is a deformed variant of the estimated mean shape, we then utilize the mean shape as a prior knowledge to guide the diffusion process of reconstructing a complete 3D image from a stack of sparse 2D slices. To evaluate the effectiveness of our proposed approach, we conduct experiments on both real brain and cardiac magnetic resonance images (MRIs). The experimental results show the superiority of SADIR over the baseline approaches, as evidenced by substantially reduced reconstruction errors. Moreover, our method successfully preserves the topology of the images during the shape-aware 3D image reconstruction process.

2   Background: Fréchet Mean via Atlas Building

In this section, we briefly review an unbiased atlas building algorithm [21], a widely used technique to estimate the Fréchet mean of group-wise images. With the underlying assumption that objects in many generic classes can be described as deformed versions of an ideal template, descriptors in this class arise naturally by matching the mean (also referred as atlas) to an input image [21,38,42,45,46]. The resulting transformation is then considered as a shape that reflects geometric changes.

Given a number of N images $\{\mathcal{Y}_1, \cdots, \mathcal{Y}_N\}$, the problem of atlas building is to find a mean or template image $\mathcal{S}$ and deformation fields $\phi_1, \cdots \phi_N$ with derived initial velocity fields $v_1, \cdots v_t$ that minimize the energy function

$$E(\mathcal{S}, \phi_n) = \sum_{n=1}^{N} \frac{1}{\sigma^2} \text{Dist}[\mathcal{S} \circ \phi_n(v_t), \mathcal{Y}_n] + \text{Reg}[\phi_n(v_t)], \tag{1}$$

where σ^2 is a noise variance and $\circ$ denotes an interpolation operator that deforms image $\mathcal{Y}_n$ with an estimated transformation ϕ_n. The Dist$[\cdot, \cdot]$ is a distance function that measures the dissimilarity between images, i.e., sum-of-squared differences [3], normalized cross correlation [2], and mutual information [40]. The Reg$[\cdot]$ is a regularizer that guarantees the smoothness of transformations.

Given an open and bounded d-dimensional domain $\Omega \subset \mathbb{R}^d$, we use Diff$(\Omega)$ to denote a space of diffeomorphisms (i.e., a one-to-one smooth and invertible smooth transformation) and its tangent space $V = T\text{Diff}(\Omega)$. A well-developed algorithm, large deformation diffeomorphic metric mapping (LDDMM) [3], provides a regularization that guarantees the smoothness of deformation fields and preserves the topological structures of objects for the atlas building framework (Eq. (1)). Such a regularization is formulated as an integral of the Sobolev norm of the time-dependent velocity field $v_n(t) \in V (t \in [0, 1])$ in the tangent space, i.e.,

$$\text{Reg}[\phi_n(v_t)] = \int_0^1 (Lv_t, v_t)\, dt, \quad \text{with} \quad \frac{d\phi_n(t)}{dt} = -D\phi_n(t) \cdot v_n(t), \tag{2}$$

where $L : V \to V^*$ is a symmetric, positive-definite differential operator that maps a tangent vector $v_t \in V$ into its dual space as a momentum vector $m_t \in V^*$. We write $m_t = Lv_t$, or $v_t = Km_t$, with K being an inverse operator of L. The operator D denotes a Jacobian matrix and $\cdot$ represents element-wise matrix multiplication. In this paper, we use a metric of the form $L = (-\alpha\Delta + \gamma\mathbf{I})^3$, in which Δ is the discrete Laplacian operator, α is a positive regularity parameter that controls the smoothness of transformation fields, γ is a weighting parameter, and $\mathbf{I}$ denotes an identity matrix.

The minimum of Eq. (2) is uniquely determined by solving an Euler-Poincaré differential equation (EPDiff) [1,29] with a given initial condition of velocity fields, noted as v_0. This is known as the *geodesic shooting* algorithm [35], which nicely proves that the deformation-based shape descriptor ϕ_n can be fully characterized by an initial velocity field $v_n(0)$. The mathematical formulation of the EPDiff equation is

$$\frac{\partial v_n(t)}{\partial t} = -K \left[(Dv_n(t))^T \cdot m_n(t) + Dm_n(t) \cdot v_n(t) + m_n(t) \cdot \text{div}\, v_n(t) \right], \tag{3}$$

where the operator D denotes a Jacobian matrix, div is the divergence, and $\cdot$ represents element-wise matrix multiplication.

We are now able to equivalently minimize the atlas building energy function in Eq. (1) as

$$E(\mathcal{S}, \phi_n) = \sum_{n=1}^{N} \frac{1}{\sigma^2} \text{Dist}[\mathcal{S} \circ \phi_n(v_n(t)), \mathcal{Y}_n] + (Lv_n(0), v_n(0)), \text{ s.t. Eq. (2)\& (3)}. \tag{4}$$

For notation simplicity, we will drop the time index in the following sections.

3   Our Method: SADIR

In this section, we present SADIR, a novel reconstruction network that incorporates shape information in predicting 3D volumes from a limited number of input 2D images. We introduce a sub-module of the atlas building framework, which enables us to learn shape priors from a given set of full 3D images. It is worth mentioning that while the backbone of our proposed SADIR is a diffusion model [16], the methodology can be generalized to a variety of network architectures such as UNet [33], UNet++ [47], and Transformer [11].

3.1   Shape-Aware Diffusion Models Based on Atlas Building Network

Given a number of N training data $\{I_n, \mathcal{Y}_n\}_{n=1}^N$, where I_n is a stack of sparse 2D images with its associated full 3D volume $\mathcal{Y}_n$. Our model SADIR consists of two submodules:

(i) An atlas building network, parameterized by θ^a, that provides a mean image $\mathcal{S}$ of $\{\mathcal{Y}_n\}$. In this paper, we employ the network architecture of Geo-SIC [39];

(ii) A reconstruction network, parameterized by θ^r, that considers each reconstructed image $\hat{\mathcal{Y}}_n$ as a deformed variant of the obtained atlas, i.e., $\hat{\mathcal{Y}}_n \triangleq \mathcal{S} \circ \phi_n(v_n(\theta^r))$. In contrast to current approaches learning the reconstruction process based on image intensities, our model is developed to learn the geometric shape variations represented by the predicted velocity field v_n.

Next, we introduce the details of our shape-aware diffusion models for reconstruction, which is a key component of SADIR. Similar to existing diffusion models [16,37], we develop a forward diffusion and a reverse diffusion process to predict the velocity fields associated with the pair of input training images and an atlas image. For the purpose of simplified math notations, we omit the index n for each subject in the following sections.

Forward Diffusion Process. Let y^0 denote the original 3D image with full volumes and τ denote the time point of the diffusion process. We assume the data distribution of y^τ is a normal distribution with mean μ and variance β, i.e., $y^\tau \sim \mathcal{N}(\mu, \beta)$. The forward diffusion of $y^{\tau-1}$ to y^τ is then recursively given by

$$p(y^\tau \mid y^{\tau-1}) = \mathcal{N}(y^\tau; \sqrt{1 - \beta^\tau} y^{\tau-1}, \beta^\tau \mathbf{I}), \tag{5}$$

where $\mathbf{I}$ denotes an identity matrix, and $\beta^\tau \in [0, 1]$ denotes a known variance increased along the time steps with $\beta^1 < \beta^2 < \cdots < \beta^\tau$. The forward diffusion process is repeated for a fixed, predefined number of time steps.

It is shown in [16] that repeated application of Eq. (5) to the original image y^0 and setting $\alpha^\tau = 1 - \beta^\tau$ and $\bar{\alpha}^\tau = \prod_{i=1}^\tau \alpha^i$ yields

$$p(y^\tau \mid y^0) = \mathcal{N}(y^\tau; \sqrt{\bar{\alpha}^\tau} y^0, (1 - \bar{\alpha}^\tau)\mathbf{I}).$$

Therefore, we can write y^τ in terms of y^0 as

$$y^\tau = \sqrt{\bar{\alpha}^\tau} y^0 + \sqrt{1 - \bar{\alpha}^\tau} \epsilon \quad \text{with} \quad \epsilon \sim \mathcal{N}(0, \mathbf{I}).$$

Reverse Diffusion Process. Given a concatenation of a sparse stack of 2D images I, an atlas image $\mathcal{S}$, and y^τ from the forward process, our diffusion model is designed to remove the added noise in the reverse process. Following the work of [41], we will now predict $y^{\tau-1}$ from the input y^τ. The joint probability distribution $p(y^{\tau-1} \,|\, y^\tau)$ is predicted by a trained neural network (e.g., UNet) in each reverse time step for all $\tau \in \{1, \cdots, T\}$, where T is the maximal time step. With the network model parameters denoted by θ^r, we can write the reverse process as

$$p_{\theta^r}(y^{\tau-1} \,|\, y^\tau) = \mathcal{N}(y^{\tau-1}; \mu_{\theta^r}(y^\tau, \tau), \Sigma_{\theta^r}(y^\tau, \tau)).$$

Similarly, we can write $y^{\tau-1}$ backward in terms of y^τ as

$$y^{\tau-1} = \frac{1}{\sqrt{\alpha^\tau}} \left(y^\tau \frac{1 - \alpha^\tau}{\sqrt{1 - \bar{\alpha}^\tau}} \epsilon_{\theta^r}(y^\tau, \tau) \right) + \sigma^t \mathbf{z},$$

where σ^τ is the variance scheme the model can learn, the component $\mathbf{z}$ is a stochastic sampling process. The model is trained with input y^τ to subtract the noise scheme $\epsilon_{\theta^r}(y^\tau, \tau)$ from y^τ to produce $y^{\tau-1}$.

The output of this reverse process is a predicted velocity field $v(\theta^r)$, which is then used to generate its associated transformation $\phi(v(\theta^r))$ to deform the atlas $\mathcal{S}$. Such a deformed atlas is the reconstructed image $\mathcal{Y} = \mathcal{S} \circ \phi(v(\theta^r))$.

An overview of the proposed SADIR network architecture is shown in Fig. 1.

3.2   Network Loss and Optimization

The network loss function of our model, SADIR, is a joint loss of the atlas building network and the diffusion reconstruction network. We first define the atlas building loss as

$$\mathcal{L}(\theta^a) = \sum_{n=1}^{N} \frac{1}{\sigma^2} \|\mathcal{S}(\theta^a) \circ (\phi_n(v_n)) - \mathcal{Y}_n\|_2^2 + (Lv_n, v_n) + \text{reg}(\theta^a), \tag{6}$$

where $\text{reg}(\cdot)$ denotes a regularization on the network parameters.

We then define the loss function of the diffusion reconstruction network as a combination of sum-of-squared differences and Sørensen−Dice coefficient [10] loss (for distinct anatomical structure, e.g., brain ventricles or myocardium) between the predicted reconstruction and ground-truth in following

$$\mathcal{L}(\theta^r) = \sum_{n=1}^{N} \|\mathcal{S} \circ \phi_n(v_n(\theta^r)) - \mathcal{Y}_n\|_2^2 + \eta \left[1 - \text{Dice}(\mathcal{S} \circ \phi_n(v_n(\theta^r)), \mathcal{Y}_n)\right] + \text{reg}(\theta^r), \tag{7}$$

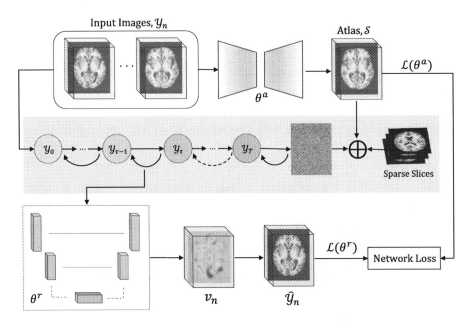

Fig. 1. An overview of our proposed 3D reconstruction model SADIR.

where η is the weighting parameter, and $\mathrm{Dice}(\hat{y}, \mathcal{Y}_n) = 2(|\hat{\mathcal{Y}}| \cap |\mathcal{Y}_n|)/(|\hat{\mathcal{Y}}| + |\mathcal{Y}_n|)$, considering $\hat{\mathcal{Y}}_n \overset{\Delta}{=} \mathcal{S} \circ \phi_n(v_n(\theta^r))$. Defining λ as a weighting parameter, we are now ready to write the joint loss of SADIR as

$$\mathcal{L} = \mathcal{L}(\theta^a) + \lambda \mathcal{L}(\theta^r).$$

Joint Network Learning with an Alternative Optimization. We use an alternative optimization scheme [31] to minimize the total loss $\mathcal{L}$ in Eq. (3.2). More specifically, we jointly optimize all network parameters by alternating between the training of the atlas building and diffusion reconstruction network, making it end-to-end learning. A summary of our joint training of SADIR is presented in Algorithm 1.

4   Experimental Evaluation

We demonstrate the effectiveness of our proposed model, SADIR, for 3D image reconstruction from 2D slices on both brain and cardiac MRI scans.

3D Brain MRIs: For 3D real brain MRI scans, we include 214 public T1-weighted longitudinal brain scans from the latest released Open Access Series of Imaging Studies (OASIS-III) [23]. All subjects include both healthy and disease individuals, aged from 42 to 95. All MRIs were pre-processed as $256 \times 256 \times 256$, $1.25\,\mathrm{mm}^3$ isotropic voxels, and underwent skull-stripped, intensity normalized,

---

Algorithm 1: Joint Training of SADIR.

Input : A group of N input images with full 3D volumes $\{\mathcal{Y}_n\}$ and a stack of sparse 2D images $\{I_n\}$.

Output: Generate mean shape or atlas $\mathcal{S}$, initial velocity fields v_n, and reconstructed images $\hat{\mathcal{Y}}_n$

1 **for** i = 1 to p **do**

    /* Train geometric shape learning network   */

2   Minimize the atlas building loss in Eq. (6)

3   Output the atlas $\mathcal{S}$

    /* Train diffusion network   */

4   Minimize the diffusion reconstruction loss in Eq. (7)

5   Output the initial velocity fields $\{v_n\}$ and the reconstructed images $\hat{\mathcal{Y}}_n$

6 **end**

7 **Until convergence**

---

bias field corrected and pre-aligned with affine transformation. To further validate the performance of our proposed model on specific anatomical shapes, we select left and right brain ventricles available in the OASIS-III dataset [23].

3D Cardiac MRIs: For 3D real cardiac MRI, we include 215 publicly available 3D myocardium mesh data from MedShapeNet dataset [24]. We convert the mesh data to binary label maps using 3D slicer [13]. All the images were pre-processed as $222 \times 222 \times 222$ and pre-aligned with affine transformation.

4.1 Experimental Settings

We first validate our proposed model, SADIR, on reconstructing 3D brain ventricles, as well as brain MRIs from a sparse stack of eight 2D slices. We compare our model's performance with three state-of-the-art deep learning-based reconstruction models: 3D-UNet [9]; DDPM, a probabilistic diffusion model [16]; and DISPR, a diffusion model based shape reconstruction model with geometric topology considered [37]. Three evaluation metrics, including the Sørensen-Dice coefficient (DSC) [10], Jaccard Similarity [19], and RHD95 score [18], are used to validate the prediction accuracy of brain ventricles for all methods. For brain MR images, we show the error maps of reconstructed images for all the experiments.

To further validate the performance of SADIR on different datasets, we run tests on a relatively small dataset of cardiac MRIs to reconstruct 3D myocardium.

Parameter Setting: We set the mean and standard deviation of the forward diffusion process to be 0 and 0.1, respectively. The scheduling is linear for the noising process and is scaled to reach an isotropic Gaussian distribution irrespective of the value of T. For the atlas building network, we set the depth of the UNet architecture as 4. We set the number of time steps for Euler integration in EPDiff (Eq. (3)) as 10, and the noise variance $\sigma = 0.02$. For the shooting, we use a kernel map valued $[0.5, 0, 1.0]$. Besides, we set the parameter $\alpha = 3$ for the operator L. Similar to [37], we set the batch size as 1 for all experiments. We utilize the cosine annealing learning rate scheduler that starts with a learning rate of $\eta = 1e^{-3}$ for network training. We run all models on training and validation images using the Adam optimizer and save the networks with the best validation performance.

In the reverse process of the diffusion network, we set the depth of the 3D attention-UNet backbone as 6. We introduce the attention mechanism via spatial excitation channels [17], with ReLU (Rectified Linear Unit) activation. The UNet backbone has ELU activation (Exponential Linear Unit) in the hidden convolution layers and GeLU (Gaussian error Linear Unit) activation with tanh approximation. For each training experiment, we utilize Rivanna (high-performance computing servers of the University of Virginia) with NVIDIA A100 and V100 GPUs for ~18 h (till convergence). For all the experimental datasets, we split all the training datasets into 70% training, 15% validation, and 15% testing. For both training and testing, we downsample all the image resolutions to 64×64×64.

4.2   Experimental Results

Figure 2 visualizes examples of ground truth and reconstructed 3D volumes of brain ventricles from all methods. It shows that SADIR outperforms all baselines in well preserving the structural information of the brain ventricles. In particular, models without considering the shape information of the images (i.e., 3D-UNet and DDPM) generate unrealistic shapes such as those with joint ventricles, holes in the volume, and deformed ventricle tails. While the other algorithm, DISPR, shows improved performance of enforcing topological consistency on the object surface, its predicted results of 3D volumes are inferior to SADIR.

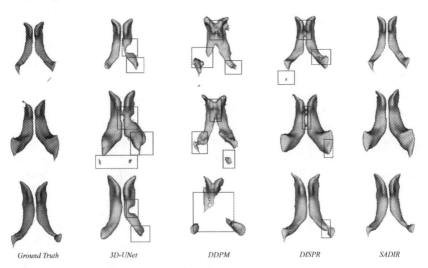

Ground Truth   3D-UNet   DDPM   DISPR   SADIR

Fig. 2. Top to bottom: examples of reconstructed 3D brain ventricles from sparse 2D slices; Left to right: a comparison of brain ventricles of all reconstruction models with ground truth.

Table 1. A comparison of 3D brain ventricle reconstruction for all methods.

Model	DSC ↑	Jaccard similarity ↑	RHD95 ↓
3D-Unet	0.878 ± 0.0128	0.804 ± 0.0204	4.366 ± 1.908
DDPM	0.731 ± 0.0292	0.652 ± 0.0365	8.827 ± 9.212
DISPR	0.918 ± 0.0097	0.861 ± 0.0158	**1.041 ± 0.130**
SADIR	**0.934 ± 0.013**	**0.900 ± 0.021**	1.414 ± 0.190

Table 1 reports the average scores along with the standard deviation of the Dice similarity coefficient (DSC), Jaccard similarity, and Hausdorff distance computed between the brain ventricles reconstructed by all the models and the ground truth. Compared to all the baselines, SADIR achieves the best performance with a 1.6%–5.6% increase in the average DSC with the lowest standard deviations across all metrics.

Figure 3 visualizes the ground truth and reconstructed 3D brain MRIs as a result of evaluating DDMP and our method SADIR on the test data, along with their corresponding error maps. The error map is computed as absolute values of an element-wise subtraction between the ground truth and the reconstructed image. The images reconstructed by SADIR outperform the DDPM with a low absolute reconstruction error. Our method also preserves crucial anatomical features such as the shape of the ventricles, corpus callosum and gyri, which cannot be seen in the images reconstructed by the DDPM. This can be attributed to the lack of incorporating the shape information to guide the 3D MRI reconstruction. Moreover, our model has little to no noise in the background as compared to the DDPM.

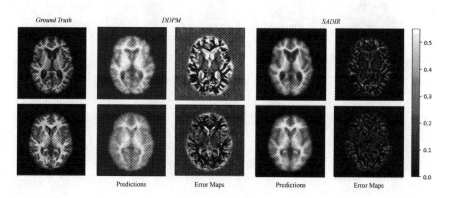

Fig. 3. Left to right: a comparison of ground truth, DDPM, and SADIR along with the error map.

Table 2 reports the average scores of DSC, Jaccard similarity, and Hausdorff distance evaluated between the reconstructed myocardium from all algorithms and the ground truth. Our method proves to be competent in reconstructing 3D volumes without discontinuities, artifacts, jagged edges or amplified structures, as can be seen in results from the other models. Compared to the baselines, SADIR achieves the best performance in terms of DSC, Jaccard similarity, and RHD95 with the lowest standard deviations across all metrics.

Table 2. A comparison of 3D myocardium reconstruction for all methods.

Model	DSC ↑	Jaccard similarity ↑	RHD95 ↓
3D-Unet	0.870 ± 0.0158	0.771 ± 0.024	0.840 ± 0.202
DDPM	0.823 ± 0.014	0.668 ± 0.019	1.027 ± 0.093
DISPR	0.950 ± 0.017	0.906 ± 0.031	0.347 ± 0.032
SADIR	**0.978 ± 0.016**	**0.957 ± 0.031**	**0.341 ± 0.023**

Figure 4 visualizes a comparison of the reconstructed 3D myocardium between the ground truth and all models. It shows that our method consistently produces reconstructed volumes that preserve the original shape of the organ with less artifacts.

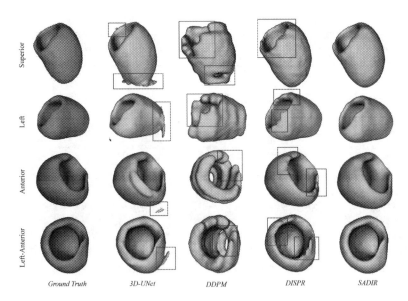

Fig. 4. A comparison of reconstructed 3D myocardium between ground truth, 3D-UNet, DDPM, DISPR, and SADIR over four different views.

Figure 5 shows examples of the superior, left, anterior and left-anterior views of the 3D ground truth and SADIR-reconstructed volumes of the myocardium for different subjects. We observe that the results predicted by SADIR have little to no difference from the ground truth, thereby efficiently preserving the anatomical structure of the myocardium.

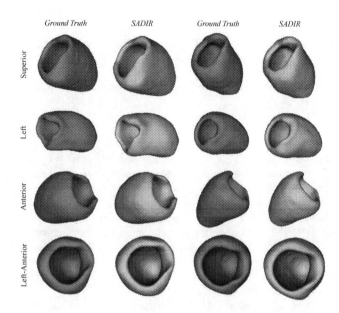

Fig. 5. 3D myocardium reconstructed from sparse 2D slices by SADIR over four different views.

5   Conclusion

This paper introduces a novel shape-aware image reconstruction framework based on diffusion model, named as SADIR. In contrast to previous approaches that mainly rely on the information of image intensities, our model SADIR incorporates shape features in the deformation spaces to preserve the geometric structures of objects in the reconstruction process. To achieve this, we develop a joint deep network that simultaneously learns the underlying shape representations from the training images and utilize it as a prior knowledge to guide the reconstruction network. To the best of our knowledge, we are the first to consider deformable shape features into the diffusion model for the task of image reconstruction. Experimental results on both 3D brain and cardiac MRI show that our model efficiently produces 3D volumes from a limited number of 2D slices with substantially low reconstruction errors while better preserving the topological structures and shapes of the objects.

Acknowledgement. This work was supported by NSF CAREER Grant 2239977 and NIH 1R21EB032597.

References

1. Arnold, V.: Sur la géométrie différentielle des groupes de lie de dimension infinie et ses applications à l'hydrodynamique des fluides parfaits. Annales de l'institut Fourier **16**, 319–361 (1966)
2. Avants, B.B., Epstein, C.L., Grossman, M., Gee, J.C.: Symmetric diffeomorphic image registration with cross-correlation: evaluating automated labeling of elderly and neurodegenerative brain. Med. Image Anal. **12**(1), 26–41 (2008)
3. Beg, M.F., Miller, M.I., Trouvé, A., Younes, L.: Computing large deformation metric mappings via geodesic flows of diffeomorphisms. Int. J. Comput. Vision **61**(2), 139–157 (2005)
4. Bruse, J.L., et al.: Detecting clinically meaningful shape clusters in medical image data: metrics analysis for hierarchical clustering applied to healthy and pathological aortic arches. IEEE Trans. Biomed. Eng. **64**(10), 2373–2383 (2017)
5. Cetin, I., Stephens, M., Camara, O., Ballester, M.A.G.: Attri-VAE: attribute-based interpretable representations of medical images with variational autoencoders. Comput. Med. Imaging Graph. **104**, 102158 (2023)
6. Chen, C., Biffi, C., Tarroni, G., Petersen, S., Bai, W., Rueckert, D.: Learning shape priors for robust cardiac MR segmentation from multi-view images. In: Shen, D., et al. (eds.) MICCAI 2019. LNCS, vol. 11765, pp. 523–531. Springer, Cham (2019). https://doi.org/10.1007/978-3-030-32245-8_58
7. Chen, L., Bentley, P., Mori, K., Misawa, K., Fujiwara, M., Rueckert, D.: Self-supervised learning for medical image analysis using image context restoration. Med. Image Anal. **58**, 101539 (2019)
8. Chung, H., Ryu, D., McCann, M.T., Klasky, M.L., Ye, J.C.: Solving 3D inverse problems using pre-trained 2D diffusion models. In: Proceedings of the IEEE/CVF Conference on Computer Vision and Pattern Recognition, pp. 22542–22551 (2023)
9. Çiçek, Ö., Abdulkadir, A., Lienkamp, S.S., Brox, T., Ronneberger, O.: 3D u-net: learning dense volumetric segmentation from sparse annotation. In: Ourselin, S., Joskowicz, L., Sabuncu, M.R., Unal, G., Wells, W. (eds.) MICCAI 2016. LNCS, vol. 9901, pp. 424–432. Springer, Cham (2016). https://doi.org/10.1007/978-3-319-46723-8_49
10. Dice, L.R.: Measures of the amount of ecologic association between species. Ecology **26**(3), 297–302 (1945)
11. Dosovitskiy, A., et al.: An image is worth 16x16 words: transformers for image recognition at scale. arXiv preprint arXiv:2010.11929 (2020)
12. Duwek, H.C., Bitton, A., Tsur, E.E.: 3D object tracking with neuromorphic event cameras via image reconstruction. In: 2021 IEEE Biomedical Circuits and Systems Conference (BioCAS), pp. 1–4. IEEE (2021)
13. Fedorov, A., et al.: 3D slicer as an image computing platform for the quantitative imaging network, November 2012
14. Feng, C.-M., Yan, Y., Fu, H., Chen, L., Xu, Y.: Task transformer network for joint MRI reconstruction and super-resolution. In: de Bruijne, M., et al. (eds.) MICCAI 2021. LNCS, vol. 12906, pp. 307–317. Springer, Cham (2021). https://doi.org/10.1007/978-3-030-87231-1_30
15. Goodfellow, I., et al.: Generative adversarial networks. Commun. ACM **63**(11), 139–144 (2020)
16. Ho, J., Jain, A., Abbeel, P.: Denoising diffusion probabilistic models. In: Advances in Neural Information Processing Systems, vol. 33, pp. 6840–6851 (2020)

17. Hu, J., Shen, L., Albanie, S., Sun, G., Wu, E.: Squeeze-and-excitation networks (2019)

18. Huttenlocher, D., Klanderman, G., Rucklidge, W.: Comparing images using the hausdorff distance. IEEE Trans. Pattern Anal. Mach. Intell. **15**(9), 850–863 (1993)

19. Jaccard, P.: Nouvelles recherches sur la distribution florale. Bull. Soc. Vaud. Sci. Nat. **44**, 223–270 (1908)

20. Jiang, J., Veeraraghavan, H.: One shot pacs: patient specific anatomic context and shape prior aware recurrent registration-segmentation of longitudinal thoracic cone beam CTs. IEEE Trans. Med. Imaging **41**(8), 2021–2032 (2022)

21. Joshi, S., Davis, B., Jomier, M., Gerig, G.: Unbiased diffeomorphic atlas construction for computational anatomy. Neuroimage **23**, S151–S160 (2004)

22. Korkmaz, Y., Dar, S.U., Yurt, M., Özbey, M., Cukur, T.: Unsupervised MRI reconstruction via zero-shot learned adversarial transformers. IEEE Trans. Med. Imaging **41**(7), 1747–1763 (2022)

23. LaMontagne, P.J., et al.: Oasis-3: longitudinal neuroimaging, clinical, and cognitive dataset for normal aging and Alzheimer disease. medRxiv (2019)

24. Li, J.: Medshapenet: a large-scale dataset of 3D medical shapes for computer vision, March 2023

25. Lin, D.J., Johnson, P.M., Knoll, F., Lui, Y.W.: Artificial intelligence for MR image reconstruction: an overview for clinicians. J. Magn. Reson. Imaging **53**(4), 1015–1028 (2021)

26. Liu, J., Aviles-Rivero, A.I., Ji, H., Schönlieb, C.-B.: Rethinking medical image reconstruction via shape prior, going deeper and faster: Deep joint indirect registration and reconstruction. Med. Image Anal. **68**, 101930 (2021)

27. Maaløe, L., Fraccaro, M., Liévin, V., Winther, O.: Biva: a very deep hierarchy of latent variables for generative modeling. In: Advances in Neural Information Processing Systems, vol. 32 (2019)

28. Maier-Hein, L., et al.: Optical techniques for 3D surface reconstruction in computer-assisted laparoscopic surgery. Med. Image Anal. **17**(8), 974–996 (2013)

29. Miller, M.I., Trouvé, A., Younes, L.: Geodesic shooting for computational anatomy. J. Math. Imaging Vision **24**(2), 209–228 (2006)

30. Nguyen, T., Hua, B.-S., Le, N.: 3D-UCaps: 3D capsules Unet for volumetric image segmentation. In: de Bruijne, M., et al. (eds.) MICCAI 2021. LNCS, vol. 12901, pp. 548–558. Springer, Cham (2021). https://doi.org/10.1007/978-3-030-87193-2_52

31. Nocedal, J., Wright, S.J.: Numerical Optimization. Springer, New York (1999). https://doi.org/10.1007/978-0-387-40065-5

32. Qin, C., Schlemper, J., Caballero, J., Price, A.N., Hajnal, J.V., Rueckert, D.: Convolutional recurrent neural networks for dynamic MR image reconstruction. IEEE Trans. Med. Imaging **38**(1), 280–290 (2018)

33. Ronneberger, O., Fischer, P., Brox, T.: U-net: convolutional networks for biomedical image segmentation. In: Navab, N., Hornegger, J., Wells, W.M., Frangi, A.F. (eds.) MICCAI 2015. LNCS, vol. 9351, pp. 234–241. Springer, Cham (2015). https://doi.org/10.1007/978-3-319-24574-4_28

34. Schlemper, J., Caballero, J., Hajnal, J.V., Price, A.N., Rueckert, D.: A deep cascade of convolutional neural networks for dynamic MR image reconstruction. IEEE Trans. Med. Imaging **37**(2), 491–503 (2017)

35. Vialard, F.-X., Risser, L., Rueckert, D., Cotter, C.J.: Diffeomorphic 3D image registration via geodesic shooting using an efficient adjoint calculation. Int. J. Comput. Vision **97**(2), 229–241 (2012)

36. von Tycowicz, C., Ambellan, F., Mukhopadhyay, A., Zachow, S.: An efficient Riemannian statistical shape model using differential coordinates: with application to the classification of data from the osteoarthritis initiative. Med. Image Anal. **43**, 1–9 (2018)

37. Waibel, D.J.E., Röell, E., Rieck, B., Giryes, R., Marr, C.: A diffusion model predicts 3D shapes from 2D microscopy images (2023)

38. Wang, J., Zhang, M.: Bayesian atlas building with hierarchical priors for subject-specific regularization. In: de Bruijne, M., et al. (eds.) MICCAI 2021. LNCS, vol. 12904, pp. 76–86. Springer, Cham (2021). https://doi.org/10.1007/978-3-030-87202-1_8

39. Wang, J., Zhang, M.: Geo-sic: learning deformable geometric shapes in deep image classifiers. In: Advances in Neural Information Processing Systems, vol. 35, pp. 27994–28007 (2022)

40. Wells, W.M., III., Viola, P., Atsumi, H., Nakajima, S., Kikinis, R.: Multi-modal volume registration by maximization of mutual information. Med. Image Anal. **1**(1), 35–51 (1996)

41. Wolleb, J., Bieder, F., Sandkühler, R., Cattin, P.C.: Diffusion models for medical anomaly detection. In: Wang, L., Dou, Q., Fletcher, P.T., Speidel, S., Li, S. (eds.) MICCAI 2022. LNCS, vol. 13438, pp. 35–45. Springer, Cham (2022). https://doi.org/10.1007/978-3-031-16452-1_4

42. Wu, N., Wang, J., Zhang, M., Zhang, G., Peng, Y., Shen, C.: Hybrid atlas building with deep registration priors. In: 2022 IEEE 19th International Symposium on Biomedical Imaging (ISBI), pp. 1–5. IEEE (2022)

43. Yang, J., Wickramasinghe, U., Ni, B., Fua, P.: Implicitatlas: learning deformable shape templates in medical imaging. In: Proceedings of the IEEE/CVF Conference on Computer Vision and Pattern Recognition, pp. 15861–15871 (2022)

44. Zelenskii, A., Gapon, N., Voronin, V., Semenishchev, E., Serebrenny, V., Cen, Y.: Robot navigation using modified slam procedure based on depth image reconstruction. In: Artificial Intelligence and Machine Learning in Defense Applications III, vol. 11870, pp. 73–82. SPIE (2021)

45. Zhang, M., Singh, N., Fletcher, P.T.: Bayesian estimation of regularization and atlas building in diffeomorphic image registration. In: Gee, J.C., Joshi, S., Pohl, K.M., Wells, W.M., Zöllei, L. (eds.) IPMI 2013. LNCS, vol. 7917, pp. 37–48. Springer, Heidelberg (2013). https://doi.org/10.1007/978-3-642-38868-2_4

46. Zhang, M., Wells, W.M., Golland, P.: Low-dimensional statistics of anatomical variability via compact representation of image deformations. In: Ourselin, S., Joskowicz, L., Sabuncu, M.R., Unal, G., Wells, W. (eds.) MICCAI 2016. LNCS, vol. 9902, pp. 166–173. Springer, Cham (2016). https://doi.org/10.1007/978-3-319-46726-9_20

47. Zhou, Z., Rahman Siddiquee, M.M., Tajbakhsh, N., Liang, J.: UNet++: a nested U-net architecture for medical image segmentation. In: Stoyanov, D., et al. (eds.) DLMIA/ML-CDS -2018. LNCS, vol. 11045, pp. 3–11. Springer, Cham (2018). https://doi.org/10.1007/978-3-030-00889-5_1

Author Index

© The Editor(s) (if applicable) and The Author(s), under exclusive license
to Springer Nature Switzerland AG 2023
C. Wachinger et al. (Eds.): ShapeMI 2023, LNCS 14350, pp. 301–302, 2023.
https://doi.org/10.1007/978-3-031-46914-5